Preface

This volume addresses the ecology and management of freshwater forests limited to within 250 km of a seacoast in the temperate and boreal zones of the Northern Hemisphere. Its chapters deal with a broad range of fields, covering topics such as physico-chemical features of soil, water, and vegetation; genetics; pathology; species diversity (including rare and endangered associated biota); geomorphology and geographic distribution; harvest and utilization; and revegetation. Parts are devoted to systems with diverse dominants in Asia and North America, and to the commercially valuable genus *Chamaecyparis* (false-cypress).

The major objectives of the book are to assess the current state of knowledge and to compare data from different geographic regions, thereby promoting collaborative research and the interchange of ideas among investigators, foresters, and land-use and water-resource managers. An equally important goal is to draw attention to systems and vegetation types that are under threat or possess untapped potential.

By correlating information on widely dispersed coastal systems, we hope to identify common features and to pinpoint aspects that are unique to each. The accumulation of comparative data promises to be of great value to researchers and resource managers working in these apparently diverse woodlands.

Two international symposia were organized to address these issues. The first was convened in "INTECOL '90," the Fifth International Congress of Ecology, Yokohama, Japan (August 9, 1990); the second was held at the School of Forestry and Environmental Studies, Yale University, New Haven, Connecticut, USA (April 18–20, 1991). This volume, drawn from those symposia, is geared to an audience that includes the lumber industry worldwide, natural resource managers, university faculty (for both instruction and research), and the growing legion of environmental consultants, as well as the ecologically concerned lay public and their legislators, and regulatory agencies at all levels of government.

Freshwater wetlands, forests, the coastal zone, natural diversity, and wetland restoration are receiving public, legislative, and research attention as never before; each has become a global priority. Simultaneously, in the past few years lumber companies in the United States, Canada, and Japan have experienced a vigorous resurgence of interest in the commercially valuable coastally restricted genus *Chamaecyparis,* a fragrant, decay-resistant tree commonly known as white and yellow cedar, or false-cypress.

The underlying purpose throughout is to aid in the application of research findings to management practice. By identifying problems in the field, this volume can promote appropriate research. By assembling data and increasing our understanding of coastal forests, it can facilitate the formulation of guidelines for the management and protection of these unique and threatened systems.

Acknowledgments

In the course of this work, I have benefited greatly from the encouragement, advice, and assistance of colleagues in many disciplines. I gratefully list them here, with the clear understanding that responsibility for omissions and errors lies with me.

Paul Alaback, John Bowers, Randy Hagenstein, Andy Mitchell, and James Weigand (Ecotrust and Conservation International coastal temperate rainforest data); George Carroll and Joel Clement for the International Canopy Network, and William Denison (old-growth canopy); Stephen Edwards (*Chamaecyparis* morphology); John Kobar (climatic data from National Climatic Data Center, Asheville, NC); Leo Hickey (influences of atmospheric and oceanic circulation on continental biota); J.L. Jenkinson and George Lottritz (Northwest USA conifers); William Libby (closed-cone pines); Paul Meyer (Morris Arboretum materials on Taiwan and China *Chamaecyparis*); Joseph Miller (literature research); Wajiro Suzuki *(Abies mariesii, Thuja orientalis)*; Wan Li Yang (Chinese translation); and Hae-Soon Yoon (Korean translation). The librarians and facilities of the Libraries of the Marine Biological Laboratory and Yale School of Forestry were a valued resource.

Many of the contributing authors generously provided additional written and oral information to me and exchanged material with other contributors as well. Each chapter was reviewed by at least two authorities in the field. Reviewers, who shared their expertise and considerably added to the value of the book, include most contributing authors of this volume and Rudolph Becking, Alice Belling, Graeme Berlyn, Elgene O. Box, Norman Christensen, Antoni Damman, John DeWitt, L.K. Mike Gantt, Steven Grossnickle, James Hamrick, Leo Hickey, Michael Hollins, Donal Hook, Satoru Kojima, Steven Leonard, Don Lester, William Libby, George Lottritz, Constance Millar, Joseph Miller, Kevin Moorehead, William Niering, Victor Rico-Gray, Peter Schoonmaker, Frank Sorenson, Robert Thorson, Daniel Ward, Robert Westfall, Dennis Whigham, Michael Wicker, and Cheng Ying. Additional in-house readers are acknowledged in various chapters.

I am particularly grateful to the Directorate of INTECOL '90 and to Yale School of Forestry Deans John Gordon and Jared Cohon for suggesting, promoting, and supporting the two preparatory conferences, and to Herbert Bormann, Richard Goodwin, Charles Remington, and David M. Smith for support, encouragement, and advice.

Many colleagues and students contributed to the smooth running of the conferences. Andre Clewell, William McWilliams, Keiichi Ohno, and Shin-Ichi Yamamoto chaired sessions; Alice Belling, Joe Hughes, W. McWilliams, and A. Clewell served as workshop panelists. At INTECOL '90 in Yokohama, S. Yamamoto and his students were essential in slide projection, audiotaping, and facilitating translation. Many students and alumni of the Yale School of Forestry assisted in 1991: Mark Jen, Anne Marsh, Nellie Tsipoura, German Andrade, Darren Eng, Richard Gill, William Kenny, Cindy McWeeney, and Anne Pernick. Jodie Chase, Sue Edwards, Denise Learned, Kenneth Metzler, Phillip Miller, Glenn Motzkin, and Robert Russo helped guide the *Chamaecyparis thyoides* field trip. Bushy Hill Nature Center in Ivoryton, Connecticut, and Camp Hazen YMCA in Chester, Connecticut, provided field facilities.

Manuscript managers, who checked and kept track of contributed chapters over many iterations were essential to our progress: Cynthia Gallo, Ginat Gutierrez, Mark Jen, Luella Thompson, Stacy Ritter, and Ted Wong. Joseph Donnette, C. Gallo, Isaiah Laderman, and Jacob Laderman lent their excellent computer skills to text and table preparation.

It has been a pleasure to work with Rachel Donnette, the editorial coordinator for the volume. Her meticulous review and editorial advice were invaluable to the progress and organization of this book; her professional acumen was crucial to its realization. Many participating authors have also expressed their gratitude for Ms. Donnette's precise and patient coordination of our efforts. Without her collaboration, it is doubtful that the volume would have been completed.

Financial and in-kind support for preparation of this work was provided by the Connecticut Forest and Park Association (Farnum Fund), Conservation and Research Foundation, A.W. Mellon Foundation, Oxford University Press, Permanone, Weyerhaeuser Company, and Yale University's Peabody Museum and School of Forestry, with other sources acknowledged in various chapters.

Woods Hole, Massachusetts   A.D.L.
October 1995

Contents

PART II. SYSTEMS WITH DIVERSE DOMINANTS
Forests not dominated by *Chamaecyparis* species

Contributors

This list includes authors of papers, abstracts, and posters presented at the preparatory symposia, chairs of sessions, and panelists at workshops, as well as those whose chapters appear in this book. This volume, and particularly the first and last chapters, is drawn from all these sources.

Henry W. Art
Biology Dept. and Center for
   Environmental Studies
Williams College
Williamstown, MA 01267 USA

Reed S. Beaman
A.F. Clewell, Inc.
Quincy, FL 32351 USA

Alice Belling
Office of Natural Resources
   New York City Department of
   Environmental Protection
New York, NY 11368 USA

Thomas W. Birch
USDA Forest Service
Northeastern Forest Experiment Station
Radnor, PA 19087 USA

Bernard T. Bormann
USDA Forest Service
Forestry Sciences Laboratory
Corvallis, OR 97331 USA

Hans B. Büchel
Botanical Institute
University of Cologne
D-50923 Köln, Germany

Marilyn A. Buford
USDA Forest Service
Southeastern Forest Experiment Station
Research Triangle Park, NC 27709 USA

Andre F. Clewell
A.F. Clewell, Inc.
Quincy, FL 32351 USA

William H. Conner
Baruch Forest Science Institute
Clemson University
Georgetown, SC 24592 USA

John W. Day Jr.
Coastal Ecology Institute
Louisiana State University
Baton Rouge, LA 70803 USA

Rachel L. Donnette
Olympia, WA 98502 USA

Glen Dunsworth
Sustainable Forestry Division
MacMillan Bloedel Ltd.
Nanaimo
B.C., Canada V9R 5H9

Rafael Durán García
Centro de Investigación Científica de
    Yucatán, A.C.
Cordemex, C.P. 97310
Mérida, Yucatán, México

Robert Eckert
Department of Natural Resources
University of New Hampshire
Durham, NH 03824 USA

Hiroko Fujita
Botanic Garden, Faculty of Agriculture
Hokkaido University
Sapporo, Japan 060

William E. Gardner
North Carolina State University
Raleigh, NC 27695 USA

Ralph E. Good
Division of Pinelands Research
Institute of Marine and Coastal Sciences
Rutgers University
New Brunswick, NJ 08903 USA

Mel Greenup
USDA Forest Service
Siskiyou National Forest
Grants Pass, OR 97526 USA

Wolfgang Grosse
Botanical Institute
University of Cologne
D-50923 Köln, Germany

E.M. Hansen
Botany and Plant Pathology
Oregon State University
Corvallis, OR 97333 USA

Mark H. Hansen
USDA Forest Service
North Central Forest Experiment Station
St. Paul, MN 55108 USA

Paul Hennon
USDA Forest Service
Region 10
Juneau, AK 99801 USA

Joseph H. Hughes
Weyerhaeuser Company
New Bern, NC 28560 USA

Mark Jen
US Environmental Protection Agency
Alaska Operations Office
Anchorage, AK 99513 USA

Peter A. Khomentovsky
Kamtchatka Institute of Ecology
Russian Academy of Sciences
Petropavlovsk-Kamtchatsky
683024 Russia

Aimlee D. Laderman
School of Forestry and Environmental
    Studies
Yale University
New Haven, CT 06511 USA
    and
Swamp Research Center
Woods Hole, MA 02543 USA

Sibylle Lattermann
Botanical Institute
University of Cologne
D-50923 Köln, Germany

Michael H. McClellan
USDA Forest Service
Forestry Sciences Lab
Juneau, AK 99802 USA

Lucinda McWeeney
Yale University Dept. of Anthropology
and School of Forestry and
Environmental Studies
New Haven, CT 06511 USA

William H. McWilliams
USDA Forest Service
Northeastern Forest Experiment Station
Radnor, PA 19087 USA

Glenn Motzkin
Forestry and Wildlife Management
University of Massachusetts
Amherst, MA 01003 USA

Keiichi Ohno
Dept. of Vegetation Science
Institute of Environmental Science and
Technology
Yokohama National University
Yokohama, 240 Japan

Ingrid C. Olmsted
Centro de Investigación Científica de
Yucatán, A.C.
Cordemex, C.P. 97310
Mérida, Yucatán, México

Robert Ornduff
Dept. of Integrative Biology
University of California, Berkeley
Berkeley, CA 94720 USA

William A. Patterson III
Forestry and Wildlife Management
University of Massachusetts
Amherst, MA 01003 USA

Anne D. Pernick
School of Forestry and Environmental
Studies
Yale University
New Haven, CT 06511 USA

Ronald W. Phillips
Extension Forestry, College of Forest
Resources
North Carolina State University
Raleigh, NC 27695 USA

John H. Russell
British Columbia Forest Service
Cowichan Lake Research Station
Mesachie Lake
B.C., Canada VOR 2NO

Charles Shaw
Forestry Science Laboratory
USDA Forest Service
Fort Collins, CO 80526 USA

Raymond M. Sheffield
USDA Forest Service
Southeastern Forest Experiment Station
Asheville, NC 28802 USA

Hisashi Shinsho
Kushiro City Museum
Kushiro, 085 Japan

Thomas Siccama
School of Forestry and Environmental
Studies
Yale University
New Haven, CT 06511 USA

Alison Specht
University of New England
Northern Rivers
Lismore NSW 2480 Australia

Robert R. Stephens
Pacific Lumber Co.
Scotia, CA 95565 USA

Dwight L. Stoltzfus
Dept. of Biological Sciences
Arkansas Tech University
Russellville, AR 72801 USA

Kenneth O. Summerville
North Carolina Division of Forest
    Resources
Clayton, NC 27520 USA

John B. Tansey
USDA Forest Service
Southeastern Forest Experiment Station
Asheville, NC 28802 USA

Fred M. White
North Carolina Forest Service
Raleigh, NC 27611 USA

Claire G. Williams
Department of Forest Science
Texas A and M University
College Station, TX 77843 USA

Shin-Ichi Yamamoto
Department of Environment and
    Resources Management
Faculty of Agriculture
University of Okayama
Okayama 700 Japan

Donald Zobel
Department of Botany & Plant
    Pathology
Oregon State University
Corvallis, OR 97331 USA

COASTALLY
RESTRICTED
FORESTS

Aimlee D. Laderman

Freshwater Forests of Continental Margins

Overview and Synthesis

A few species of forest trees are found in nature only in narrow ribbons bordering an ocean's edge. The trees and the ecosystems they inhabit present many intriguing puzzles. Despite their maritime location, these tree species are unable to survive in saline waters and soils. What do the trees have in common? What characteristics enable them to succeed in such challenging habitats? Why don't these species occur inland naturally? Our aim here is to begin to answer these questions by establishing some correlations between the facets of the environment and the characteristics of the trees. Then, testable hypotheses can be developed in an attempt to relate cause and effect.

In situations where other species do not readily colonize or thrive, the success of particular trees indicates unusual morphological, physiological, or ecological adaptations. These properties are not just of intrinsic interest to the inquiring mind. The remaining coastal forests must be preserved because they also may harbor great practical value in medicine, agriculture, water resources, and other areas as yet unforeseen.

Recognition, understanding, and respect are essential for the protection and wise management of any ecosystem. The gathering and assessment of existing data is a vital precursor and stimulant to relevant research. By identifying common as well as unique features of coastally restricted species, their ecology and properties may be better understood, and their inherent values better appreciated. This volume assembles information on widely dispersed coastal systems to achieve these ends.

There are rarely sharp, immutable boundaries in nature. The shorelines of continents and islands move, climates shift in degree and extent, leaving relict populations, outliers that belie the most carefully assembled generalizations. Before we can be sure that isolated communities are indeed not "typical," they must be recognized as belonging to an ecosystem type. Each example must be observed and described, its many facets and parameters compared to reach an understanding of the general ecosystem type. From such study, we hope to gain insight into the primary factors shaping the system. Once the driving forces are perceived, it is often necessary to go back and take measurements that were omitted earlier. Only after this laborious process— a zig-zag process of isolation and synthesis, isolation and resynthesis—does a useful, appropriate definition and delineation of an ecosystem type emerge.

Any generalization about natural systems is primarily a construct, a model useful in gaining further understanding. It must be reformulated as patterns emerge and more relevant data are accumulated. Accordingly, the boundaries selected to delimit and include forests as "coastally restricted" were chosen as a first approximation. This approximation determined what was to be included in this volume.

Contents of the Book

The ecosystems dealt with here are forest types found only within 250 km of a marine shore in the Northern Hemisphere (Fig. 1.1 and Tables 1.1 and 1.2). They are in the temperate zone, with the exception of a group in the Yucatán Peninsula, Mexico, a few degrees south of the Tropic of Cancer. In Asia, these ecosystems are native to five countries bordering the Sea of Okhotsk and the Sea of Japan: Russia, Japan, the Republic of Taiwan, South Korea, and a small region in China. In North America, the forests border the Pacific Ocean from Alaska (USA) to the northern shore of Mexico, and the Atlantic Ocean from Maine to Florida (USA) and the Mexican Yucatán.

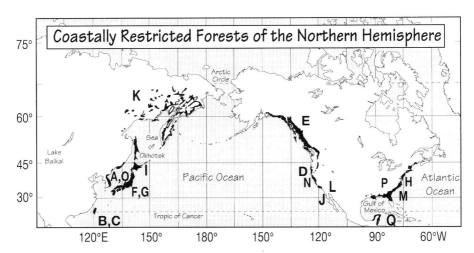

Figure 1.1. Black ink outlines the regions within which the coastal species dominate. Regional maps showing the distribution of each species are placed with the relevant text. A: *Alnus japonica.* B: *Chamaecyparis formosensis.* C: *C. taiwanensis.* D: *C. lawsoniana.* E: *C. nootkatensis.* F: *C. obtusa.* G: *C. pisifera.* H: *C. thyoides.* I: *Picea glehnii.* J: *Pinus muricata.* K: *P. pumila.* L: *P. radiata.* M: *P. serotina.* N: *Sequoia sempervirens.* O: *Thuja orientalis.* P: *Taxodium distichum/Nyssa aquatica.* Q: Yucatan species. (Map by John Cook and Fritz Heide.)

Box 1.1 Note on Tropical, Broad-leaved, and Deciduous Systems.

Three coastal freshwater-based systems are included in this volume but excluded from comparisons in this chapter because their physical and biological properties significantly differ from the others:

1. Tropical Mexican Yucatán forests (Olmsted and Durán García)[1];
2. Broad-leaved, deciduous *Alnus japonica* (Fujita); and
3. Distinctive deciduous forests co-dominated by coniferous *Taxodium* and broad-leaved *Nyssa* (Conner and Day, McWilliams et al.).

Another group of coastal trees, the mangroves and their associates, is adapted to brackish water. Their geographic confinement has an inherent logic, and the literature about them is extensive. Mangroves will not be dealt with in this volume, for their adaptations and problems are quite different from those of freshwater species.

Three chapters deal with general coastal and wetland phenomena rather than specific species or regions. They address the effects of sea-level rise (Conner and Day), the impact of hurricanes (Conner), and the ecophysiological implications of wetland tree root aeration (Grosse et al.).[2]

Information available on different species and regions varies widely. This uneven distribution is reflected in the discussion and presentation of data. The treatments are also weighted by the varied foci and degrees of attention historically accorded each tree. Trees with the greatest commercial value generally have the most detailed and diverse data sets, with substantial bodies of information on provenance and management. For example, the small amount of accumulated genetic data is concentrated on the three *Chamaecyparis* species most highly prized as timber: *C. obtusa* (hinoki cypress), *C. nootkatensis* (Alaska yellow-cedar), and *C. thyoides* (Atlantic white cedar) (see Zobel, Russell, and Eckert, respectively).

This chapter is divided into three sections. Section 1 is a comparison of the physical and biological properties of the species studied. Section 2 is a comparison of coastal forest environments, including implications for species selection. Section 3 is a synthesis of the species-environment overviews. Comparisons here concentrate on the nondeciduous needled conifers of the North Temperate zone. (See Box 1.1, a note on tropical, broad-leaved, and deciduous systems.)

PROPERTIES OF THE DOMINANT TREES

In this section, we examine and compare the morphological, physiological, and ecological properties of coastal tree species. Separating the properties common to all the coastally restricted species from those that they do not share allows the emergence of the key adaptations that restrict them—or fit them—to marine shores. To ascertain the key adaptations that determine their ecological perimeters, we examine the distri-

Table 1.1 Coastally Restricted Forests: Site Descriptions

Species	Country	Site Description [a]	Information Sources [b]
Alnus japonica Japanese alder	China, Japan, Korea, Russia	Swamps, water edges, peatlands, low moors, fog zone	Fujita; Ohno (pers. comm.)
Chamaecyparis formosensis Taiwan cypress	Taiwan	Montane, moderately moist coniferous forests (often pure stands, in clouds)	Li 1963; Zobel
Chamaecyparis taiwanensis Taiwan red cypress	Taiwan	Higher montane, well-drained coniferous forests (often pure stands, in clouds)	Li 1963; Zobel
Chamaecyparis lawsoniana Port-Orford-cedar	USA (W)	Dry and wet, often poor, soils, sand dunes, stream margins, bogs; fog common	FNA 1993; Greenup; Zobel
Chamaecyparis nootkatensis Yellow-cypress	Canada (W), USA (W)	Poor soils, bogs, disturbed sites; subalpine (S) to sea level (Alaska); high humidity	Dunsworth; FNA 1993; Hennon; Russell; Zobel
Chamaecyparis obtusa Hinoki cypress	Japan	Xeric sites, middle to upper slopes, ridges; soil: podzol, serpentine; warm temperate to subalpine	Yamamoto; Zobel
Chamaecyparis pisifera Sawara cypress	Japan	Wet or mesic sites, lower montane slopes, rocky depressions; thin, dry soils on ridges	Yamamoto; Zobel
Chamaecyparis thyoides Atlantic white cedar	USA (E)	Low-elevation acid bogs, fluctuating high water table; peat over sand soils	Eckert; Laderman 1989; Stoltzfus
Picea glehnii Sakhalin spruce	Japan (NE), Kurile Is. (S), Sakhalin Is. (S)	Boreal, subalpine, raised bogs, transitional moor margins; cool summer fogs	Fujita (pers. comm.); Fujita et al. 1994; Nakamura 1988; Nakamura et al. 1994; Ohno (pers. comm.); Shinso (pers. comm.)

Pinus muricata Bishop pine	Mexico (W), USA (W)	Dry ridges to coastal windshorn forests; steep slopes in fog belt; in or near bogs	FNA 1993; Libby, pers. comm. 1994, 1995; Libby et al. 1968; Perry 1991
Pinus pumila Dwarf stone pine	China, Japan, Korea, Kurile Is, Russia, Sakhalin Is.	Severe sites, subalpine to subarctic, poor soils, colonizes dunes, volcanic ash; montane to sea level, tundra	Khomentovsky
Pinus radiata Monterey pine	Mexico (W), USA (W)	Rocky or volcanic ridges and slopes; fog belt	FNA 1993; Libby, pers. comm. 1994; Libby et al. 1968; Perry 1991
Pinus serotina Pond pine	USA (E)	Flatwoods, bogs, savannahs, barrens, pocosins; high water table	Bramlett 1990; FNA 1993
Sequoia sempervirens Coast redwood	USA (W)	Canyon slopes, river flats, not waterlogged (deep soils); prolonged summer fog	FNA 1993; Ornduff
Thuja orientalis Oriental arbor-vitae	China, Iran?, Japan, S. Korea (E)	Poorly drained rock gaps in mountains	Bailey & Bailey 1947; Lee 1980; Suzuki, pers. comm. 1992
Taxodium distichum/Nyssa aquatica Bald cypress/water tupelo	USA (SE)	Brownwater rivers, lake margins, alluvial swamps; soil saturated to 60 cm standing water	Conner; FNA 1993; McWilliams

? = Information contradictory.

[a]*Site description; distribution.* All entries refer to the natural condition in the native range, not to protected or horticultural situations. Most of the species will grow in a broad range of conditions if protected from competition.

[b]*Information sources.* Citations without dates refer to chapters in this volume and to personal communications from their authors. Climatic data was also obtained from Great Britain Meteorological Office (1980) and NOAA (1992).

Table 1.2 Coastally Restricted Forests: Distribution[a]

Species	°N. Latitude Min–Max	°N. Latitude Total Range	Altitude (m) Min–Max	Altitude (m) Total Range	Distance from Coast (km) Min–Max	Restricted to Coast? (< 250 km)	Discontinuous Distribution[b]
A. japonica	30° – 46°	16°	0–1400	1400	0.1–100 avg. 50	Y	?
C. formosensis	22° 28' – 24° 49'	2° 21'	850–2900	2050	8–75	Y	Y
C. taiwanensis	22° 55' – 24° 46'	1° 51'	1200–2800	1600	12–75	Y	Y
C. lawsoniana	40° 50' – 43° 35'	2° 45'	0–1950	1950	0–209	Y	Y
C. nootkatensis	41° 25' – 61° 07'	19° 42'	AK: 0–1000 BC: 0–1600 OR, WA: 600–2300 (2500 rare)	1000 1600 1700	0–200 Outliers 425	Y	Y
C. obtusa	30° 15' – 37° 10'	6° 55'	10–2200	2190	0–150	Y	Y
C. pisifera	32° 48' – 39° 32'	6° 44'	110–2590	2480	0–150	Y	Y
C. thyoides	29° 14' – 44° 20'	15° 6'	0–457 avg. < 50	457	0–200 (250) avg. < 50	Y	Y
P. glehnii	42° 28' – 46° 48' [Outlier: 39° 31' Honshu]	4° 20'	5–1800	1795	0.3–100	Y	Y
P. muricata	31° 20' – 41° 05'	9° 45'	USA: 0–300 Mexico: 30–50	300 20	0–20 avg. < 10	Y	Y
P. pumila	35° – 71°	36°	5 to > 3100	3100	Close and along mountains to Lake Baikal	N (extends to Lake Baikal)	Y
P. radiata	Mainland: 35° 30' – 37° Island: 28° 15' – 29°	1° 30' 45'	M: 0–400 I: 275–1200	M: 0–400 I: 925	0–10 avg. 2	Y	Y
P. serotina	27° – 38°	11°	0–200	200	0–275 avg. 100	Y	Y
S. sempervirens	35° 41' – 42° 09'	6° 28'	0–1000 avg. < 300	1000	0.2–72	Y	Y
T. orientalis			700–1400	700			Y
T. distichum/ N. aquatica	29° – 39°	10°	avg. < 30	< 30	1–250 (0–1000 km along the Mississippi River)	Y (except along Mississippi River)	N

Notes: Y = Yes. N = No. Avg. = Average. Blank = Information not available.
[a] See Table 1.1. [b] Discontinuous distribution: "Y" indicates that the species is along a few, more separated sites; "N" indicates continuous.

Box 1.2 Properties Common to Coastally Restricted Nondeciduous Trees[a]

1. Native only to within 250 km of a marine coast (by definition)
2. Sensitivity to saline groundwater and salt-laden winds
3. High aerial moisture requirement
4. Broad-niche potential (wide ecological amplitude; capacity to thrive in a great range of environments)
5. Poor competitive ability
   (a) outside the favored niche
   (b) in favorable or "easy" environments
   (c) in much of its potential range
6. Become dominants in special circumstances wherein they are better adapted to one or more restrictive factors
7. Discontinuous distribution and therefore contain the genetic and evolutionary effects found in ecological islands
   (a) founder effect
   (b) high susceptibility to change in the environment
   (c) greater variation between stands than from tree to tree within stands, due to reproductive isolation
8. Often, but not exclusively, monotypic stands
9. Do best in protected sites; do not thrive in exposed windy sites
10. Moisture-conserving features (sclerophylly)
    (a) waxy cuticle
    (b) needles
    (c) thick bark
11. Bark deeply fissured
12. Those in saturated soil have shallow roots, no tap root
13. Thrive in protected plantations or amenity plantings far from the natural range and under very different conditions
14. Wind pollinated
15. Seed transported by wind
16. Paleoendemic

[a]These properties refer to temperate zone trees in the northern hemisphere. Coastal deciduous species and non-evergreens in the temperate zone *(A. japonica, T. distichum, and N. aquatica)* and southern hemisphere (Weigand et al. 1992) and tropical trees (Olmsted and Durán García) differ significantly in many properties from those compared here.

---

bution, longevity, reproduction and growth habit, morphology, ecological strategies, and evolutionary patterns of coastally restricted trees.

Characteristics common to all coastally restricted nondeciduous conifers are listed in Box 1.2. These adaptations are most probably essential to success in coastal freshwater forests, or are caused by coastal conditions. Characteristics common to most of the species are listed in Box 1.3. These are probably adaptations closely related to, but not necessarily essential for, the coastal situation. Characteristics in which the

---

Box 1.3 Properties Common to Most Coastally Restricted Trees[a,b]

1. Stomata small, with thick "lips"
2. Needles very small
3. "Drip-tip" features
4. Distinct assemblage in each type of environment
5. Possess secondary chemicals and therefore
   (a) resist decay
   (b) one or more parts toxic to some animal species
6. High seed production
7. Mycorrhizal fungi

[a] See footnote Box 1.2.
[b] For properties common to all coastally restricted nondeciduous trees, see Box 1.2.

---

species vary widely are listed in Box 1.4; these are probably not closely related to the coastal environment.

Biogeography

All the coniferous coastal trees studied here have a markedly discontinuous distribution; they are absent in many apparently suitable habitats situated between extant stands.

Many of these species are distributed in a paleoendemic pattern indicating that their environments have changed greatly, and/or the population units have been separated for a long period. Using detailed comparative morphological analysis, Edwards (1992) determined that the closest living relative of *C. nootkatensis,* a native of western North America, is *C. obtusa* in Japan. The closest relatives of Atlantic coastal *C. thyoides* grow thousands of kilometers away near the Pacific shore. Some characteristics of the southern variant, *C. thyoides* var. *henryae,* appear to be more similar to those of west coast *Chamaecyparis* species than to *C. thyoides* itself (Angus Gholson, personal communication 1980; Clewell and Ward 1987; see "Evolution" later in this

---

Box 1.4 Properties Not Shared by Most Coastally Restricted Trees[a]

1. Longevity
2. Pace of initial seedling growth
3. Shade tolerance of seedlings and saplings
4. Capacity for vegetative reproduction
5. Height, girth at maturity
6. Epiphytes
7. One dominant family in associated species

[a] See footnote Box 1.2.

---

chapter for further discussion). Plate tectonics, paleobotany, and genetics may provide clues to the significance of the great distances between related species.

Longevity, Reproduction, and Growth

The maximum age of the trees varies widely, from the relatively short-lived *C. thyoides* with a maximum lifespan of less than 300 years (Clewell and Ward 1987), to *Taxodium distichum* and *Sequoia sempervirens* with over 1,000 and 1,200 years, respectively (Flora North America [FNA] 1993), to the Methusalahs of the forest, *C. lawsoniana* and *C. nootkatensis,* which may live longer than 3,000 years (Zobel).

All but the broad-leaved *Nyssa* are monoecious (possess both male and female flowers on a single tree). All the species are wind-pollinated; the deciduous *Taxodium* and *Nyssa* are insect-pollinated as well. Seeds of all the species are transported by wind; those of approximately one third of the trees are also water-borne. *Pinus muricata* and *P. pumila* seed are reported to be carried by animals as well.

The dependence on vegetative reproduction in nature varies widely, from the recumbent *P. pumila,* which regularly produces intertwined coppices (Khomentovsky), to *C. thyoides,* which rarely resprouts even after being repeatedly browsed.

Initial seedling growth also varies widely from species to species and generally responds to edaphic (soil), nutrient, and surface conditions, the water and light regimen, weather, and competition from other trees and shrubs. There is no discernable common pattern among coastal species' responses to each of these factors. However, all the trees seem capable of colonizing open sites.

The size at maturity varies among these species as greatly as within any other group, ranging from the prostrate *P. pumila* with a diminutive maximum height of six meters (Khomentovsky) to *S. sempervirens'* average 61 m and a towering maximum of 115 m (Ornduff).

Morphology: Leaves, Drip-tip, and Bark

Almost all the 17 coastal tree species examined have narrow evergreen needles. Of the 17, only *Alnus japonica, Nyssa aquatica,* and *T. distichum* are deciduous; only *A. japonica* and *N. aquatica* are broad-leaved; and only one, *Alnus,* is neither needle-leaved nor evergreen. The evergreen conifer needles have waxy surfaces with sunken stomata, and the cones are woody or fleshy.

All the evergreens have a "drip-tip" feature, which is more pronounced in some genera than in others. The needle, branchlet, and stem arrangements promote collection, condensation, and retention of ambient moisture, allowing for maximal absorption of both fluid and nutrient along the stems. Water gathers at the outer branch tips and drips off, providing moisture to the root zone even when there is no precipitation.

Fissured or furrowed thick bark is another feature common to almost all the species (the exception being the deciduous broad-leaved *Nyssa*). The crenelation slows and channels moisture, maximizing absorption during throughflow. Thick, insulating bark also protects mature trees from the heat of brief fires.

This cluster of morphological aspects presents a picture of species well adapted to conserving both nutrients and water.

Dominance and Competition

The canopy is often codominated by other conifers in the higher-latitude coastally restricted forests, but at lower latitudes (in the southern United States), deciduous and coniferous species often share the canopy. Conifer dominance may be correlated with temperature, soils, and other variables not dependent on a marine coast. The shrub layer also tends to be dominated by plants with water-conserving features.

The success of coastally restricted trees is probably dictated by their interactions with other regionally dominant plants. Conditions that stress other regionally dominant trees appear to favor the coastally restricted species. Thus, coastally restricted trees are poor competitors in favorable or easy environments, but in special circumstances where they are better adapted to one or more restrictive factors, they become the dominants. Zobel reviewed and analyzed this concept for the genus *Chamaecyparis*. Specific adaptations are discussed below.

Ecological Strategies and Properties

Mutualism and properties such as evergreenness, sclerophylly, and the synthesis of secondary chemicals (plant products not necessary to basic metabolism), are trophic strategies of plants living in nutrient-deficient environments (Specht 1979). Other hallmarks of dominant species in stressful habitats are expressed in their competitive capacity, their specialist/generalist qualities, and their role as ecological determinants. Their existence in ecological islands and remnant patches strongly influences the rarity of both the dominants and their associated species.

Mutualism

The majority of coastally restricted tree species are known to harbor either mycorrhizae (fungi intertwined in their roots) or epiphytes (plants growing upon them). Mycorrhizal fungi transfer nutrients to their plant hosts, as, for instance, noted by Molina (1994) in the redwood forests. Such a handy cost-free nutrient supply is of obvious benefit, particularly in nutrient-poor situations. Epiphytes (plants physically supported by, but not metabolically interdependent with, other vegetation), which are often abundant in moist forests, may also benefit their hosts. For example, it has been found that *Lobaria organa,* a lichen of the old-growth temperate coastal rainforest canopy, contributes up to 50% of nitrogen added to the rainforest soil (Pike et al. 1977, Denison, personal communication 1995).

Evergreenness

As previously noted, the majority of the coastally restricted trees studied are evergreens and are often (but not exclusively) found in infertile soils. It is thought that evergreenness may be an effective adaptation to nutrient deficiency (Monk 1966, Small 1972, reviewed in Chabot and Hicks 1982, and Jonasson 1989). The reasoning is that when all leaves are not lost each year, each molecule of photosynthesized carbon is retained longer, so that nutrients are more efficiently used (Specht 1979). The major contributing factors are thought to be the long life span of each leaf, the reabsorption of leaf nutrients from aging leaves, and the translocation (transfer) of nutrients directly from aging to newly expanding young leaves (Shaver 1981, Jonasson and Chapin 1985).

The evergreen-nutrient correlation is further strengthened by the fact that *Taxodium*, the only deciduous conifer studied, has a generally rich substrate, whereas all but one of the evergreen species grow predominantly in poor soils. In a much-quoted study, Small (1972) found that evergreen species reabsorbed a higher percentage of leaf nutrients than deciduous species in the same environment. Jonasson (1989) observed that abundant nutrients favored fast-growing deciduous plants; a moderately reduced supply favored evergreens with short-lived leaves; and severe nutrient deprivation favored long-lived evergreen leaves.

The picture is extremely complex, for the rates and ratios of nutrient removal and translocation vary from species to species (Chapin 1980, Schulze and Chapin 1987, Chapin and Shaver 1989, and Jonasson 1989). Further variation in plant biochemistry is observed when the same species grows in different environments (Whigham and Richardson 1988). Research is required to clarify the correlation between soil nutrient availability, the evergreen habit, and coastal restriction.

Sclerophylly: A Paradox

Vegetation of dry regions is often evergreen and is typically sclerophyllous, having thick, leathery, tough leaves with recessed stomata. However, these qualities are also characteristic of coastally restricted trees. Sclerophylly prevents desiccation, providing a great advantage in desert-like environments, but it is a very strange capacity for plants living in areas of high humidity. Why would plants possess water-conserving strategies in water-logged, foggy, or rainy environments? Are these actually techniques for maintaining a physiological balance of salts and water in an environment with frequently salt-laden air? Or for conserving nutrients in oligotrophic situations? These are expensive strategies that may partially explain the poor competitive ability of coastally restricted species outside their zones of greatest adaptation.

The apparent anomaly may be more readily understood if one adopts the assumption suggested repeatedly in *Forested Wetlands* (Lugo et al. 1990a): In stressful sites, it is not the average situation, but the rare, erratically occurring catastrophe that determines the vegetational profile.

The paradox of water-conserving properties in wetland plants has intrigued scientists for decades. Ironically, an internal water debt may be created by one property that allows many species to survive in a low-nutrient environment: evergreenness. Evergreens continue to photosynthesize and respire at low temperatures. At below-freezing temperatures water uptake is low or non-existent, but transpiration water is still lost due to photosynthesis. Hence, water that is not replaced is lost by evergreens in cold weather. This does not, however, explain sclerophylly in frost-free areas (Brown 1990, Lugo et al. 1990b).

Physiological drought may also be induced by a peat substrate, which, even when wet, does not readily release water. For example, plants that do not wilt in sand until saturation is reduced to between 1.5 and 7.8% will wilt in peat at 50% saturation (Lugo et al. 1990b). Water conservation may also be of advantage when there are toxic levels of elements such as humic or other acids (as in peat and podzol), or aluminum (as in serpentine soil and volcanic ash).

The root pattern of trees growing in a normally high-moisture environment may be another key to the utility of conservation strategies. Such trees would be expected to be shallow-rooted for two reasons: First, in saturated soils with no water flow, anoxia and chemically toxic conditions develop, killing deeper roots. Second, in un-saturated soil, the availability of abundant moisture in the air generally encourages surface root development. During the rare but recurring drought periods, these plants adopt a strategy of tolerance. They cannot adapt by allocating sufficient energy for rapid root growth when water tables drop below the shallow rhizosphere (root zone); their water-conserving properties allow them to tolerate the stress.

These characteristics and processes are significant at two levels. At the species level, water stress that is catastrophic for their competitors allows survival of adapted species. At the ecosystem level, water loss is minimized during drought.

Competitive Capacity and Niche

All coastally restricted tree species normally fill only a limited subset of the environments in which they can grow. They have the capacity to thrive in a variety of situations (i.e., they are all broad-niche forms), but they do not compete well in much of their potential range.

Zobel observed that, in nature, all the *Chamaecyparis* species occur in unusual forest communities, often limited to particular topo-edaphic conditions within a range of climatically controlled regional forest types. Although the evidence is incomplete, this appears to be true of other coastal forests as well: for *A. japonica* (Fujita), *P. glehnii* (Fujita et al. 1994; Fujita, personal communication 1993; Shinsho (personal communication 1993); *P. muricata* and *P. radiata* (Libby, personal communication 1994), *P. pumila* (Khomentovsky), *S. sempervirens* (Ornduff), and *Taxodium-Nyssa* forests (McWilliams et al.).

Specialists and Generalists

Generalists (called r-selected species) are defined as forms that are phenotypically plastic or adaptable: they can change to fit different environments. For example, *C. nootkatensis* does not form shoot buds (i.e., it has indeterminate growth) (Russell). Growth ceases if there is a severe drought or unusually cold weather during the growing season, and growth resumes without damage when warmer or wetter conditions return. In other words, it is a generalist for these traits. It has been shown that forms with a preponderance of r-selected traits tend to resist extinction (Laurance 1991).

Species that are good dispersers (i.e., are r-selected for dispersal ability) may be good survivors, because the demographic and genetic contributions of immigrants can bolster small populations and provide a buffer against extinction (Laurance 1991). All the coastal trees studied are poor dispersers and lack this buffer.

Specialists (called K-selected species) are forms with a narrow capacity to use habitats or resources. An example is again *C. nootkatensis*, which has many subpopulations for seedling morphology and physiology, with each subpopulation being classed as a specialist in these traits (Russell). Laurance (1991) theorizes that a preponderance of K-selected, specialist traits are significant predictors of extinction-proneness. All the coastal trees have many specialist traits.

It appears that many coastally restricted species may follow the pattern of *C. nootkatensis,* in that it is a specialist in some characteristics and a generalist in others. Russell (chapter 5 and 1993) examines this problem in detail.

Few researchers have systematically explored the ecological implications of the genetics of coastal forest trees. Pioneers in this area include Eckert, Edwards (1992), Russell, and Zobel for all *Chamaecyparis* genera; Libby et al. (1968) for *P. muricata, P. radiata,* and related closed-cone pines; and Brunsfeld et al. (1994) for the families Taxodiaceae and Cupressaceae.

Remnant Patches, Rarity, and Extinction

Although the tree species here are all dominants or codominants (by definition locally abundant), they are in great danger of extinction because their habitats are so specialized and distributed as islands over the landscape. Tilman et al. (1994) developed a model describing how even moderate habitat destruction of patches can lead to the extinction of the dominant competitor. As habitat loss increases, they predict that associated species will be lost, with the best competitors going first. Furthermore, they expect that the more fragmented the ecological islands, the greater the extinctions that additional destruction will cause. This phenomenon first affects those best adapted to the specialized environment, the endemics, those that tend to be rare outside that habitat. The poor dispersal ability exhibited by many coastally restricted species exacerbates the situation.

The extinction is a delayed effect, particularly among species with a long life span. Tilman et al. (1994) describe this as an extinction debt, a future ecological cost of current habitat destruction. As these authors suggest for all systems, field work is required to determine how the extinction debt model applies to coastally restricted systems, and to see the spatial and temporal scales to which it applies.

Habitat-restricted species are almost by definition rare or absent in most environments, although they may be abundant in their native sites. From a conservation viewpoint, the existence of coastally restricted forests in ecological islands is highly significant. The preservation of "sample" or representative stands will not save the taxonomic or genetic diversity of these systems.

Ecological Determinants

Many of the coastally restricted trees appear to create conditions that dictate, dominate, or determine their associates by shaping their environment. The associated species, not surprisingly, frequently possess many of the same ecological characteristics as the dominant (Box 1.2, 1.3): wide ecological amplitude, acid tolerance, and water-conserving features. Although many of these "constant companions" (in ecosociological parlance) are competent competitors in a broad range of situations, others, like their determinant tree, do not compete well outside their native habitat.

Secondary Chemicals

The restrictive nature of many coastal sites appears to encourage the production of secondary chemicals, which are plant substances not essential to self-maintenance. All conifers contain resins, secondary chemicals occurring in canals in the bark and elsewhere (Prance and Prance 1993). About half the coastally restricted trees are known to resist decay, which may also reflect the production of secondary chemicals. Some conifer needles also harbor symbiotic endophytic fungi that produce insect-repelling alkaloids (Carroll 1988, 1991; Barbosa et al. 1991). Such chemicals are often toxic and possess pharmaceutical or antibiotic properties that may be useful in medicine, agriculture, and industry.

Evolution: Paleoendemism and Neoendemism

It is postulated that plants growing in a restrictive environment probably have evolved in one of two ways (Stebbins 1942, Terrill 1951, and Brooks 1987), resulting in two population types.

The paleoendemic, relict, or depleted population type is the remnant of species that were once highly successful. These plants currently have a widespread but disjunct distribution and are confined to isolated stands. As the environment changed, paleoendemics failed to adapt. Descendants of the original species were restricted to progressively smaller, more remote populations. Their closest relatives may now be geographically far removed. The greater the environmental change, and the longer the population units remain separated, the more probable it is that they will be separated by greater distances. At the same time, the subpopulations become more distinct.

The other type (termed a neoendemic island population) was derived relatively recently from plants growing in neighboring, more moderate habitats. When some seeds of the moderate-habitat plants were able to germinate and survive in the harsh environment, their "tolerant" genes were progressively selected for, and "intolerant" genes were weeded out of the population. The restrictive environment forced rapid evolution, ultimately producing a new biotype capable of thriving under harsh conditions. Their closest relatives are generally found nearby. The speed with which the biotype may evolve is related to the generation time of the plant: some herbaceous species can evolve quite rapidly under harsh conditions. For example, new grass strains tolerant to mine wastes can develop within 30 years (Antonovics et al. 1971). Tolerant and intolerant biotypes can coexist in the moderate environment, with the neoendemic plants superficially bearing a close resemblance to their relatives lacking the capacity to thrive in the harsher habitat.

The evidence presented by the contributors to this volume seems to indicate that the evergreen coastally restricted trees are paleoendemics.

Catastrophe Dependence

Dependence on catastrophic events appears to be an opportunistic strategy for the use of limited resources. Four types of catastrophes encountered in coastally restricted forests are discussed below: fire, harvest, hurricane, and drought.

Fire

Organisms use two basic strategies to successfully survive stress: avoidance and accommodation. Both strategies may be illustrated by the mature *S. sempervirens* (Ornduff) and *C. lawsoniana* (Greenup, Zobel). Their thick fire-resistant bark can "avoid" and hence survive most forest fires. Their seed "accommodates" by sprouting vigorously in the newly cleared gaps in the canopy. Another example of accommodation: *C. thyoides* is killed by hot fires, but only about one third of any year's seed crop germinates each year. The rest of the seed, protected in the forest floor, remains to opportunistically germinate after a devastating fire. The seed bank benefits from the fertilized, cleared, sunlit seedbed created by the conflagration. *Pinus serotina* uses a combination of strategies: it is not only highly fire-resistant, but it goes beyond accommodation to take advantage of, and actually require, fire to release seed from its fire-resistant cones. This trait enables the species to colonize whenever a fire clears the seedbed: it has a fresh store of seed ready for planting, released by the agent that destroyed its competitors.

Harvest

Under certain circumstances, clearcut harvest may have an effect similar to a hot fire for some species. If the forest floor is left in a state that encourages seed germination, or if conditions promote stump-sprouting for those species capable of it, pure stands can regenerate. Of course, restoration of biodiversity depends on many factors and is often impossible to replicate. In old-growth forests, the complex interactive living web that requires decades or centuries to develop will be destroyed even when an apparently healthy stand replaces the cut forest.

Hurricane

As with other catastrophic disturbances, hurricanes present opportunities as well as problems. The effect on biodiversity and productivity can appear more like renewal than disaster for species capable of adapting in some way to the rare, erratically occurring major storm. Conner discusses what little is known of this process and describes research efforts under way.

Drought

In riverine forests, or those well supplied by adequate flow, water is seldom limiting due to the continual inflow to the root zone. Therefore, there is little selection for water conservation (Lugo et al. 1990b). Isolated basin wetlands, which have no inflow or outflow, however, rely primarily on precipitation for their water. They are therefore dependent on the vagaries of the weather. Water stress may not occur often, but it is the extreme condition, not the average, that determines the plant community. The minimal data available on transpiration rates of forested wetlands indicate that transpiration is lower in basin wetlands than in riverine wetlands even when standing water is present (Brown 1981). This supports the hypothesis that where water stress is a recurring (but possibly rare) challenge, the trees that dominate such sites are those that have been selected to survive that challenge.

ENVIRONMENTAL DIVERSITY

This section examines the environmental parameters of coastal forest habitats. The environments are compared, and the implications of common factors are explored. In segregating the habitat qualities that are common to all the forests from those they do not share, the key environmental factors that select for coastal species should emerge. Characteristics of coastally restricted forests are summarized in Tables 1.1–1.6. Properties shared by all the ecosystems are listed in Box 1.5; properties that vary among the systems are listed in Box 1.6.

Biogeography

Although all the systems considered here are native only to a narrow coastal band (Table 1.1), it is evident that each dominant tree species occupies a wide range of environments (Table 1.2). Environmental range is expressed by some in north-south geographic extent and by others in altitude, soil type, nutrient status, or atmospheric temperature range. Some species exhibit an extreme degree of confinement (e.g., less than 1° in latitude or 30 m in altitude), while others cover a range hundreds of times as great.

Clearly, the native environments (the expressed niche breadths) vary greatly among the species of coastally restricted forests.

Range

The native ranges of coastally restricted forest species in the temperate zone are indicated in Tables 1.1 and 1.2 and Figure 1.1. Three geographic aspects are discussed here: latitude, altitude, and cultivated situations.

Chamaecyparis nootkatensis has the greatest breadth of latitudinal range among coastally restricted forest species, extending 19.4° from Alaska to northern Oregon (Dunsworth, Hennon). In contrast, *P. radiata* covers only 0.45° in its island distribu-

Box 1.5 Environmental Diversity: Common Factors—Characteristics of the native habitats of all coastally restricted forests[a]

1. Less than 250 km from marine coast (by definition)
2. Wide range of environmental types for each species
3. Includes but not restricted to stressful environments
4. Low-nutrient substrate is common, never mandatory
5. Salt- and nutrient-laden fogs and aerosols
6. Markedly discontinuous sites
7. Often limited to particular topoedaphic conditions within a range of climatically controlled regional forest types
8. Frequent fogs
9. High humidity
10. Irregularly recurring (or rare) major disturbances (drought, fire, hurricane, and clearcut)
11. Geographic range previously greater than at present

[a] See footnote Box 1.2.

tion, and 1° 30′ on the continent (Libby et al. 1968; Libby, personal communication 1994). The Taiwan species *C. formosensis* and *C. taiwanensis* (Li 1963, Zobel) and the North American *C. lawsoniana* (FNA 1993, Zobel) extend little more, between two and three degrees latitude. There is a nearly 20-fold difference between species with the greatest and smallest north-south range. *Pinus pumila* has the greatest vertical range, extending from sea level to over 3,000 m (Khomentovsky). The two Japanese *Chamaecyparis* species, *C. pisifera* and *C. obtusa,* follow with ranges of 2,480 and 2,190 m respectively (Yamamoto, Zobel). In contrast, the deciduous *T. distichum-*

Box 1.6 Environmental Diversity: Divergent Factors—Varying characteristics of the native habitats of coastally restricted forests[a]

1. Environment type (e.g., montane vs. sea level; flat vs. hilly; or saturated vs. mesic soil)
2. Expressed niche breadth of native habitat
   (a) latitudinal range
   (b) temperature range
   (c) altitudinal range
3. Soil type (e.g., sand, peat, podzol, or serpentine)
4. Soil hydration (dry, saturated, or submerged)
5. pH of standing water, if and when it is present
6. Soil nutrient status

[a] See footnote Box 1.2.

Table 1.3 Coastally Restricted Forests: Physical Features of the Ecosystem

Species	Soil		Water			Regional Climate			
	Soil Type[a]	Nutrient	Saturated	F = Flowing S = Standing	pH	Rainfall (mm/yr)	Relative Humidity	Temperature Mean (°C)[b]	
								Minimum	Maximum
A. japonica	P alluvial soil	mesotrophic, eutrophic	Y	S	3.5–6.4	abundant	25–95% avg. 60–75%	cool–temperate	warm–temperate
C. formosensis	shallow to moderately deep, moist mineral soils		N	none	N.R.	3000–4800	72–96% local clouds		
C. taiwanensis	shallow, often well-drained mineral soils		N	none	N.R.	3000–4800	72–96% local clouds		
C. lawsoniana	L, M, P, S, Se varies widely	all	N/Y	F, S, or none	4.2–7	1000–2400 1275 avg.	high at night; foggy; 77–89%	[−3]	33
C. nootkatensis	P, Po	poor neutral–acidic	N/Y	occasional or none		1300–3300	71–91%	AK*:[−29]–[−17] [−9]–3.5 BC*:[−26.7] WA*:[−24.4]–[−28.9] CA*:[−18.9]	29–32 15–18 33.3 32.9 43.9
C. obtusa	M, Po, Se	poor	N	none	N.R.	1400–4100; most in growing season		[−9.2]	30

C. pisifera	M, P		N.R.	none	N (sometimes Y)	1400–4100		[−9.2]	30
C. thyoides	P over S	poor	3.1–7	F, S	Y	1186	53–86%	[−27.8]	41.7
P. glehnii	P, S, V lithosol	mesotrophic, eurotrophic	N.R.	none	some Y some N	1000–1200	local heavy fog in summer	cool	cool, humid summer very high
P. muricata	L, M, Po, S, diatomaceous		N.R.	none	N?	< 500	mist and fog	frost	
P. pumila	P, S, V all types with good drainage	poor to mesotrophic	4.5–6	none or F	N	300–2000	70–80%	[−26]([−51*])	21(33*)
P. radiata	S, SL, V sedimentary, metamorphic	pH 4–6	N.R.	none	N	< 300	mist and fog	freezing	high
P. serotina	L, M, P, S			S?	Y	1200	53–87%	[−23]	43
S. sempervirens	L deep, moist	rich; pH neutral-slightly acidic	N.R.	none	N	640–3100	68–89%	[−5.6]	38.9
T. orientalis				S, some sites					
T. distichum/ N. aquatica	L, M, P, S, other sandy clays, silts	rich	5.5–8.4	S, F	Y	1270–1330	53–89%	[−21.1]	40.6

Notes: Y = Yes. N = No. Avg. = Average. N.R. = Not relevant. Blank = Information not available. ? = Information not available. * = Information contradictory. * = Extreme temperature reported from any year at weather stations nearest the species' range.

a Soil type. L = loam, M = mineral, P = peat, Po = podzol, S = sand, Se = serpentine, and V = volcanic.

b Temperature. Average mean daily temperatures (except as noted) at weather stations in species' region. There are generally no weather stations at the colder high-altitude locations; no reliable data are available on conditions within the stands. Brackets [ ] denote negative numbers.

Table 1.4 Coastally Restricted Trees: Reproduction and Growth

Species	Monoecious	Pollination	Seed Transport	Vegetative Reproduction[a]	Init. Growth	Colonize Open Sites	Shade Tolerant
A. japonica	Y	wind	wind	Y	fast	Y	N
C. formosensis	Y	wind	wind	N?		Y	N
C. taiwanensis	Y	wind	wind	N?		Y	Y
C. lawsoniana	Y	wind	wind, water	Y	slow	Y	Y
C. nootkatensis	Y	wind	wind	Y	slow–moderate	Y	N (varies)
C. obtusa	Y	wind	wind	N	slow	Y	Y
C. pisifera	Y	wind	wind	Y	slower than C. obtusa	Y	Y
C. thyoides	Y	wind	wind, water	N	varies	Y, fast	N
P. glehnii	Y	wind	wind, animal	Y	slow	Y	Y
P. muricata	Y	wind	wind, (water)	N	fast	Y	N
P. pumila	Y	wind	birds, mammals	Y	slow	Y	N
P. radiata	Y	wind	wind, (water), some animal	N	fast	Y	N
P. serotina	Y		wind	Y	slow	Y	N
S. sempervirens	Y	wind	wind	Y	very fast	Y	moderately
T. orientalis	Y						
T. distichum/ N. aquatica	Y/N	wind, insect	water, wind		fast/ moderate		Y/N

Notes: Y = Yes. N = No. Blank = Information not available. ? = Information contradictory.
[a]*Vegetative reproduction.* "Y" indicates that this is a common strategy in nature; it does not refer to the success of rooted cuttings or greenhouse layering.

N. aquatica forests cover only 30 m in altitude. Island populations of *P. muricata* extend over only 20 m; populations of *P. glehnii* extend for 140 m. There is an almost 3,000-fold difference between species with greatest and least altitudinal range.

Although the natural range of coastally restricted trees is geographically very limited, many of these species can thrive under cultivation far from their natural range and under very different conditions. For example, *S. sempervirens* is, in nature, most successful in the Central Valley of California and at low elevations in the Sierra Nevada Mountains (Ornduff). Its dominance is limited to a well-defined area 725 km N-S and at most 72 km inland from the Pacific Ocean. However, in New Zealand, France, and Spain, it is readily cultivated far from ocean coasts. Similarly, although *C. thyoides* has been grown in gardens and wetlands worldwide—in Hawaii, Great

Table 1.5 Coastally Restricted Trees: Morphology and Other Properties

| Species | Size at Maturity | | | Bark | Leaf | | Epiphytes | Mycorrhizae | Resists Decay |
| | Height (m) | | DBH (m) | Furrowed | | | | | |
	Average	Maximum	Max.	Fissured	Needle	Evergreen			
A. japonica	17	30		Y	N	N		Y	N
C. formosensis		60–65	5.8–7	Y	Y	Y			Y
C. taiwanensis		40	3	Y	Y	Y			Y
C. lawsoniana	45	65–73	3.8–4.8	Y	Y	Y	Y	Y	Y
C. nootkatensis	25–35	40–50	2–3.7	Y	Y	Y	Y	Y	Y
C. obtusa		40–52	1.5–2	Y	Y	Y	N	Y	Y
C. pisifera		46–50	1.5–2	Y	Y	Y	N		Y
C. thyoides	28	36	1.2–2.1	Y	Y	Y	Y	N	Y
P. glehnii		50	1.5	Y	Y	Y	N	N	
P. muricata	15–20	30	0.9	Y	Y	Y	Y	Y	
P. pumila	vertical: 0.5–6 (prostrate form)	20[a]	0.1–0.3	slightly Y	Y	Y	Y	Y	
P. radiata	25	60	0.9	Y	Y	Y	Y few	Y	
P. serotina	20.1	29	0.35	Y	Y	Y	Y		
S. sempervirens	61	115	3–4.6	Y	Y	Y	Y	Y	Y
T. orientalis	25	25	1	Y	Y	Y			Y
T. distichum/ N. aquatica	25/24	50/34	4/	Y/N	Y/N	N	Y	Y	Y/N

Notes: See Notes, Table 1.4
[a] *P. pumila* size *maximum*. The tree is prostrate; this refers to length of trunk.

Table 1.6 Coastally Restricted Forests: Characteristics of Associated Species

Species	Distinct Assemblage in Each Type of Environment	One Dominant Family	Rares/ Endemics	Comments
A. japonica	Y	N	N	
C. formosensis	many ferns few conifer spp.			
C. taiwanensis	many ferns few conifer spp.			
C. lawsoniana	N	N	Y, several	high variability; seed crops not regionally synchronized; roots intermingled, shallow; oils toxic
C. nootkatensis	Y, several	Y, Ericaceae		regeneration limited by deer browse, poor seed germination; epiphytes: lichens, liverworts, mosses
C. obtusa	many conifers & woody angiosperms			
C. pisifera				
C. thyoides	Y	Y, Ericaceae	Y	
P. glehnii	Y evergreen conifers	Y, Ericaceae	Y *Rhododendron parvifolium*	
P. muricata	Y, often ericaceous	Y, Ericaceae		seed dispersal: some closed cones via water
P. pumila	?	N	N	prostrate; cone mass increases in open, not extremely cold sites
P. radiata	Y	Y (Mainland) Anacardiaceae	few (Monterey)	seed dispersal: some closed cones via water, some by birds, squirrels; endemic
P. serotina	Y	N		fire-resistant, fire-dependent: requires fire to release seed from cones
S. sempervirens	Y	Y Coniferae		leaves on lower branches differ from upper branches in shape, size, and array
T. orientalis				
T. distichum/ N. aquatica	Y	N		dependent on fluctuation in water level; cannot germinate under water; Nyssa: low and relatively constant N:P ratio

Notes: See Table 1.4 notes.

Britain, Australia, Russia, and many other countries—it is not known to naturalize outside its native range.

Climate

The range of climatic parameters varies widely, reflecting the longitudinal and altitudinal perimeters of the species being studied. Moreover, information on temperature, precipitation, and humidity is difficult to quantify and reliably evaluate at present, for localized conditions frequently differ markedly from those of the nearest weather station. Accurate multi-year weather records have been maintained for relatively small portions of the local habitats of these species.

Precipitation

Precise rainfall data are generally unavailable for determining the microclimate affecting a particular species. However, some useful general observations may be offered. Many coastally restricted species occur in regions of high rainfall, as, for example, all those *Chamaecyparis* species classified as coastal temperate rainforest (CTRF): *C. lawsoniana* and *C. nootkatensis* in the American northwest, and *C. obtusa* and *C. pisifera* of Japan (Weigand et al. 1992). Following the monograph "Coastal Temperate Rain Forests" (Weigand et al. 1992), it appears that other systems addressed here also fall within the CTRF zone: Taiwan's *C. taiwanensis* and *C. formosensis;* Japan's *P. glehnii; P. pumila* near the Sea of Okhotsk of Japan, China, and the Russian Kamchatka Peninsula; and many *S. sempervirens* stands in California and Oregon, USA.

By definition the CTRF belt has a precipitation minimum of 2,000 mm (Weigand et al. 1992). Some species not in the temperate rainforest, such as the closed-cone pines *(P. muricata, P. radiata)* and some *S. sempervirens,* are subject to regular seasonal drought (Libby et al. 1968, Ornduff). This apparent anomaly may be offset by high humidity in the immediate microclimate of individual stands. A common requirement of all coastal species is an ample supply of water, either in the atmosphere as rainfall, fog, or clouds, or in the substrate, as we will see below.

Soils

The soils supporting coastally restricted trees vary widely, including rich loam, sand, organic peat, podzol, serpentine, and volcanic ash. They range from saturated to well drained, and may be eutrophic, mesotrophic, or dystrophic (i.e., from rich to impoverished). Where there is standing water, its pH ranges from 3.1 to 8.4. Many of these soils, especially pure peat, podzol, serpentine, and volcanic ash, are recognized as difficult or impossible habitat for most other regional vegetation.

More than half the coastally restricted forest trees may be found on peats: *A. japonica, C. lawsoniana, C. nootkatensis, C. pisifera, C. thyoides, P. glehnii, P. pumila, P. serotina, T. distichum,* and *N. aquatica.* Peat soils consist of organic matter in various states of structural breakdown ranging from fibrous to mucky (Moore and

Bellamy 1974, Allaby 1992). Pure peat contains few nutrients available to plants, due to its high cation exchange capacity (CEC), the near absence of oxygen in its waters, and the binding of nutrients within undecomposed organic compounds (Bear 1964, Moore and Bellamy 1974). Cation exchange results in acidification and the removal of available nutrients from the surrounding water and soil. Available phosphorus, potassium, and nitrogen are often extremely low or entirely absent. In acid peat, there is virtually no nitrification (conversion of ammonia to nitrate) due to the absence of the necessary aerobic (oxygen-requiring) bacteria (Moore and Bellamy 1974). The lack of oxygen also effectively prevents the bacterial breakdown of organic matter into forms usable by plants.

Most often, peat is mixed with components such as clay, muck, and other mineral matter containing varying amounts of nutrients available to plants. Hence, the soil's nutrient status cannot be inferred solely from the identification of peat on a site. However, as peat reflects a saturated or periodically submerged substrate, particular site qualities and capacities of the plant population may be inferred from its presence.

Podzol supports *C. nootkatensis, C. obtusa,* and *P. muricata.* Podzols are acidic, infertile soils with an ashlike upper layer depleted of colloids, iron, and aluminum, and a dark lower layer where these substances have accumulated in toxic concentrations.

Serpentine supports *C. lawsoniana, C. obtusa,* and *P. glehnii.* In contrast, *S. sempervirens* is conspicuously absent from serpentine outcrops that are common near redwood stands. Serpentine soils, weathered from ultramafic rocks, are composed primarily of silicates high in magnesium, iron, and aluminum (Bear 1964). Where *C. lawsoniana* grows over ultramafic rock, the rock has weathered to produce a dense layer of fine clay, creating a perched water table with consistent seepage (Zobel). Serpentines hold a suite of plants differing from their surrounding habitats in both species composition and superficial aspect. The plants are a xeromorphic (desert-like) group with marked vegetational discontinuities, impoverished in both species and numbers of individuals (Brooks 1987).

Volcanic ash supports *P. glehnii, P. radiata,* and *P. pumila.* Ash weathers to amorphous aluminosilicate gels (allophane, halloysite, and gibbsite) that are infertile primarily due to their high permeability and the subsequent leaching of essential nutrients (Bear 1964). Like peat, volcanic ash has a high CEC, which, combined with high aluminum and iron activity, results in the virtual absence of essential phosphate.

Implications of Environmental Factors

Influence of Oceans on Nearby Land

Oceans influence continental and island margins in many ways (Box 1.7). The sea functions as a giant heat sink (in other words, it stores heat) due to the high latent heat of water. It moderates the extremes of temperature in summer and winter, and produces a longer fall and a later spring than are found further inland. The sea also strongly influences wind and storm patterns, which in turn will limit and shape animal communities and vegetation.

As hurricanes often sweep through many coastal areas, they are a recurring and,

Box 1.7 Environmental Factors Resulting from Being Adjacent to a Marine Coast

1. Giant heat sink effects
    (a) moderates winter and summer temperature extremes
    (b) longer fall, later spring
        (i) encourages extending range of both southern and northern species
        (ii) encourages growth of regionally rare species
2. Higher salinity in air (as aerosols and fogs), which creates
    (a) stress via leaves and soil*
    (b) high aerial nutrient supply
3. High humidity
4. Variable hydroperiod
    (a) increased stress for non-adapted species*
5. High winds
    (a) increased danger of desiccation*
    (b) increased windthrow*
6. High rate of violent storms*
    (a) controls ecosystems via catastrophe*
    (b) favors colonizer species*
7. Ecotone effects
    (a) accumulation and transfer of nutrients
    (b) atmospheric nutrient content increased
8. Frequent fogs

*Factor that inhibits competitors.

therefore, a possibly determining aspect of coastal forest development (discussed in detail by Conner).

Materials produced in a system tend to accumulate at the boundary zone (ecotone) and are frequently transformed there as well. The land-water ecotone especially is recognized as a principal route for the transport of nutrients across landscapes via physical and biological means. High-intensity winds produce salt-spray aerosols, which form as air bubbles burst and force droplets to spray up from the sea surface (Boyce 1951, 1954; Kientzler et al. 1954). The transfer of gases and aerosols from oceans restores biogeochemical components to the land in the form of gases, precipitation, and windborne materials. This forms a major source of inorganic nutrients for coastal habitats (Art et al. 1974). In any ecosystem, the degree of the influence of meteorologic components depends on local weather and the distance from the sources of airborne materials (Gorham 1958, Gambell and Fisher 1966), in this case, the sea. Applying the nutrient cycling model of Bormann and Likens (1967) to systems restricted to ocean borders, Henry Art (1971, 1976) suggested that the sea itself may not only directly affect, but actually control, the function and structure of coastland ecosystems.

Relatively little is known of the role of the sea in controlling the distribution and development of coastal ecosystems, beyond the toxic effects of salt spray on vegetation (Conner and Day). The major portion of the available data concerns the mineral content of coastal precipitation, fogs, and dusts. The nearer a system is to the sea, the higher the concentration of chemicals in precipitation—for example, magnesium, sodium, chloride, sulfate, and often calcium and potassium (Boyce 1951).

The forests most strongly influenced by atmospheric nutrients are probably those with minimal geological and biological inputs (Art et al. 1974). Such circumstances are found where surface water inflow is low or absent, and where the parent rock is resistant to weathering. When biodegradation is inhibited by toxins, low temperature, or the lack of oxygen, atmospheric contributions are important components of the mineral supply.

The accumulation and transfer of materials in this complex fashion provides multiple benefits to adapted species. It acts both as a stressor eliminating those competitors unable to cope, and as a source of wealth available to the survivors.

Influence of Fog

All the coastally restricted species share one environmental feature: they thrive in very humid environments. All the species are affected by frequent fogs or shrouding by clouds (Libby et al. 1968, Fujita et al. 1994, Paul Meyer, personal communication, and all authors in this volume). The *Chamaecyparis* found in the region with least rainfall, *C. thyoides,* grows in nature only in or immediately adjacent to wetlands. Fog-drip, in many cases prevalent in all seasons and sometimes localized as low clouds, makes rainfall data in the coastal region extremely difficult to interpret relative to tree requirements.

Some striking examples of the correlation between coastal species and fog come from the eastern shore of the Pacific Ocean: Libby et al. (1968) observed mildly foggy days on Cedros Island off Baja, California, when each *P. radiata* was covered by a separate cloud, while the desert between the pine groves was exposed to clear sky. The presence of localized fog seems critical for the growth of pines on Cedros and Guadalupe Islands. Ornduff reports that *S. sempervirens* fog-drip may add more than 25 cm to the effective summer precipitation at the root zone.

Catastrophe as a Controlling Element

Catastrophic events play an important role in determining the characteristics of coastally restricted ecosystems. These events include abrupt yet often impermanent changes in hydrology and geomorphology such as those caused by drought, storm, fire, beaver, or harvest. These disasters can be seen as essential agents favoring the ecosystems' characteristic biota.

Effects of Stress

In humid environments, stresses that permanently affect the hydrological regime, such as draining, diking, damming, and road or levee construction tend to have a more severe, adverse, permanent impact on the vegetation than lumbering or most natural disasters (Lugo et al. 1990b). The canopy dominants native to the regions under study are adapted to survive or regenerate after fire, violent storms, temporary floods, and certain types of harvest. They appear to compete best under conditions that eliminate their competitors. Such occasional catastrophes may determine the vegetational structure of coastally restricted systems. Again, it is the extreme condition, not the average, that determines the plant community.

Drought in Wet Sites   Lack of water is a major environmental stress in any environment. In normally humid or wet regions, drought has peculiar evolutionary, ecological, and physiological implications. The influence of rare but erratically recurring droughts was considered in the first section of this chapter, under Properties of the Dominant Trees.

Regeneration after Disturbance

Both in nature and under management, some coastally restricted species are known to regenerate successfully and rapidly into monotypic stands after catastrophic events that completely kill the standing mature trees. For example, under certain conditions, *C. thyoides* and *S. sempervirens* are successful colonizers from seed after clearcut harvest, fire, or prolonged flooding (Little 1950, Laderman 1989, Stephens 1991, Ornduff). These forests may be classed as perturbation-dependent ecosystems (Vogl 1980). In his study of hurricanes, Conner concludes that natural disasters are vital to the ecosystem dynamics of coastal systems.

Selection by Fire   Fire, a strong agent for species selection, functions as an erratic but recurring stress, affecting ecosystems in a way similar to drought. Fire is an important thinning agent (Schlesinger 1978) that increases landscape diversity (McKinley and Day 1979, Christensen et al. 1981) and hence maintains and helps determine species diversity. It reduces the importance of some species, and enhances the dominance of others (Ewel and Mitsch 1978).

Although there are still many gaps in our understanding of the role of fire in forest systems, several effects are recognized. Fire clears the surface of most vegetation and debris, provides nutrients, exposes the forest floor to sunlight and warmth, and scarifies the surface. These multiple actions provide an ideal seedbed for the germination of some species. In wetlands, due to local variations in the wetness of the upper layers of the forest floor, fire resculpts the surface into crenelated hummocks and hollows. Hollows quickly fill with water. Burning also forms an impermeable skin on peat, creating or increasing poor drainage (Tallis 1983:322). This situation perpetuates and amplifies the saturation of the soil, which gives the advantage to wetland-adapted species.

Global Climate Change

There are climatic considerations at a different scale that should be factored into this analysis. Global climate change, which influences the distribution of all vegetation, affects coastal species most directly via alterations in sea level (Conner and Day).

This form of disturbance either regionally extirpates the coastally restricted forms, or it induces inland shifts of entire coastal system types as oceans rise. The success of a system in surviving such large-scale change depends on (1) the capacity of all ecosystem components to "travel" fast enough to keep pace with the encroachment of salt water and (2) the existence of suitable terrain at the landward margin. A major factor now limiting the successful translocation of ecosystems with such stringent environmental parameters is the heavy development of so many coastal areas, which has already eliminated many potential habitats.

In the scenario of a rising sea level, a variety of species mixes would develop, reflecting the variations in the capacities of each species existing in a specific over-washed area to migrate, colonize, and thrive in the precise conditions of adjacent inland sites. This variety is indeed what is now seen in many ecological island habitats dominated by coastally restricted trees.

Quite frequently, the stringent site requirements cannot be fulfilled, and a coastal forest ecosystem is locally extirpated. This may help to explain the presence of numerous range gaps and relict stands, as well as macrofossil evidence of coastally restricted species in sites where they do not currently grow.

CORRELATIONS BETWEEN PROPERTIES OF THE TREES AND THEIR ENVIRONMENTS

We can now attempt to relate the capacities of coastally restricted species to the limitations of the habitats they dominate. Figure 1.2 is a first approach to matching the species-habitat parameters. The next step is to devise experimental and analytical procedures to test the validity of the suggested relationships.

It may be fruitful to explore the following connected hypotheses. The factors that appear most closely tied to each other (as listed in Fig. 1.2) are the key aspects that limit these species to the marine coastal environment. They are those properties shared by all the species, and in turn by all the habitats.

Coastal ecosystems have been shaped by the co-occurrence of the multiple factors resulting from the atmospheric conditions of a marine coast. The high humidity, salt-laden fogs and aerosols, giant heat sink effects, and violent storms combine to encourage the growth of species that are poor competitors in more moderate environs. The keys to their survival and dominance in isolated areas along the world's coasts are their mechanisms to withstand or to take advantage of stress. Avenues of research to test these hypotheses are suggested in chapter 21, Coastal Forest Management and Research.

Success through Failure

The strategy of coastally restricted species is that of outlasting others, a tortoise strategy, not in duration, but spatially, in niche breadth. Coastally restricted trees occupy those parts of the sites where it is hardest to subsist. Most of their predecessors were repeatedly weeded out with all other species when successive disasters struck.

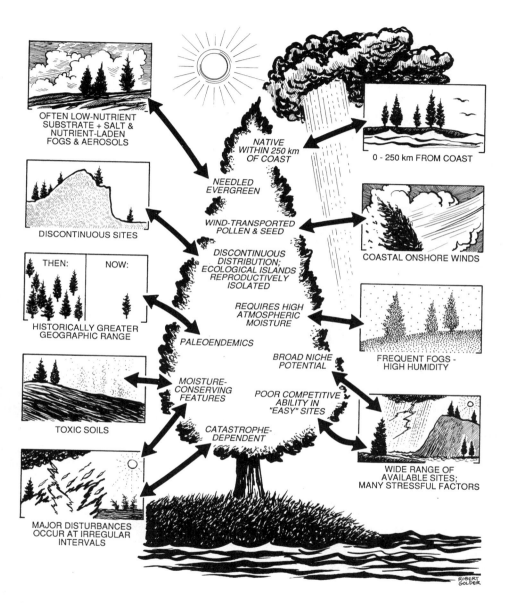

Figure 1.2. Correlations between coastally restricted trees and their habitats. (Original illustration by Robert Golder.)

It is not the average situation prevailing on each coast, but rather the rare catastrophe that selects for coastally restricted trees. Over the long haul, such species depend on major disturbances for their survival.

The principle of their success is the same as that taken advantage of by microbiologists who use selective media to encourage an increase in desired bacterial traits.[2] Here, environmental stress has selected for those characteristics that allowed survival after successive disasters by progressively eliminating all those individuals unable to cope. Eventually, progeny appeared with suites of genes adapted to multiple stressors. Similar characteristics may have also entered the genome of other species, but these traits would have been swamped by all those that gave an advantage on more moderate sites, preventing the development of multi-resistance genomes. It is their very failure, their inability to compete on nonstressful sites, that provides an advantage to coastally restricted species. The quality of persistence under adversity encourages the selection of traits useful in a variety of disastrous situations.

A related hypothesis is worth further study: The combination of a high moisture requirement with sclerophylly may be the major factor in the development of coastally restricted species. Prime candidates to explain the apparent paradox of the existence of xeromorphic species in a humid environment are: (1) the accumulation of toxic conditions in the soil combined with (2) salinity and nutrients in the atmosphere and (3) the erratic recurrence of drought. Only those species capable of withstanding each of these multiple stressors could succeed in such habitats.

In summary, coastally restricted trees are endemics, poor dispersers, poor competitors restricted to becoming dominants in increasingly fragmented, specialized ecological islands. Each of these properties indicates that such species are prime candidates for local extirpation. As each habitat island is destroyed, the environmental debt (Bormann 1990 and 1994) and the extinction debt become greater (Tilman et al. 1994). Certain catastrophic events favor the survival of coastally restricted trees. These stresses eliminate competition from canopy codominants, and actually stimulate renewal of species with adaptations that create advantage from disaster.

Notes

1. Unless otherwise indicated, all citations without dates refer to chapters in this volume.
2. An example of the use of selective media: An antibiotic is placed in the food material, or medium, of the bacteria to kill all but the few preexisting antibiotic-resistant cells. Each resistant cell gives rise to a colony identical to it; the entire next generation then consists of drug-resistant bacteria.

References

Allaby, M., ed. 1992. The concise Oxford dictionary of botany. Oxford Univ. Press, New York.
Antonovics, J., A.D. Bradshaw, and R.G. Turner. 1971. Heavy metal tolerance in plants. Adv. Ecol. Research 7:1–85.
Art, H.W. 1971. Atmospheric salts in the functioning of a maritime forest ecosystem. Ph.D. dissertation. Yale University, New Haven.

————. 1976. Ecological studies of the Sunken Forest, Fire Island National Seashore, New York. Natl. Park Serv. Sci. Monogr. Series #7.

Art, H.W., F.H. Bormann, G.K. Voigt, and G.M. Woodwell. 1974. Barrier island forest ecosystem: role of meteorologic nutrient inputs. Science **184**:60–62.

Bailey, L.H., and E.Z. Bailey. 1947. Manual of cultivated plants. Macmillan, New York.

Barbosa, P., V.A. Krischik, and C.G. Jones, eds. 1991. Microbial mediation of plant-herbivore interactions. Wiley Interscience, New York.

Bear, F.E., ed. 1964. Chemistry of the soil. Reinhold, New York.

Bormann, F.H. 1990. The global environmental deficit. BioScience **40**:74.

————. 1994. Landscape restoration and ecosystem health. A series of talks at Yale School of Forestry and Environmental Studies, New Haven. Unpublished.

Bormann, F.H., and G.E. Likens. 1967. Nutrient cycling. Science **155**:424–429.

Boyce, S.G. 1951. Source of atmospheric salts. Science **113**:620–621.

————. 1954. The salt spray community. Ecol. Monogr. **24**:29–67.

Bramlett, D. 1990. *Pinus serotina*. Pages 470–474 in R.M. Burns and B.H. Honkala, eds. Silvics of North America. Vol. 1. Conifers. U.S. Dept. Agric. For. Serv. Agric. Handb. 65.

Brooks, R.R. 1987. Serpentine and its vegetation. Dioscorides Press, Portland, OR.

Brown, S. 1981. A comparison of the structure, primary productivity and transpiration of cypress ecosystems in Florida. Ecol. Monogr. **51**:403–427.

————. 1990. Structure and dynamics of basin forested wetlands in North America. Pages 171–199 in Lugo et al., 1990a.

Brunsfeld, S.J., P.S. Soltis, D.E. Soltis, P.A. Gadek, C.J. Quinn, D.D. Strenge, and T.A. Ranker. 1994. Phylogenetic relationships among the genera of Taxodiaceae and Cupressaceae: evidence from rbcL sequences. Sys. Bot. **19**:253–262.

Carroll, G.C. 1988. Fungal endophytes in stems and leaves: from latent pathogen to mutualistic symbiont. Ecology **69**:2–9.

————. Fungal associates of woody plants as insect antagonists in leaves and stems. Pages 253–271 in Barbosa et al., 1991.

Chabot, B.F., and D.J. Hicks. 1982. The ecology of leaf life spans. Ann. Rev. Ecol. Syst. **13**:229–259.

Chapin, F.S., III. 1980. The mineral nutrition of wild plants. Ann. Rev. Ecol. Syst. **11**:233–260.

Chapin, F.S., and G.R. Shaver. 1989. Differences in growth and nutrient use among arctic plant growth forms. Funct. Ecol. **3**:73–80.

Christensen, N., R.B. Burchell, A. Liggett, and E.L. Simms. 1981. The structure and development of pocosin vegetation. Pages 43–61 in C.J. Richardson, ed. Pocosin wetlands. Hutchinson Ross, Stroudsburg, PA.

Clewell, A.F., and D.B. Ward. 1987. White cedar in Florida and along the northern Gulf Coast. Pages 69–82 in A.D. Laderman, ed. Atlantic white cedar wetlands. Westview Press, Boulder.

Edwards, S. 1992. Foliar morphology of *Chamaecyparis* and *Thuja*. Four Seasons **9**:4–29.

Ewel, K.C. and Mitsch, W.J. 1978. The effects of fire on species composition in cypress dome ecosystems. Fla. Sci. **41**:25–31.

Flora of North America Editorial Committee (FNA). 1993. Flora of North America North of Mexico, Vol. 2. Pteridophytes and Gymnosperms. Oxford Univ. Press, New York.

Fujita, H., S. Kojima, and M. Nakata. 1994. Sakhalin spruce *(Picea glehnii)* swamp forest in eastern Hokkaido. Proceedings, VI International Congress of Ecology, Manchester, England.

Gambell, A.W., and D.W. Fisher. 1966. Chemical composition of rainfall, eastern North Carolina and southeastern Virginia. U.S. Geol. Surv. Water-Supply Paper 1535-K.

Gorham, E. 1958. The influence and importance of daily weather conditions in the supply of chloride, sulphate, and other ions to fresh waters from atmospheric precipitation. Royal Soc. Lond. Phil. Trans. Ser. **B241**:147–178.

Great Britain, Meteorological Office. 1980. Tables of temperature, relative humidity, precipitation, and sunshine for the world. Part 1: North America and Greenland. Part 5. Asia. Her Majesty's Stationery Office, London.

Jonasson, S. 1989. Implications of leaf longevity, leaf nutrient reabsorption and translocation for the resource economy of five evergreen plant species. Oikos **56**:121–131.

Jonasson S., and F.S. Chapin, III. 1985. Significance of sequential leaf development for nutrient balance in the cottonsedge, *Eriophorum vaginatum* L. Oecologia (Berl.) **67**:511–518.

Kientzler, C.F., A.B. Arons, D.C. Blanchard, and A.H. Woodcock. 1954. Photographic investigation of the projection of droplets by bubbles bursting at the water surface. Tellus **6**:1–7.

Laderman, A.D., ed. 1989. The ecology of Atlantic white cedar wetlands: a community profile. U.S. Fish Wildl. Serv. Biol. Rep. **85**(7.21).

Laurance, W.F. 1991. Ecological correlates of extinction proneness in Australian tropical rain forest mammals. Conservation Biology **5**:79–87.

Lee, Chang Bok. 1980. Illustrated flora of Korea. Hyang Moon Comp., Seoul.

Li, H. 1963. Woody flora of Taiwan. Morris Arboretum. Livingston Publishing, Narberth, PA.

Libby, W.J., M.H. Bannister, and Y.B. Linhart. 1968. The pines of Cedros and Guadalupe Islands. J. For. **66**:846–853.

Little, S. 1950. Ecology and silviculture of white cedar and associated hardwoods in southern New Jersey. Yale Univ. Sch. For. Bull. **56,** 103 pp.

Lugo, A.E., S. Brown, and M.M. Brinson, eds. 1990a. Forested wetlands. Ecosystems of the world, Vol. 15. Elsevier, New York.

———. 1990b. Concepts in wetland ecology. Pages 53–85 *in* Lugo et al., 1990a.

McKinley, C.E., and F.P. Day. 1979. Herbaceous production in cut-burned, uncut-burned, and control areas of a *Chamaecyparis thyoides* (L.) BSP, (Cupressaceae) stand in the Great Dismal Swamp. Bull. Torrey Bot. Club **106**:20–28.

Molina, R. 1994. The role of mycorrhizal symbioses in the health of giant redwoods and other forest ecosystems. Pages 78–81 *in* P.S. Aune, coordinator. Proceedings of the symposium on giant sequoias. U.S. Dept. Agric. Gen. Tech. Rep. PSW-GTR-151. For. Serv., Pacific Southwestern Resch. Sta., Albany, CA.

Monk, C.D. 1966. An ecological significance of evergreeness. Ecology **47**:504–505.

Moore, P.D., and D.J. Bellamy. 1974. The peatlands. Springer-Verlag, New York.

Nakamura, Y. 1988. Subalpine Nadelwalder. Pages 307–313 *in* A. Miyawaki, ed. Vegetation of Japan, Hokkaido. Shibundo, Tokyo. (As seen in Nakamura et al. 1994:145,150)

Nakamura, Y., M.M. Grandtner, and N. Villeneuve. 1994. Boreal and oroboreal coniferous forests of eastern North America and Japan. Pages 121–154 *in* A. Miyawaki, K. Iwatsuki, and M.M. Grandtner, eds. Vegetation in Eastern North America. Univ. of Tokyo Press, Japan.

NOAA (National Oceanic and Atmospheric Administration). Meteorological Data for 1992. US Dept. of Commerce, National Climatic Data Ctr., Asheville, NC.

Perry, J.P., Jr. 1991. The Pines of Mexico and Central America. Timber Press, Portland, OR.

Pike, L.H., R.A. Rydell, and W.C. Denison. 1977. A 400 year old Douglas fir tree and its epiphytes: biomass, surface area, and their distributions. Can. J. For. Res. **7**:680–699.

Prance, G.T., and A.E. Prance. 1993. Bark. Timber Press, Portland, OR.

Russell, J.H. 1993. Genetic architecture, genecology and phenotypic plasticity in seed and seedling traits of yellow-cedar (*Chamaecyparis nootkatensis* [D. Don] Spach). Dissertation. Univ. of British Columbia, Canada.

Schlesinger, W. 1978. Community structure, dynamics and nutrient cycling in the Okefenokee Cypress Swamp forest. Ecol. Monogr. **48**:43–65.

Schulze, E.D., and F.S. Chapin III. 1987. Plant specialization to environments of different resource availability. Pages 120–148 *in* E.D. Schulze and H. Zwolfer, eds. Potentials and limitations of ecosystem analysis. Springer-Verlag, Berlin.

Shaver, G.R. 1981. Mineral nutrition and leaf longevity in an evergreen shrub, *Ledum palustre* ssp. *decumbens*. Oecologia (Berl.) **49**:362–365.

Small, E. 1972. Photosynthetic rates in relation to nitrogen recycling as an adaptation to nutrient deficiency in peat bog plants. Can. J. Bot. **50**:2227–2233.

Specht, R.L., ed. 1979. Heathlands and related shrublands of the world. Ecosystems of the world, Vol. 9A. Elsevier, Amsterdam.

Stebbins, G.L. 1942. The genetic approach to rare and endemic species. Madrono **6**:241–272.

Stephens, R.R. 1991. Regeneration and management of coast redwoods *(Sequoia sempervirens)*. Abstract *in* Symposium: Coastally Restricted Forests (April 9–11, 1991), Yale Univ. School of Forestry, New Haven. Unpublished.

Tallis, J.H. 1983. Changes in wetland communities. Pages 311–347 *in* A.J.P. Gore, ed. Mires: swamp, bog, fen, and moor. Ecosystems of the world, Vol. 4A. Elsevier, Amsterdam.

Terrill, W.B. 1951. Some problems of plant range and distribution. J. Ecol. **39**:205–227.

Tilman, D., R.M. May, C.L. Lehman, and M.A. Nowak. 1994. Habitat destruction and the extinction debt. Nature **371**:65.

Vogl, R.J. 1980. The ecological factors that produce perturbation dependent ecosystems. Pages 63–94 *in* J. Cairns, Jr., ed. The recovery process in damaged ecosystems. Ann Arbor Sci. Publ., Ann Arbor.

Weigand, J., P.B. Alaback, A. Mitchell, and D. Morgan. 1992. Coastal temperate rain forests. Ecotrust, Portland, OR. Unpublished report.

Whigham, D.F., and C.J. Richardson. 1988. Soil and plant chemistry of an Atlantic white cedar wetland on the Inner Coastal Plain of Maryland. Can. J. Bot. **66**:569–576.

Part I

CHAMAECYPARIS
(FALSE-CYPRESS)
STUDIES

2

Donald B. Zobel

Chamaecyparis Forests

A Comparative Analysis

The conifer genus *Chamaecyparis* Spach (family Cupressaceae) includes seven taxa, two each in Japan, Taiwan, and western North America and one in eastern North America (Table 2.1). All grow in coastal environments. These trees have similar, scalelike foliage. Their woods are decay resistant, relatively strong, easy to work, light colored, take finishes well, and have all been valuable for local use. The primary market for *Chamaecyparis* wood now is Japan, however, where the high value of the wood depends on its aesthetic properties.

These superficially similar species differ in their maximum size and age, form, habitat, and the composition and structure of the forests where they grow. The purpose of this chapter is to survey and contrast the characteristics of native forests in which *Chamaecyparis* species are important.

Chamaecyparis species vary in size (Table 2.1). In general, *C. thyoides* (Atlantic white cedar) is the smallest, and *C. formosensis* (Taiwan cypress) and *C. lawsoniana* (Port-Orford-cedar) are the largest. Maximal height of smaller *Chamaecyparis* species is shorter than most associated conifers and even some associated hardwoods. Several associated coniferous species equal or exceed the potential height of *C. lawsoniana*, the tallest species. In contrast, *C. formosensis* is the largest conifer in eastern Asia (Li and Keng 1954). *Chamaecyparis formosensis* and *C. nootkatensis* (Alaska yellow-cedar) are reported to live more than 3,000 years (Lee 1962, Waring and Franklin 1979).

Ranges

Chamaecyparis obtusa (Hinoki cypress) and *C. pisifera* (Sawara cypress) both grow in central to southern Japan (Table 2.1), generally in mountainous terrain and sometimes together. In Taiwan, ranges of *Chamaecyparis* species also overlap substantially in their high-elevation montane habitat. In western North America, *C. lawsoniana* has

Table 2.1 Nomenclature, Ranges, and Maximum Reported Sizes of *Chamaecyparis* spp.

| Species[a] | Location | Range[b] | | Max. Size[c] | |
		Latitude (°N)	Elevation (m)	Height (m)	Diameter (m)
C. formosensis Matsumura	Taiwan	22°28′–24°49′	850–2600	60–65	5.8–7
C. lawsoniana (A. Murray) Parlatore	W. North America	40°50′–43°35′	0–1950	65–73	3.8–4.8
C. nootkatensis (D. Don) Spach	W. North America	41°25′–61°07′	0–2300	40–50	2–3.7
C. obtusa (Siebold and Zuccarini) Endlicher	Japan	30°15′–37°10′	10–2200	40–52	1.5–2
C. pisifera (Siebold and Zuccarini) Endlicher	Japan	32°48′–39°32′	110–2400	46–50	1.5–2
C. taiwanensis[d] Masamune and Suzuki	Taiwan	22°55′–24°46′	1200–2800	40	3
C. thyoides[e] (Linnaeus) Britton, Sterns and Poggenberg	E. North America	29°14′–44°20′	0–457	28–36	1.2–2.1

[a]Nomenclature of Dallimore and Jackson (1966) and Liu (1966).
[b]From Hayashi (1951), Li (1953), Japan For. Tech. Assoc. (1964), Liu (1966), Viereck and Little (1975), Antos and Zobel (1986), Laderman (1989), Harris (1990), and Zobel (1990).
[c]Range of three largest values found in the literature.
[d]Also classified as *C. obtusa* var. *formosana* (Hayata) Rehder (Li and Keng 1954).
[e]Some of the southern populations have been segregated into *C. thyoides* var. *henryae* Li (Laderman 1989).

a very restricted geographic range, but grows in a variety of habitats. It co-occurs with *C. nootkatensis* in only two known stands. *Chamaecyparis nootkatensis* has an extended latitudinal range; it is subalpine in the south, but descends to sea level in Alaska. It has outliers 400 and 425 km inland, far east of its usual range. *Chamaecyparis thyoides* occurs over a wide range of latitude and up to 285 km inland (Little 1971), but grows primarily in low-elevation bogs, often at or near sea level.

Environment

Weather stations within the native range of *Chamaecyparis* forests have mean annual temperatures from less than 5°C to greater than 20°C (Fig. 2.1). Although *C. nootkatensis* grows at the lowest temperatures, values for the other species overlap substantially. Precipitation exceeds 1,000 mm for all stations and exceeds 3,000 mm for three species. It is generally lowest for *C. thyoides* and highest in Taiwan (Fig. 2.1).

Climates of the species are more clearly differentiated by their seasonality (Fig. 2.2). In western North America, summer (June to August) rainfall is low; the proportion of total annual rainfall approaches 25% only in Alaska. For Asian species, precipitation is concentrated during summer. Seasonal temperature variability is greatest in Japan and for *C. thyoides*, and is least in Taiwan. These two seasonal climatic criteria clearly differentiate the environments of the five regions where the genus occurs (Fig. 2.2).

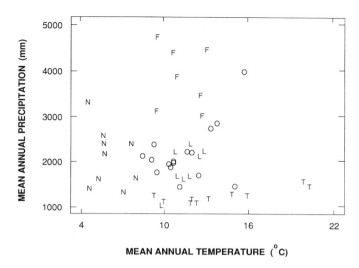

Figure 2.1. Mean annual temperature and mean annual precipitation at weather stations within the ranges of *Chamaecyparis* spp. F: range of Taiwanese spp. L: range of *C. lawsoniana.* N: range of *C. nootkatensis.* O: range of Japanese spp. T: range of *C. thyoides.* Data are primarily from Maeda (1951), Sato (1974), Hawk (1977), Willmott et al. (1981), and U.S. National Oceanographic and Atmospheric Administration summaries.

Local microclimates of *Chamaecyparis* stands may differ somewhat from weather station data. For example, understory temperatures in *C. lawsoniana* stands have annual means that vary from being the same as for weather stations to almost 5°C below; annual ranges vary similarly both in stands and at weather stations (Zobel et al. 1985).

Chamaecyparis species occupy a wide variety of substrates. *Chamaecyparis thyoides* usually grows on organic soils in wetlands (Laderman 1989), and *C. nootkatensis* is common in such sites in the northern part of its range. *Chamaecyparis nootkatensis* may also occupy rocky, shallow ridgetop soils in the Cascade Range, associated with xerophytic species characteristic of semi-arid woodland and shrub-steppe (Antos and Zobel 1986). Its inland outlier populations are in drier, more continental regions than its usual marine coastal climate. At least three species grow on ultramafic rocks (which contain high concentrations of magnesium and iron), sometimes dominating there. *Chamaecyparis obtusa* and *C. lawsoniana,* in particular, grow on a variety of substrates (Sato 1974, Zobel et al. 1985). Availability of soil moisture in summer limits the distribution of *C. lawsoniana* (Zobel et al. 1985).

The members of both species pairs in Asia are separated similarly in their topo-edaphic occurrence. In Japan, *C. obtusa* occurs on upper slopes and ridges, and *C. pisifera* exists on more gentle topography with moist soils (Sato 1974). In Taiwan, *C. taiwanensis* (Taiwan red cypress, recognized by many as *C. obtusa* var. *formosana*)

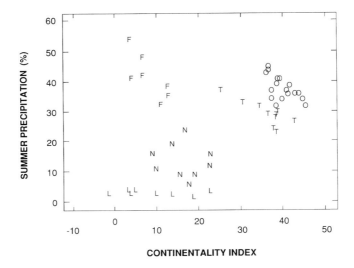

Figure 2.2. Seasonality of climate at weather stations within the ranges of *Chamaecyparis* spp. "Summer Precipitation" is the percent of annual total received in June through August. "Continentality" (C) is the mean annual temperature range (A) (A = mean of warmest month − mean of coldest month) adjusted for latitude and elevation (C = [1.7A/sin (f + 10 + 9 h)] −14, where f = latitude and h = elevation in km; Tuhkanen 1980, p. 39, eq. 74). Data sources and symbols are as in Figure 2.1.

occurs at higher elevations and topographic positions and on better-drained sites than does *C. formosensis* (Hawk 1977). In both areas, there is substantial overlap of ranges and habitat requirements, and mixed stands occur.

Associated Flora

The flora associated with forests of each species was determined from descriptions of well-developed forest stands (>45% tree cover) that include a substantial component of *Chamaecyparis* in the overstory (≥15% of tree importance; Tables 2.2 and 2.3). The total flora with which the species grow may be much larger, as data given here exclude forest types where the canopy is open or *Chamaecyparis* is rare, and nonforested areas with occasional trees or shrub-form individuals. For example, Laderman's (1989) species list for *C. thyoides* includes over 560 species, compared to the 192 analyzed here, and Antos and Zobel (1986) report 387 species with only a part of the range of *C. nootkatensis*. Estimates of species richness increase as the number and geographic dispersion of available stand descriptions increase; richness will depend partially on the sources of data available and those selected for use. Comparisons are made using relative importance of large taxonomic categories, which are not confined to one region as genera or even families may be. Importance

Table 2.2 Sources of Stand Descriptions Used to Compile Floristic Lists (Table 2.3)

Species	References
C. formosensis and *C. taiwanensis*	Chang (1961) Chang (1963) *Hawk (1977) [2 types for each species]
C. lawsoniana	*Atzet and Wheeler (1984) [4 associations] *Hawk (1977) [7 types]
C. nootkatensis	Antos, J. A. *(unpublished data)* *Antos and Zobel (1986) [5 stand groups] *Atzet and Wheeler (1984) [1 association] Banner et al. (1983) *Brooke et al. (1970) [2 associations, 1 variant] *Franklin et al. (1988) [3 associations] *Frenkel (1974) [1 mean for "dry" forest in a disjunct stand] *Guiguet (1953) [1 transect] Hennon (1986) *Ver Hoef (1985) [2 stand groups (based on soils)]
C. obtusa and *C. pisifera*	*Maeda (1951) [5 *C. obtusa* types] *Maeda and Yoshioka (1952) [1 *C. pisifera* association, 1 *C. obtusa* association] Nunotani (1978, 1979) Sato (1974) *Yamanaka (1957) [2 *C. obtusa* types—on and off serpentine]
C. thyoides	Collins et al. (1964) Ehrenfeld (1986) Eleuterius and Jones (1972) Harper (1926) Harshberger (1916) Hull and Whigham (1987) *Kologiski (1977) [1 mean for all stands] *Laney et al. (1990) [1 mean for a local area] Little (1951) Lynn (1984) McKinley and Day (1979) Niering (1953) Sipple and Klockner (1980) Sorrie and Woolsey (1987) *Stoltzfus (1990) [1 mean for mature forests]

*References used in ordination analysis, using growth forms, indicating the number of vegetation units for which data were extracted.

of a taxonomic type was determined as the number of species associated with forests of each *Chamaecyparis* species and then displayed as a percent of species present. Exclusion of *C. pisifera* from Table 2.3 was necessitated by limited vegetation data. Data for Taiwanese species are similar, and they are discussed below as one unit.

Composition of the flora varies among *Chamaecyparis* species (Table 2.3). *Chamaecyparis obtusa* grows with many conifers and woody angiosperms of the Dilleniidae and Hamamelidae families, but few monocots. In Taiwan, many ferns and few

Table 2.3 Summary of Taxa in the Flora of Forests of *Chamaecyparis* spp.[a]

Taxon	*C. formosensis*	*C. lawsoniana*	*C. nootkatensis*	*C. obtusa*	*C. taiwanensis*	*C. thyoides*
Dicotyledons						
Magnoliidae	10	8	7	5	11	7
(Lauraceae, Ranunculaceae)						
Hamamelidae	11	5	1	14	9	7
(Fagaceae)						
Caryophyllidae	2	2	5	0	2	0
Dilleniidae	16	14	13	25	19	20
(Theaceae, Violaceae, Ericaceae)						
Rosidae	19	22	35	19	15	14
(Saxifragaceae, Rosaceae, Aquifoliaceae, Umbelliferae)						
Asteridae	11	14	18	6	10	14
(Scrophulariaceae, Caprifoliaceae, Compositae)						
Monocotyledons						
Arecidae	0	1	<1	0	0	2
Commelinidae	5	4	9	5	4	19
(Cyperaceae, Gramineae)						
Liliidae	3	15	10	2	3	8
(Liliaceae, Orchidaceae)						
Conifers	4	10	7	18	3	4
Ferns	18	6	6	5	22	5
Other Vascular Cryptogams	2	0	2	1	2	2
Total Species (No.)	134	163	273	131	130	192

[a]Data are percentages of vascular species in each taxon. The best-represented families within each subclass of angiosperms are listed. The subclass classification for angiosperms of Cronquist (1988) is used.

Sources of data are listed in Table 2.2.

conifers are characteristic, and Magnoliidae are most important. In western North America *(C. lawsoniana* and *C. nootkatensis),* Rosidae, Asteridae, and Liliidae are more important than elsewhere, and Dilleniidae and Hamamelidae are less so. In addition, Caryophyllidae are more important with *C. nootkatensis* than elsewhere. In forests of *C. thyoides,* Commelinidae and Dilleniidae are important, but conifers are not. In general, the frequency of monocot species is greater in American *Chamaecyparis* forests than in Asian; an important exception is the large grasses that dominate many Asian stands. In Japan and western North America, *Chamaecyparis* species co-occur with four to five other members of the Cupressaceae; other species co-occur with only one.

In North America, most species are herbaceous, especially so with *C. nootkatensis;* in Asia, most species are woody (Table 2.4). In Asia, most woody species are trees;

Table 2.4 Summary of Growth Form Characteristics of Woody Species in Five Areas of *Chamaecyparis* Forest

Characteristic	Species				
	C. formosensis & C. taiwanensis	C. obtusa	C. lawsoniana	C. nootkatensis	C. thyoides
Woody Species					
(% of total)	61	85	39	24	45
Trees					
(% of woody spp.)	62	56	44	24	30
Evergreen					
(% of woody spp.)	66	50	61	48	38
Lianas					
(% of woody spp.)	8	1	3	3	7
Leaf Size* (% of woody spp.)					
Lentophyll (25)	3	6	8	12	1
Nanophyll (225)	4	15	24	26	6
Microphyll (2025)	29	19	26	32	26
Mesophyll (18225)	62	50	35	23	61
Macrophyll (164,025)	2	8	6	7	6
Megaphyll	0	0	0	0	1

*Following Raunkaier (1934); the leaf size (value) in parentheses is the area (mm2) of the largest leaf in that size class.

in North America, most are shrubby. With *C. lawsoniana* and in Taiwan, most woody species are evergreen (Table 2.4); with *C. thyoides,* most are deciduous. Lianas are most common with *C. thyoides* and in Taiwan. Large, woody grasses (including bamboos) are common in Asia, rare with *C. thyoides,* and absent from western North America.

Leaf sizes of woody species vary (Table 2.4). Species in western North America have a range of smaller leaf sizes, whereas in Taiwan and with *C. thyoides* there are few small-leaved species, and a majority of woody species are mesophyllous. These leaf size differences probably reflect primarily the drier, sunnier growing season conditions (Givnish 1987) in western North America.

Regional Vegetation

Chamaecyparis species often occur in unusual forest communities, often limited to particular topo-edaphic conditions within a range of climatically controlled regional forest types. In Japan, *C. obtusa* ranges from warm temperate, broadleaved evergreen forests in the south to deciduous forests and subalpine evergreen forests (Sato 1974). Taiwanese *Chamaecyparis* forests are primarily within the upper *Quercus* (oak) vegetation zone, but often extend from the lower *Quercus* zone into the higher *Tsuga–Picea* (Hemlock–Spruce) zone (Su 1984: Fig. 2.3). They occur in "montane mixed coniferous forest" with other coniferous types, on sites that are moister than the montane deciduous broadleaved forest, but drier than the montane evergreen broadleaved forest (Su 1984: Fig. 2.2).

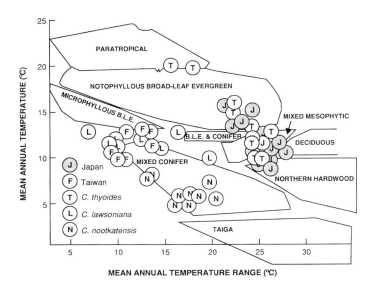

Figure 2.3. Mean annual temperature and mean annual temperature range at weather stations within the ranges of *Chamaecyparis* spp., superimposed on Wolfe's (1979) classification of mesic forest types in Asia. B.L.E.: broadleaved evergreen. Leaf sizes (mm²): microphyll 225–2025, notophyll 2025–4500. Data sources are as in Figure 2.1.

In western North America, *C. nootkatensis* occurs within mixed coniferous forests of the Pacific coastal mountains (Franklin 1988). It is part of the moister subalpine forests dominated by *Tsuga mertensiana* (mountain hemlock) in the southern part of its range, often being most important in wet or open sites and sometimes as a shrub in the understory. Farther north, *C. nootkatensis* is associated with the *Picea sitchensis–Tsuga heterophylla* (Sitka spruce–western hemlock) forests, often near sea level. Within its limited range, *C. lawsoniana* grows with all major types of coniferous forest, from sea level to subalpine, as well as in mixed evergreen forests with substantial broadleaved dominance (Zobel et al. 1985). It is most dominant in moist areas on ultramafic rocks.

Chamaecyparis thyoides, an edaphic specialist, is confined to alluvial and nonalluvial wetlands that are associated with forest types from boreal and mixed deciduous forest in the north (Laderman 1989) to a complex mixture of coniferous and broadleaved forests on the coastal plain of southeastern North America (Christensen 1988).

Based on temperature data, the type of vegetation expected (Wolfe 1979) in a mesic forest climate varies substantially for weather stations within *Chamaecyparis* species' ranges (Fig. 2.3). There is substantial concentration of the stations at ecotonal temperature combinations, especially for species in Taiwan and for *C. lawsoniana.* For *C. thyoides* and in Japan, stations represent several major forest types. Only *C. nootkatensis* is confined primarily to one major vegetation type (Fig. 2.3).

Forest Physiognomy

In order to compare forests of such variable composition, vegetation data were summarized by growth form. Each species in the vegetation descriptions was assigned to one of nine major growth forms: (1) trees in Cupressaceae, (2) trees in Pinaceae and Taxodiaceae, (3) evergreen angiosperm trees, (4) deciduous angiosperm trees, (5) evergreen shrubs, (6) deciduous shrubs, (7) lianas, (8) herbaceous plants, and (9) bamboos and other robust grasses. From available vegetation descriptions, 44 data sets from mature forests were selected (Table 2.2) that included species importance values for all vascular plants present. Thirty-nine data sets ("types") represent summary values for a given association, community-type, or similar classificatory unit identified by the investigator. Three sets represent a single transect or plot, and two others represent a summary of multiple samples in one small area.

The 44 sets of growth-form importance data were grouped by the species of *Chamaecyparis* present. A multiresponse permutation procedure (Zimmerman et al. 1985, McCune 1987) was used to determine that forests of different *Chamaecyparis* species differed ($P < 10^{-6}$) when the importance of all growth forms was considered. A Kruskal–Wallis test (Zar 1984) was used to compare importance of individual growth forms among forests of the seven *Chamaecyparis* species. Importance of herbs and deciduous shrubs did not differ among *Chamaecyparis* species, but importance of other growth forms did (Table 2.5).

Vegetation relationships among *Chamaecyparis* forest types were summarized by ordination of data from the 44 forest types, using detrended correspondence analysis (DCA; Hill and Gauch 1980). The first two axes (Fig. 2.4) separated western North American from other forests. There was some overlap between *C. lawsoniana* types with high broadleaf evergreen cover and Asian types with little or no bamboo. The one major exception was a *Tsuga*-dominated stand with 15% *C. obtusa* cover that

Table 2.5 Mean Percent Cover by Growth Form for Forest Types Including *Chamaecyparis* spp.

		Growth Form[a,b]								
		Trees				Shrubs				
Chamaecyparis spp.	*n*	Cupres.	Pinac.	Ever.	Decid.	Ever.	Decid.	Liana	Herb	Bamboo
C. formosensis	2	52abc	10c	66a	0bc	42a	14a	2a	31a	39ab
C. lawsoniana	11	46ab	53a	7bc	6b	45a	19a	0.3b	28a	0c
C. nootkatensis	16	27c	47ab	0d	0c	6b	30a	0b	42a	0c
C. obtusa	9	59ab	28bc	21abc	8a	21a	19a	1b	35a	12b
C. pisifera	1	65	0	0	0	30	4	8	56	0
C. taiwanensis	2	54abc	25bc	24ab	0bc	51a	28a	2a	22a	49a
C. thyoides	3	63a	4c	18abc	18a	52a	29a	13a	6a	0c

Note: C. pisifera was not included in the multiple comparison tests; *n* = no. of data sets used.

[a] Abbreviations: Cupres. = Cupressaceae, Pinac. = Pinaceae, Ever. = evergreen, Decid. = deciduous. For trees, "ever." and decid." refer to angiosperms.

[b] Values in a column that share a letter do not differ statistically (Kruskal-Wallis multiple comparison test, $P < .01$; Zar 1984).

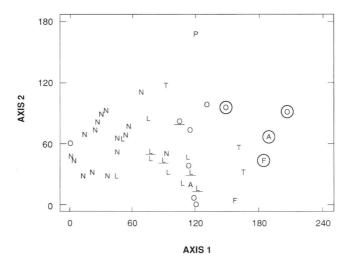

Figure 2.4. Results of ordination (Axes 1 and 2) of 44 *Chamae-cyparis* forest types using detrended correspondence analysis of type descriptions based on importance of nine growth forms. A: *C. taiwanensis.* F: *C. formosensis.* L: *C. lawsoniana.* N: *C. noot-katensis.* O: *C. obtusa.* P: *C. pisifera.* T: *C. thyoides.* Underlined symbols are types that occur primarily on ultramafic rock. Cir-cled symbols include >30% cover of bamboo. Sources of data are given in Table 2.2.

was grouped with *C. nootkatensis* stands. The first two axes also separated most *C. lawsoniana* forests from most of *C. nootkatensis.* Asian and *C. thyoides* forests over-lapped broadly on the first two axes (Fig. 2.4), except that the single *C. pisifera* type was extreme for all types on the second axis. Forests including all other *Chamaecy-paris* species were broadly distributed over the second axis. The third axis (Fig. 2.5) separated most *C. lawsoniana* forests (in the lower half) from Asian and *C. thyoides* forests (in the upper two-thirds). Ordinations of growth-forms produced by DCA (Fig. 2.6) indicated that the axes represent changes as follows: Axis 1—dominance by Pinaceae and deciduous shrubs to dominance by evergreen angiosperms and bamboo; Axis 2—evergreen angiosperm trees and shrubs to deciduous angiosperm trees with lianas; and Axis 3—Pinaceae trees with evergreen shrubs to angiosperm trees with deciduous shrubs.

Environmental patterns cannot definitely be assigned to Axis 1, except for the western North American/Asian dichotomy; Axis 1 is clearly not associated with either the mean or seasonal variation for either temperature or precipitation. Variation within forests of a given *Chamaecyparis* species shows no clearly geographical pattern on any axis, except for a north-to-south gradient among the *C. thyoides* stands from low to high values of Axis 1 and the negatively correlated changes for Axes 2 and 3. Much of the difference among types appears to be associated with undefined local

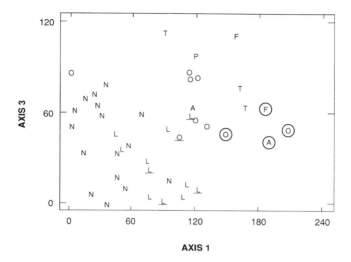

Figure 2.5. Results of ordination (Axes 1 and 3) of 44 *Chamae-cyparis* forest types using detrended correspondence analysis of type descriptions based on importance of nine growth forms. Symbols are as in Figure 2.4. Underlined symbols are types that occur primarily on ultramafic rock. Circled symbols include >30% cover of bamboo. Sources of data are given in Table 2.2.

habitat variability, rather than with regional climatic gradients. Forest types regularly on ultramafic rocks were concentrated in the center of Axis 1 (Figs. 2.4 and 2.5), but were not clearly separated from types on other substrates.

The Habitat of *Chamaecyparis* Forests

Chamaecyparis forests occupy a wide range of environments. Some species occur across climatic ranges that support several different regional forest types. All are more-or-less coastal and receive substantial precipitation. Lack of water seems to limit the generic distribution, even though two species occupy summer-dry climates, and several occur on apparently dry montane sites. Dry-site occurrence of *C. lawsoni-ana,* however, can usually be accounted for by microclimatic or edaphic compensa-tion (Zobel et al. 1985). In addition, *C. thyoides,* with the lowest overall precipitation and highest temperatures, dominates only on wetlands. For *Chamaecyparis,* then, temperature, substrate, excessive moisture, and snowpack, which may limit other tree species, seem generally less important as limiting factors than low availability of water.

The abiotic requirements of *Chamaecyparis* seem to differ from those of trees that dominate regional vegetation types. No *Chamaecyparis* species dominates what could be called zonal vegetation, widespread enough to represent a biome or even a sub-stantial part of one. *Chamaecyparis* forests are usually fragmented, and the species

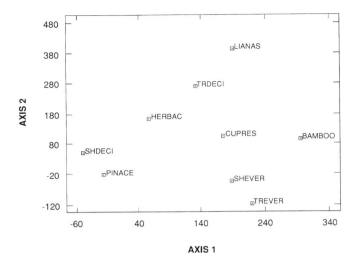

Figure 2.6. Results of ordination (Axes 1 and 2) of nine growth forms in *Chamaecyparis* forests, based on their importance in 44 forest types, using detrended correspondence analysis. PINACE: trees of Pinaceae. CUPRESS: trees of Cupressaceae. TREVER: evergreen broadleaf trees. TRDECI: deciduous broadleaved trees. SHEVER: evergreen shrubs. SHDECI: deciduous shrubs. HERBAC: herbaceous plants. Data from sources in Table 2.2.

are often confined to extreme edaphic conditions or to subordinate status in forests on productive soils. Three species that occupy narrow ranges of climate are ecotonal between physiognomically well-defined regional forest types (Fig. 2.3). Although *Chamaecyparis* forests vary greatly in physiognomy among and within species, the major axis of vegetational variation (Fig. 2.4, Axis 1) does not parallel any clear climatic gradient.

Chamaecyparis is distributed as are most smaller conifer genera (Li 1953, Florin 1963); it is concentrated in moist habitats in mountainous regions at mid-latitudes on the margins of the Pacific basin. In these locations on ancient land masses, temperature fluctuations are muted, substantial relief has allowed species to migrate up- and down-slope as temperatures changed, and a moist habitat has been available consistently during the Cenozoic climatic changes. Climatic reconstructions from fossils suggest that, from about 50 million to 10 million years ago, North American *Chamaecyparis* grew with a wide variety of floristic associates at temperatures within the range now occupied by the genus, although sometimes with lower (but seasonally uniform) rainfall (Edwards 1983).

Based on these generalities and detailed studies of some species (e.g., Little 1950, Maeda 1951, Yamanaka 1957, Sato 1974, Hawk 1977, Zobel et al. 1985, Antos and Zobel 1986), it seems that any general explanation of *Chamaecyparis* forest distribution must emphasize limitation by water. Within areas with adequate moisture, however, control of species importance is probably dominated by interactions with other biota, particularly the regionally dominant trees. Success of *Chamaecyparis* seems limited to situations that reduce the importance of the major dominant trees, and usually only to a limited subset of those azonal environments.

References

Antos, J.A., and D.B. Zobel. 1986. Habitat relationships of *Chamaecyparis nootkatensis* in southern Washington, Oregon and California. Can. J. Bot. **64**:1898–1909.

Atzet, T., and D.L. Wheeler. 1984. Preliminary plant associations of the Siskiyou Mountain Province. U.S. Dept. Agric. For. Serv., Pac. Northwest Region.

Banner, A., J. Pojar, and G.E. Rouse. 1983. Post-glacial paleoecology and successional relationships of a bog woodland near Prince Rupert, British Columbia. Can. J. For. Res. **13**:938–947.

Brooke, R.C., E.B. Peterson, and V.J. Krajina. 1970. The subalpine mountain hemlock zone. Ecol. West. N. Amer. **2**:147–349.

Chang, L.-M. 1961. Ecological studies on the vegetation of Mt. Ta-Yuan. Taiwan For. Res. Inst. Bull. 70 (in Chinese, with English summary).

———. 1963. Ecological studies of *Chamaecyparis formosensis* and *Chamaecyparis taiwanensis* mixed forest in Taiwan. Taiwan For. Res. Inst. Bull. 91 (in Chinese, with English summary).

Christensen, N.L. 1988. Vegetation of the southeastern Coastal Plain. Ch. 11. Pages 317–363 *in* M.G. Barbour and W.D. Billings, eds. North American terrestrial vegetation. Cambridge Univ. Press, Cambridge.

Collins, E.A., C.D. Monk, and R.H. Spielman. 1964. White-cedar stands in northern Florida. Q. J. Fla. Acad. Sci. **27**:107–110.

Cronquist, A. 1988. The evolution and classification of flowering plants. Second Ed. New York Bot. Gard., Bronx, NY.

Dallimore, W., and A.B. Jackson. 1966. A handbook of Coniferae and Ginkgoaceae. Revised by S.G. Harrison. Edward Arnold, London.

Edwards, S.W. 1983. Cenozoic history of Alaskan and Port Orford *Chamaecyparis* cedars. Dissertation. Univ. California, Berkeley.

Ehrenfeld, J.G. 1986. Wetlands of the New Jersey pine barrens: the role of species composition in community function. Am. Midl. Natur. **115**:301–313.

Eleuterius, L.N., and S.B. Jones. 1972. A phytosociological study of white-cedar in Mississippi. Castanea. **37**:67–74.

Florin, R. 1963. The distribution of conifer and taxad genera in time and space. Acta Horti Bergiani **20**:121–312.

Franklin, J.F. 1988. Pacific northwest forest. Chapter 4. Pages 103–130 *in* M.G. Barbour and W.D. Billings, eds. North American terrestrial vegetation. Cambridge Univ. Press.

Franklin, J.F., W.H. Moir, M.A. Hemstrom, S.E. Greene, and B.G. Smith. 1988. The forest communities of Mount Rainier National Park. U.S. Dept. Inter. National Park Serv. Sci. Monogr. Ser. 19.

Frenkel, R.E. 1974. An isolated occurrence of Alaska-cedar (*Chamaecyparis nootkatensis* [D. Don] Spach) in the Aldrich Mountains, central Oregon. Northw. Sci. **48**:29–37.

Givnish, T.J. 1987. Comparative studies of leaf form: assessing the relative roles of selective pressures and phylogenetic constraints. New Phytol. **106** (Suppl.):131–160.

Guiguet, C.J. 1953. An ecological study of Goose Island, British Columbia, with special reference to mammals and birds. British Columbia Prov. Mus. Occas. Pap. 10.

Harper, R.M. 1926. A middle Florida cedar swamp. Torreya **26**:81–84.

Harris, A.S. 1990. *Chamaecyparis nootkatensis* (D. Don) Spach. Pages 97–102 *in* R.M. Burns and B. H. Honkala, tech. coords. Silvics of North America. Vol. 1, Conifers. U.S. Dept. Agric. For. Serv. Agric. Handb. 654. Washington, DC.

Harshberger, J.W. 1916. The vegetation of the New Jersey pine-barrens. An ecologic investigation. Christopher Sower, Philadelphia.

Hawk, G.M. 1977. A comparative study of temperate *Chamaecyparis* forests. Dissertation. Oregon State Univ., Corvallis.

Hayashi, Y. 1951. The natural distribution of important trees indigenous to Japan. Conifers. Report I. Bull. Gov. Forest Exper. Sta. 48 (in Japanese, with English summary).

Hennon, P.E. 1986. Pathological and ecological aspects of decline and mortality of *Chamaecyparis nootkatensis* in southeast Alaska. Dissertation. Oregon State Univ., Corvallis.

Hill, M.O., and H.G. Gauch, Jr. 1980. Detrended correspondence analysis: an improved ordination technique. Vegetatio. **42**:47–58.

Hull, J.C., and D.F. Whigham. 1987. Vegetation patterns in six bogs and adjacent forested wetlands on the inner Coastal Plain of Maryland. Pages 143–173 in A. Laderman, ed. Atlantic white cedar wetlands. Westview Press, Boulder.

Japan Forest Tech. Assoc. 1964. Illustrated important forest trees of Japan.

Kologiski, R.L. 1977. Phytosociology of the Green Swamp, North Carolina. Dissertation. North Carolina State Univ., Raleigh.

Laderman, A. 1989. The ecology of Atlantic white cedar wetlands: A community profile. U.S. Fish Wildl. Serv. Biol. Rep. 85(7.21).

Laney, R.W., R.E. Noffsinger, and J.H. Moore. 1990. Personal communication.

Lee, S.-C. 1962. Taiwan red- and yellow-cypress and their conservation. Taiwania **8**:1–13.

Li, H.-L. 1953. Present distribution and habitats of the conifers and taxads. Evolution **7**:245–261.

Li, H.-L., and H. Keng. 1954. *Icones Gymnospermum Formosanarum.* Taiwania **5**:25–83.

Little, E.L., Jr. 1971. Atlas of United States trees. 1. Conifers and important hardwoods. U.S. Dept. Agric. For. Serv. Misc. Publ. 1146.

Little, S., Jr. 1950. Ecology and silviculture of white cedar and associated hardwoods in southern New Jersey. Yale Univ. Sch. For. Bull. 56.

Little, S. 1951. Observations on the minor vegetation of the pine barren swamps in southern New Jersey. Bull. Torrey Bot. Club **78**:153–160.

Liu, T. 1966. Study on the phytogeography of the conifers and taxads of Taiwan. Bull. Taiwan For. Res. Inst. 122 (in Chinese, with English summary).

Lynn, L.M. 1984. The vegetation of Little Cedar Bog, southeastern New York. Bull. Torrey Bot. Club **111**:90–95.

Maeda, T. 1951. Sociological study of *Chamaecyparis obtusa* forest and its Japan-sea elements. Enshurin **8**:21–47 (in Japanese, with English summary).

Maeda, T., and J. Yoshioka 1952. Studies on the vegetation of Chichibu Mountain forest (II). The plant communities of the temperate mountain zones, with plates II–III. Bull. Tokyo Univ. Forests **42**:129–150 (in Japanese, with English summary).

McCune, B. 1987. Multivariate analysis on the PC-ORD system. Butler Univ., Holcombe Res. Inst. Rep. 75.

McKinley, C.E., and F.P. Day, Jr. 1979. Herbaceous production in cut-burned, uncut-burned and control areas of a *Chamaecyparis thyoides* (L.) BSP (Cupressaceae) stand in the Great Dismal Swamp. Bull. Torrey Bot. Club **106**:20–28.

Niering, W.A. 1953. The past and present vegetation of High Point State Park, New Jersey. Ecol. Monogr. **23**:127–150.

Nunotani, T. 1978. Ecological structure and dynamics in forest floor vegetation. I. Age structure and distribution in Ooe hinoki *(Chamaecyparis obtusa)* forest. Osaka Mus. Nat. Hist. Bull. **31**:1–11 (in Japanese, with English summary).

———. 1979. Ecological structure and dynamics in forest floor vegetation. II. Age distribution and structure in Kamigamo hinoki *(Chamaecyparis obtusa)* forest. Osaka Mus. Nat. Hist. Bull. **32**:19–30 (in Japanese, with English summary).

Raunkaier, C. 1934. The life forms of plants and statistical plant geography. Clarendon Press, Oxford.

Sato, K. 1974. *Chamaecyparis obtusa* in Japan. Partial translation by M.E. Hale. Oregon State Univ., Dept. of Botany and Plant Pathology, Corvallis.

Sipple, W.S., and W.A. Klockner. 1980. A unique wetland in Maryland. Castanea **45**:60–69.

Sorrie, B.A., and H.L. Woolsey. 1987. The status and distribution of Atlantic white cedar in Massachusetts. Pages 135–142 *in* A. Laderman, ed. Atlantic white cedar wetlands. Westview Press, Boulder.

Stoltzfus, D.L. 1990. Development of community structure in relation to disturbance and ecosystem fragmentation in Atlantic white cedar swamps in the Pinelands National Reserve, New Jersey. Dissertation. Rutgers Univ., New Brunswick.

Su, H.-J. 1984. Studies on the climate and vegetation types of the natural forests in Taiwan. II. Altitudinal vegetation zones in relation to temperature gradient. Quart. J. Chinese For. **17**:57–73.

Tuhkanen, S. 1980. Climatic parameters and indices in plant geography. Acta Phytogeogr. Suecica 67.

Ver Hoef, J.M. 1985. Vegetation patterns of two areas in southeast Alaska's forests. Masters' thesis. Univ. of Alaska, Fairbanks.

Viereck, L.A., and E.L. Little, Jr. 1975. Atlas of United States trees. 2. Alaska trees and common shrubs. U.S. Dept. Agric. For. Serv. Misc. Publ. 1293.

Waring, R.H., and J.F. Franklin. 1979. Evergreen coniferous forests of the Pacific Northwest. Science **204**:1380–1386.

Willmott, C.J., J.R. Mather, and C.M. Rowe. 1981. Average monthly and annual surface air temperature and precipitation data for the world. 1. The Eastern Hemisphere. 2. The Western Hemisphere. Publica. Climatol. 34, C.W. Thornthwaite, Elmer, NJ.

Wolfe, J.A. 1979. Temperature parameters of humid to mesic forests of eastern Asia and relation to forests of other regions of the Northern Hemisphere and Australasia. U.S. Geol. Surv. Prof. Pap. 1106.

Yamanaka, T. 1957. On the *Chamaecyparis obtusa* forest in Shikoku, southern Japan. Jap. J. Ecol. **6**:149–157 (in Japanese, with English summary).

Zar, J.H. 1984. Biostatistical analysis. Second edition. Prentice-Hall, Englewood Cliffs, NJ.

Zimmerman, G.M., H. Goetz, and P.W. Mielke, Jr. 1985. Use of an improved statistical method for group comparisons to study effects of prairie fire. Ecology **66**:606–611.

Zobel, D.B. 1990. *Chamaecyparis lawsoniana* (A. Murr.) Parl. Port-Orford-cedar. Pages 88–96 *in* R.M. Burns and B.H. Honkala, tech. coords. Silvics of North America. Vol. 1, Conifers. U.S. Dept. Agric. For. Serv. Agric. Handb. 654.

Zobel, D.B., L.F. Roth, and G.M. Hawk. 1985. Ecology, pathology and management of Port-Orford-cedar *(Chamaecyparis lawsoniana)*. USDA Forest Serv. Gen. Tech. Rep. PNW-184.

3

Paul E. Hennon, Charles G. Shaw III, & Everett M. Hansen

Reproduction and Forest Decline of *Chamaecyparis nootkatensis* (Yellow-Cedar) in Southeast Alaska, USA

Chamaecyparis nootkatensis (D. Don) Spach (yellow-cedar, Alaska yellow-cedar) is an ecologically important and extremely valuable tree species in coastal Alaska. Its ecology, silvics, and methods of management, however, are poorly understood (Harris 1990, Hennon 1995). *Chamaecyparis nootkatensis,* which is now dying on more than 200,000 hectares (ha) of unmanaged forest in southeast Alaska, is suffering from the most severe forest decline in western North America (Fig. 3.1). In addition, this species has limited successful reproduction in many declining and healthy unmanaged forests in Alaska; numerous searches have failed to locate seedlings (Harris 1990). Forest managers are particularly concerned because *C. nootkatensis* is not frequently regenerating on sites harvested by logging.

In 1981, we initiated the first detailed studies on reproduction, community relationships, and etiology and epidemiology of *C. nootkatensis* decline in southeast Alaska. This chapter briefly reviews the resource and silvics of *C. nootkatensis* and summarizes our recent studies.

Value and Uses

On a per-unit-volume basis, *C. nootkatensis* is the most valuable tree grown in Alaska and consistently commands the highest price (see Fig. 4.1 in Dunsworth, Chapter 4, this volume). Its bright yellow, aromatic wood has narrow annual rings, extreme decay resistance, and excellent strength (Harris 1971, Hennon 1995). Currently, its wood is used for boat building and other marine purposes, interior molding, cabinets, furniture, fence posts, bleacher seats, saunas, musical instruments, and carving (Harris 1971, Frear 1982). Most of the harvested wood is exported to countries along the Pacific Rim, especially Japan, where it is used in home construction

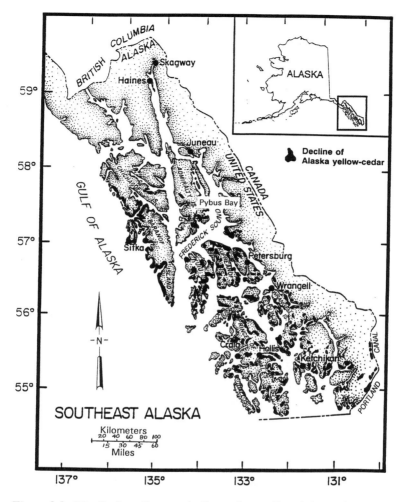

Figure 3.1. Distribution of severe decline and mortality of *C. nootkatensis* in southeast Alaska.

and is prized for making ceremonial boxes and restoring temples and shrines (Frear 1982).

Natural Distribution

Chamaecyparis nootkatensis has a natural distribution from near Port Wells in Prince William Sound in Alaska, south through southeast Alaska and British Columbia, to near the Oregon–California border (Fig. 3.2; Harris 1971, Harris 1990). It occurs from tidewater to timberline in southeast Alaska (Harris and Farr 1974), where it can

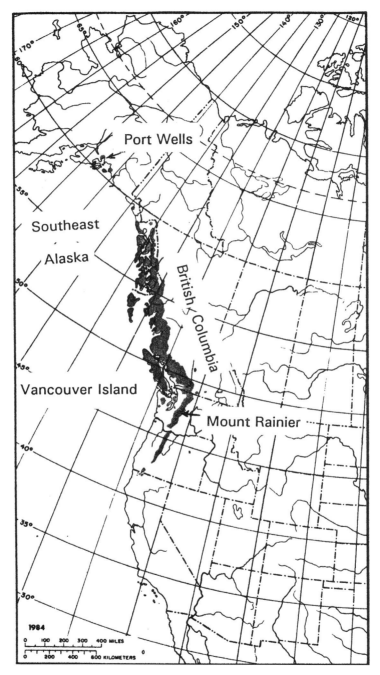

Figure 3.2. Natural distribution of *C. nootkatensis* (modified from Harris 1990).

grow in nearly pure stands. More commonly, it exists in scattered groups or as individual trees mixed with other conifers (Ruth and Harris 1979). In its natural range south of Vancouver Island, *C. nootkatensis* is restricted to higher elevations (e.g., 600–2,300 m). It grows within 200 km of the Pacific Coast, except for several isolated, disjunct stands that occur to the east in British Columbia and Oregon.

Throughout its range, *C. nootkatensis* is confined to a cool, humid climate with a relatively short growing season and winter temperatures that are not exceptionally cold. It apparently can grow on extremely poor soils if there is an abundant supply of moisture (Anderson 1959), and may be uncommon on more productive sites because of its inability to compete with faster growing tree species (Harris 1971). It occurs on dry sites in the southern portion of its range (Antos and Zobel 1986).

Chamaecyparis nootkatensis is among conifers with the greatest longevity; trees more than 1,000 years old are common (Franklin and Dyrness 1973). It can also be one of the slowest growing conifers in North America. Radial growth rates of 2 rings/mm have been recorded (Harris 1971), and we examined trees with 12 rings/mm. The largest known *C. nootkatensis* was found in Mount Rainier National Park, Washington; it had a diameter at breast height of 243 cm and was 40.2 m tall (Harris 1970).

Southeast Alaska

Southeast Alaska, the location of our studies, consists of a thin strip of mainland dominated by mountains and bisected by fjords and glaciers, with a scattering of large and small islands to the west. The cool, moist climate has an annual precipitation ranging from 150 to 500 cm (Harris 1990). Winters have relatively moderate temperatures, although brief cold periods occur. Summers are without prolonged dry periods and lightning is rare; thus fire is not an important factor in forest succession (Harris 1990). Windthrow and landslides are common disturbances (Harris 1990); diseases and insects are probably also important factors in mortality. Poorly drained soils, which are highly organic and shallow to deep, generally occur on sites without a steep slope and overlay unfractured bedrock or compact glacial till. Bogs and scrub forests occur on these poorly drained sites, but their patterns of plant succession are not well understood.

This region is dominated by undisturbed old-growth forests of *Tsuga heterophylla* (western hemlock) and *Picea sitchensis* (Sitka spruce), which account for 89% of the volume of commercial forests. By comparison, *C. nootkatensis, Thuja plicata* (western redcedar), *Tsuga mertensiana* (mountain hemlock), and *Pinus contorta* var. *contorta* (shore pine) compose far less volume. *Chamaecyparis nootkatensis* accounts for about 4% of the volume on commercial forests in southeast Alaska (Hutchinson and LaBau 1975), but if lower volume forests, which are numerous and expansive in some areas, were included, then the percentage of the total volume would be greater. Despite a relatively small volume, the high value of its wood makes *C. nootkatensis* a commercially important tree species.

Community Relationships

Many results reported in this chapter on reproduction, community relationships, and decline of *C. nootkatensis* are from data collected during ground surveys in declining forests on Baranof and Chichagof Islands in southeast Alaska (Hennon et al. 1990b). Surveys were also conducted in declining forests on Prince of Wales and Wrangell Islands near the northern limit of the natural distribution of *T. plicata* to determine if this species is also suffering from decline.

Along with data on overstory trees (Hennon et al. 1990b), the abundance of 55 understory plant taxa, including regenerating *C. nootkatensis,* was recorded from 280 variable radius plots on 21 transects. These data were used in an ordination analysis (DECORANA; Hill and Gauch 1980) to indicate relationships of plant communities with forest trees, reproduction, and decline.

Ordination of understory plant taxa produced only one important axis, the first axis. Field observations on the distribution of these taxa suggest that this axis represents a gradient from bog communities to those with better drainage (Hennon et al. 1990c). The first axis of DECORANA also produced ordination scores for each plot that represent the same gradient. Based on the average basal areas of live trees from these plots, *C. nootkatensis* was the dominant overstory component along most of the vegetation gradient (Fig. 3.3). It was rarely missing on plots from this intermediate zone. *Pinus contorta* was common in the open bogs where total conifer basal area was at a minimum. At the other extreme (better drainage) where total conifer basal

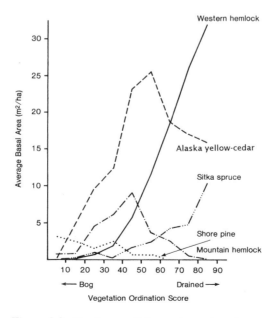

Figure 3.3. Basal area of *C. nootkatensis* and other conifers in plots along the ordination gradient from bog to better drainage.

area was greatest, *T. heterophylla* was predominant. Although less dominant, *C. nootkatensis* attained its greatest dimensions of height and diameter on these productive sites. Thus, the occurrence of *C. nootkatensis* appears to be relatively unrestricted by the major factor (soil drainage) that controls the distribution of plants in these forests.

Reproduction

Antos and Zobel (1986) discussed natural reproduction of *C. nootkatensis* in Washington and Oregon, but its reproduction in southeast Alaska had not been previously studied. To evaluate its natural reproduction in undisturbed cedar forests, we recorded the number of seedlings of *C. nootkatensis* and the presence or absence of prostrate growing patches of *C. nootkatensis* from plots during the surveys already discussed. We observed the occurrence of seedlings, saplings, and prostrate patches of *C. nootkatensis* at numerous locations in bogs, mature cedar stands, and clearcut harvest sites throughout southeast Alaska. In addition, we spoke to U.S. Forest Service land managers responsible for reforestation about the incidence of natural reproduction on logged sites.

Vegetative Reproduction

Chamaecyparis nootkatensis frequently occurs as a low-growing, prostrate form in southeast Alaska. Patches typically have several to many small upright stems that sweep downslope or, on flat terrain, in many directions. They have short (e.g., 1 m tall), intermingled crowns. The upright stems are usually more than 100 years old, although they are often less than 5 cm in diameter. Excavations of their roots (Hennon 1986) show that stems within single patches are typically connected to the same root system, and that lower branches from these stems frequently root adventitiously. These patches actively spread by means of vegetative reproduction, sometimes called layering. Stems become independent, yet genetically similar, plants when the "root-stem" connecting them dies and eventually decays.

The incidence of prostrate patches of cedar is strongly associated with bog understory plants and becomes less common and eventually absent along the gradient to plant communities on sites with better drainage (Fig. 3.4). This is not surprising; layering occurs in bogs where trees are short and lower branches are in contact with the ground. Layering would presumably be infrequent in closed-canopy forests on well-drained sites where lower limbs are usually far above ground. Thus, asexual reproduction cannot be relied upon to regenerate logged sites, which typically are well-drained and lack prostrate patches of *C. nootkatensis*. Outside of our study areas, we have found layering in avalanche and landslide sites.

Perhaps due to drier soil characteristics or escaping heavy snows, some upright stems on wet sites slowly acquire conical form and grow into small trees. Many mature trees of *C. nootkatensis* growing on wet sites may have originated from this form of vegetative reproduction.

Some small, isolated, and upright stems of cedar were emergent in bogs. These small cedars clearly developed from seedlings; their root systems could be followed

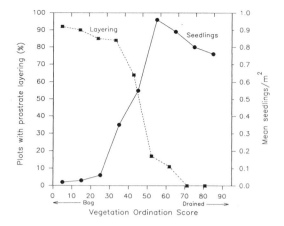

Figure 3.4. Density of seedlings (closed circles) and occurrence of prostrate patches of *C. nootkatensis* (closed squares) in 3 m² plots along the gradient from bog to better drainage.

distally to fine root-ends and did not attach to other root systems, as was the case with layering. Some of these smaller seedlings had juvenile foliage, indicating recent origin from seed. Some older seedlings had adventitious roots along lateral branches and were apparently developing vegetatively into patches of prostrate cedar. Other patches of prostrate cedar could be traced back to the adventitious rooting of branches on fallen or broken trees. Thus, these prostrate patches of *C. nootkatensis* may originate from the stem of a fallen or broken cedar, a lower limb from an existing patch, or a seedling establishing in a bog.

Sexual Reproduction

Seedlings of *C. nootkatensis* were never frequent, and their density (Fig. 3.4) closely paralleled the amount of live *C. nootkatensis* basal area of mature trees (Fig. 3.3). Seedlings were uncommon in bogs, reached peak densities (0.1 seedlings/m²) in communities with moderate drainage, and became less common on sites with better drainage dominated by *T. heterophylla.* The distance of seed dispersal for *C. nootkatensis* is normally short (less than 120 m; Harris 1990), perhaps because of its small seed wings (Owens and Molder 1984), and may account for the close association of cedar seedlings and live cedar basal area.

Most seedlings that we found were in the germinant stage (e.g., up to three years old with cotyledons recognizable). They were distinguished from other conifers in this region by having only two cotyledons (Franklin 1961), except for germinants of *T. plicata.* In the case of *T. plicata,* which has two cotyledons, foliage morphology was used to discriminate among seedlings. Larger, established seedlings of *C. nootkatensis,* but still with juvenile foliage, were less common than germinants. Seedlings with mature foliage were rare, except where found occasionally establishing on bogs. Sapling-stage *C. nootkatensis* were rare or absent in forests with moderate to high basal area.

Chamaecyparis nootkatensis appears to be out-competed by faster growing *Picea sitchensis* and *Alnus rubra* (Sitka alder) on revegetating earthslides and creek bottoms. Seedling establishment near windthrown trees could be important to stand development in unmanaged stands, but no sizable cedar seedlings were observed at these sites.

Thus, seedlings are not establishing where seed source and germinants are the most abundant—in mature, relatively high-volume stands of *C. nootkatensis.* Poor reproduction from seed of *C. nootkatensis* is independent of decline. Factors limiting successful establishment of seedlings of *C. nootkatensis* on most non-bog sites are not understood, but a requirement for light, soil temperature, and perhaps for soil disturbance is suggested.

Forest managers report that *C. nootkatensis* infrequently reproduces by seed on some clearcut sites, even when it is a common component of the previous old-growth stand. It is conceivable that its seed dormancy, which often takes two years to break (B. Pawuk, personal communication), puts *C. nootkatensis* at a disadvantage relative to competing species that germinate quickly following logging or species, such as *T. heterophylla,* that occur as understory in old-growth forests and respond to logging with rapid growth.

Planting nursery-grown seedlings of *C. nootkatensis* may be necessary to replace it where this valuable tree species is harvested. Planted seedlings and cuttings appear to have high survival and good growth performance. In a regeneration study on Etolin Island (Hennon 1992), planted nursery-grown seedlings of *C. nootkatensis* on burned and unburned sites have greater than 90% survival and show rapid early height and diameter growth. Browsing by deer, shoot blight by the fungus *Apostrasseria* sp., and competing vegetation (on unburned sites) appear to be the primary factors limiting the success of planted seedlings.

Forest Decline

Distribution

Besides the problem of natural reproduction, *C. nootkatensis* is experiencing a severe and widespread decline in southeast Alaska. Figure 3.1 portrays the distribution of concentrated cedar decline. It is based on detailed sketchings made on 1:250,000 scale maps during aerial surveys that we conducted annually during the 1980s and 1990s. Some 200,000 ha of decline have been mapped in southeast Alaska from the British Columbia–Alaska border near Portland Canal to the northwest side of Chichagof Island. Decline is either absent or not severe farther south in British Columbia and around the Gulf of Alaska near the northwest limits of the cedar's range.

Aerial Photographs and Ground Surveys

Analysis of aerial photographs and numerous ground surveys (Hennon et al. 1990b) were conducted to determine in which forest communities decline occurs, which tree species are dying, and if mortality has been spreading over short or long distances. Such information should aid in evaluating potential causal factors.

Each dead *C. nootkatensis* encountered on ground surveys was placed into one of six snag classes based on degree of foliage, twig, or branch retention, and deterioration of its bole. Average time since death is estimated for five of these snag classes (Table 3.1).

Poorly Drained Sites

Our aerial observations and study of recent color aerial photographs confirm that cedar decline is strongly associated with forests on poorly drained sites. Some open bogs extend, contiguously or in chains, for many kilometers along fairly flat terrain at lower elevations. Mortality is consistently associated with the edges of these bogs. Decline is also severe in scrub forests on large areas without open bogs, but the understory flora and stature of trees suggest that these sites are influenced by moderately poor drainage. Within the general distribution of severe mortality, low-elevation forested areas without concentrated mortality generally lack bogs and are dominated by high-volume hemlock forests. Data from ground surveys indicate that the incidence of mortality is significantly more common ($P = 0.05$) in bog communities and is progressively less common in communities with better drainage (Hennon et al. 1990b).

Tree Species Dying

Chamaecyparis nootkatensis accounts for 74% of the dead basal area in stands with a high proportion of mortality (Hennon et al. 1990b). Dead *Tsuga* accounts for 17% of the mortality in declining stands; other species have negligible dead basal area.

Table 3.1 Average Time since Death for Dead *Chamaecyparis nootkatensis* Trees (Snags)

Snag Class	Appearance	Time since Death (yrs)
I	Foliage retained	4
II	Twigs retained	14
III	Secondary branches retained	26
IV	Primary branches retained	51
V	No branches retained	81
VI	Bole broken and deteriorated	not dated

Source: Hennon et al. 1990c.

Since *C. nootkatensis* is the principal species that is dead or dying in most declining stands, the percent basal area dead for each species provides a better measure of which species have been affected. Sixty-five percent of *C. nootkatensis* basal area is dead in these stands, nearly twice the percentage of any other species. *Chamaecyparis nootkatensis* is also dying at disproportionate levels in stands with *T. plicata:* on Prince of Wales Island, 34% of *C. nootkatensis* basal area is dead compared to 9% for *T. plicata;* on Wrangell Island, 54% of *C. nootkatensis* is dead compared to only 3% of *T. plicata.*

Mortality Spread

Maps delineating the extent of mortality at seven sites made from aerial photographs taken in 1927, 1948, 1965, and 1976 clearly show that peripheral boundaries of mortality have expanded over the last 60 years at all sites (Hennon et al. 1990b). In 1927, however, the mortality apparent on each site covered a large portion of the area where trees are now dead and dying; thus, local spread since 1927 accounts for only a small proportion of the total area of decline. Subsequent mortality has rarely extended more than 100 m beyond the 1927 boundary.

Results from ground surveys also indicate local spread at most surveyed sites. In such stands, dying and recently killed cedars often surround areas containing the old snags. This mortality spread, which commonly occurs upslope and in all cardinal directions, has been along the gradient from bog to better drainage, as evidenced by the ordination of understory plants. Snags with no limbs (class V) predominate in bog and semibog plant communities, and plots with more recently killed snags (classes IV, III, II, and I) support progressively better-drained plant communities. Snags in the longest-dead class VI (deteriorating boles) are uncommon, are not associated with severe mortality, and are not confined to bogs, as are class V snags. Thus, spread of mortality within any one site has occurred as a slow advance along an established ecological gradient that is often related to slope. The common upslope spread of mortality results from mortality originating in bogs and semibogs and spreading upslope along the gradient to better-drained communities.

Ecological Effects of Decline

Different forest conditions have developed in the areas of earliest mortality where the long-dead (class V) snags are present. A new stand of vigorous-appearing trees has grown up beneath the bark-free, white snags at some sites. *Chamaecyparis nootkatensis, T. heterophylla,* and to some extent *T. mertensiana,* are the dominant tree species in these areas and appear as a green zone from a distance or on color aerial photographs. Many of these trees are older than 100 years and were probably present as understory conifers during the initial mortality or, in the case of *Tsuga,* are surviving overstory trees. Others have regenerated after death of the overstory cedars.

In other stands with long-dead snags, continued mortality of smaller *C. nootkatensis* trees has apparently prevented development of this green zone. Reasons for the recurrence of mortality in some stands, but not in others, are unclear. On extremely boggy sites, live trees have not "released" after old mortality, perhaps because factors limiting the growth of live trees (e.g., anaerobic or cold soils) are not improved by the death of *C. nootkatensis*. Once the dominant overstory of *C. nootkatensis* dies, the reduced transpiration causes a degeneration to the site that may cause the soils to become wetter, affecting the survival or growth of other conifer species. Interestingly, southeast Alaskan tree species suspected of being relatively intolerant to excessive moisture (Ruth and Harris 1979) suffered higher rates of death (35% of *P. sitchensis* basal area and 29% for *Tsuga*) than have the tolerant *P. contorta* (6%) on sites of mortality. *Chamaecyparis nootkatensis* is the exception; the species is reportedly well adapted to wet sites, but it suffers the greatest mortality (65%) in declining stands.

Snags of *C. nootkatensis* that lack limbs (class V) were present on all declining sites and constituted at least 8% (range = 8–60%) of all snags on 23 sites with heavy mortality that we surveyed. More recently killed trees (i.e., classes I–IV) were also present at all transects, indicating that mortality has continued at all locations since initiation. Our general reconnaissance revealed only one site that had only long-dead class IV and V snags and lacked recent mortality. No sites, however, had recently killed trees in the absence of long-dead class V snags. Thus, we have no evidence of site-to-site spread of decline.

Dating the Onset

To determine how long *C. nootkatensis* decline has been occurring in Alaska, we examined old aerial photographs, dated the death of standing dead trees, and inspected historical accounts of botanical expeditions (Hennon et al. 1990c).

The earliest available aerial photographs of southeast Alaska, taken in 1926 and 1927 by the U.S. Navy, represent one of the first efforts anywhere to photograph large areas of forest (Sargent and Moffit 1929). These photographs are now of variable quality, but, on both vertical and oblique prints with good contrast, cedar mortality clearly appears as patches of white snags. Mortality of *C. nootkatensis* was already widespread by 1927.

Estimates of the time since death for *C. nootkatensis* trees in snag classes (Table 3.1) were determined by counting annual rings of hemlock trees growing under large cedar snags and by counting annual rings in callus growth on partially killed stems of *C. nootkatensis* ("rope trees") that were interspersed among cedar snags (Hennon et al. 1990c). Rope trees have a dead top (snag class I–V) and one narrow strip of live tissue, consisting of callusing bark and sapwood, that connects roots to one live, bushy branch cluster. The cause of this condition is not known; we hypothesize that these trees were severely injured, but not completely killed, by whatever caused nearby cedars to die. Rope trees were not injured during one sudden incident such as an extreme climatic event; the cambiums among the trees that we sampled did not die during the same year. Also, their slow decline in growth many years prior to bole death does not support a sudden event as the cause of tree injury.

We consider that class V snags (boles intact, but no primary limbs retained) are the original extensive mortality. These trees died an average of 81 years ago (Hennon et al. 1990c), as estimated by the rope tree method, and are present and common at all mortality sites examined. Because numerous class V snags died before the average of 81 years ago, some *C. nootkatensis* probably began to die before the turn of the century. The early aerial photographs confirm the widespread occurrence of dead trees in 1927. The more deteriorated snags in class VI with broken-off and decayed boles were infrequently encountered in surveys and were not associated with distinct mortality sites (Hennon et al. 1990b). These latter trees probably died prior to the onset of extensive mortality and may represent the nonepidemic or background level of mortality.

The appearance of abundant dead *C. nootkatensis* around 1900 is also supported by historical observations. Sheldon (1912) was the first observer to report extensive mortality and noted dead *C. nootkatensis* near Pybus Bay on Admiralty Island in 1909, stating "vast areas are rolling swamp, with yellow-cedars, mostly dead." Numerous botanical expeditions to Sitka and other areas in southeast Alaska where cedar decline is now extensive were conducted prior to 1880; most report the occurrence of *C. nootkatensis,* but none mentions dead or dying trees (Hennon et al. 1990c).

The excessive level of cedar mortality on sites of decline (65% has died in the last 100 years), the time required to establish mature *C. nootkatensis* trees (nearly all are well over 100 years old), and inadequate replacement by natural regeneration suggest that the population of *C. nootkatensis* in these forests is diminishing and that intensive mortality could not have been an ongoing phenomenon for centuries.

Biotic Factors

The cause of decline has been variously attributed to bark beetles, root disease, and winter injury, but these suggestions were based on brief observations. The causes of decline did not receive detailed investigation until 1981 when we began to study the symptoms of dying trees, organisms associated with symptomatic tissues, and the ability of these organisms to incite disease.

Symptoms of dying trees suggest a root or below-ground problem (Hennon et al. 1990d). Crowns of declining trees die suddenly or slowly as a unit, rather than as individual dying branches. Slowed radial growth precedes crown symptoms. We excavated the root systems of 62 *C. nootkatensis* in various stages of decline and made systematic observations of symptoms (Shaw et al. 1985, Hennon et al. 1990d). The initial symptom of decline is death of the fine root system. As the crown begins to show symptoms of thinning and off-color foliage, small diameter roots die and larger roots develop necrotic cambial lesions. In final stages of death, these necrotic lesions spread vertically up the boles of trees.

Phloeosinus cupressi (bark beetles) and the fungus *Armillaria* are frequently found on declining cedars, but only attack in late stages of tree decline (Shaw et al. 1985). None of the 50 taxa of fungi that we isolated from the symptomatic tissues or collected from *C. nootkatensis* (Hennon 1990) was consistently associated with dying or dead trees. Of the ten most common fungi isolated from symptomatic tissues, none demonstrated the ability to kill unstressed seedlings in inoculation trials (Hennon et al. 1990d). Four genera of nematodes (parasitic worms) were found in declining forests, but their low populations and association with healthy forests indicate that they do not cause decline (Hennon et al. 1986). Five species of *Pythium,* known as rootlet feeding fungi, were recovered from beneath dying trees, but their recovery rates were not associated with mortality (Hamm et al. 1988). Viruses and mycoplasmas, which are unlikely threats to conifers, are currently under investigation.

We made a concentrated effort to determine whether any species of the fungal genus *Phytophthora* were present in declining forests, because necrotic lesions on roots and root collars of dying cedars suggested similarity to the serious disease of *C. lawsoniana* (Port Orford cedar) in southwest Oregon caused by *P. lateralis* (Roth et al. 1972). The occurrence of dying *C. nootkatensis* on wet sites and patterns of local spread also suggested *Phytophthora* involvement. *Phytophthora lateralis* was not recovered, but we successfully isolated another species of *Phytophthora* from baits placed below four of 69 *C. nootkatensis* (Hansen et al. 1988, Hennon et al. 1990d). We never isolated any *Phytophthora* directly from cedar tissues even though selective media were used. We recently identified the fungus as *P. gonapodyides,* a fungus that has been found in healthy, nondeclining forests in Oregon (Hansen and Hamm 1988). The fungus lacked strong pathogenicity on *C. nootkatensis* (Hansen et al. 1988). *Phytophthora gonapodyides* is not the cause of decline, and *C. nootkatensis* may not even be a host to the fungus.

Basal scars are common on many cedar trees in declining stands—nearly one-half (49%) of the *C. nootkatensis* sampled in several areas on Chichagof and Baranof Islands had either fresh or old, callusing scars (Hennon et al. 1990a). Fresh scars consistently have teeth or bite marks and are almost certainly the result of feeding by *Ursus arctos* (Alaskan brown bears; Hennon et al. 1990a). A smaller number of scars are caused by Alaska Native people stripping bark from cedar trees (Hennon et al. 1990a). Regardless of cause, basal scars are not the primary cause of cedar decline.

Abiotic factors

Epidemiological evidence and the lack of aggressive biotic factors indicate that some abiotic factor is the primary cause of decline. We must now begin to investigate the most likely factors. One hypothesis for an abiotic cause of tree death is that bogs, for climatic or other reasons, are advancing onto the adjacent semibog sites where so many trees are dying (Klinger 1988). The development from forest to bog requires the waterlogging of the forest floor, which may result from the proliferation of *Sphagnum* moss, the development of poor drainage, or both. This process, paludification, may lead to the death of forest trees as sufficient oxygen or nutrients become less available in the wet soil. Whether there is a general successional direction for forests in southeast Alaska, from forest to bog or bog to forest, is presently unresolved.

Our observations, however, suggest that if bogs are advancing on forests, then the rate of advancement is imperceptible. Bogs observable on the 1927 aerial photographs have not noticeably expanded, yet cedar mortality has been substantial during this period. If bogs had enlarged during the last 100 years, one might expect to see some evidence of rapidly expanding *Sphagnum* mats or invasion of other bog plants into forests. In recent observations of understory flora growing beneath several hundred dying cedars at four sites in southeast Alaska, more than 75% of trees had no

Sphagnum spp. within a 1-m radius of tree boles (Hennon, unpublished data). Even if bogs are expanding, the rate of expansion is probably too slow to explain the widespread decline of *C. nootkatensis*. Also, the relatively high rate of mortality for *C. nootkatensis*, perhaps one of the Alaska conifers best adapted to bogs (aside from *P. contorta*), contradicts the simple hypothesis that expanding bogs kill cedars.

Soil toxicity, which presumably could affect *C. nootkatensis* more than other coni-fers, may explain tree death. Perhaps toxic, organic compounds result from anaerobic decomposition in these wet, highly organic soils (Hennon and Shaw 1995).

Another hypothetical explanation for cedar death around bogs and on wet sites is the poor protection from atmospheric events that these sites offer. Cedar trees on such sites are open-grown and probably more vulnerable to extreme weather events (e.g., hard frosts, droughts) than cedars grown within protective canopies. It is con-ceivable that if some trees on bog edges died following such a hypothetical weather event, then trees in adjacent forest stands might lose protection and become vulnera-ble to damage. Such action might lead to a slow, local spreading of mortality from open stands in bogs and semibogs.

Subtle variations in climate may be responsible for triggering some forest declines. Because climate is not static, vegetation may be constantly adjusting to reflect its new environment. The long-lived *C. nootkatensis* may have difficulty adapting to a changing environment, particularly if cedar is reproducing primarily asexually (Hen-non 1986). Interestingly, a warming trend has occurred in most of Alaska since the late 1800s (Hamilton 1965), marked by the end of the Little Ice Age in 1870, which nearly coincides with the onset of *C. nootkatensis'* decline (Hennon and Shaw 1995). A slight increase in average winter temperatures would change some precipitation from snow to rain, reducing the snowpack at low elevations. Perhaps the primary cause of cedar decline is frost damage to root systems (fine root necrosis is apparently the initial symptom of dying cedars; Hennon et al. 1990d) during periods when cold continental air moves over the region and roots are inadequately protected by snow. Trees growing in wet soils at low elevations, where decline has been severe, would be susceptible due to their shallow root systems and poor insulation offered by satu-rated soils. Less decline on wet sites at higher elevations could be explained by the persistent winter snowpack at those elevations even in today's warmer climate.

Summary

The cumulative results from our studies do not support the hypothesis that an organ-ism is the primary cause of mortality. The specificity of mortality to *C. nootkatensis* and evidence of local spread seem to suggest a pathogen; however, no new sites of mortality appear to have developed since the nearly simultaneous onset of *C. nootka-tensis* decline some 100 years ago at numerous locations throughout southeast Alaska. Snags that represent the original extensive mortality were present at every sampling location and observed on all good-quality 1927 aerial photographs of sites where cedars are currently dying. It is difficult to imagine a pathogen that is capable of initiating and continuing to cause the level of mortality that occurs on remote and dispersed islands in isolated wilderness, but not capable of reinitiating the problem on other similar bog and semibog sites. The pattern of local spread, along an existing moisture gradient, also suggests that a pathogen is not involved. Detailed studies on fungi, insects, nematodes, and bears reveal that none of them is the primary cause. Because decline is not contagious and does not spread to new sites, forest managers can attempt to regenerate *C. nootkatensis* on healthy, productive sites, such as most logged sites, without fear that the decline problem will affect their plantings.

Chamaecyparis nootkatensis' decline appears to be a unique and outstanding example of a naturally induced forest decline. The extreme decay resistance of this species has allowed us the rare opportunity to reconstruct the onset and development of decline. The occurrence of extensive mortality before 1900 in countless remote, undisturbed sites without nearby sources of pollution argues against atmospheric pollution as the cause of decline. Additionally, no introduced exotic pathogen or insect (another potential form of anthropogenic activity) was found to be associated with decline. Whether or not a general warming trend, perhaps caused by human activities, has affected the onset and development of decline is an open question. Except for the probable abiotic cause, the specific, primary stresses in the etiology of *C. nootkatensis* decline remain a mystery.

This spectacular forest decline has reduced the population of *C. nootkatensis* on more than 200,000 ha in southeast Alaska. In addition, timber harvesting has further reduced *C. nootkatensis* populations. With limited natural reproduction on sites of decline and on harvested areas, artificial regeneration will become necessary to reestablish and perpetuate this valuable tree species.

References

Anderson, H.E. 1959. Silvicultural characteristics of Alaska-cedar. U.S. Dept. Agric. For. Serv., Alaska For. Res. Cent. Sta. Pap. No. 11.
Antos, J.A., and D.B. Zobel. 1986. Habitat relationships of *Chamaecyparis nootkatensis* in southern Washington, Oregon, and California. Can. J. Bot. **64**:1898–1909.
Franklin, G.F. 1961. A guide to seedling identification for 25 conifers of the Pacific Northwest. U.S. Dept. Agric. For. Serv., Pac. Northwest For. and Range Exp. Sta., Portland, OR.
Franklin, G.F., and C.T. Dyrness. 1973. Natural vegetation of Oregon and Washington. U.S. Dept. Agric. For. Serv., Pac. Northwest For. and Range Exp. Sta., Portland, OR. Gen. Tech. Rep. PNW-8.
Frear, S.M. 1982. What's killing the Alaska yellow-cedar? Am. For. **88**(11):41–43,62–63.
Hamilton, T.D. 1965. Alaskan temperature fluctuations and trends: an analysis of recorded data. Arctic **18**:105–117.
Hamm, P.B., E.M. Hansen, P.E. Hennon, and C.G. Shaw III. 1988. *Pythium* species from forest and muskeg areas of southeast Alaska. Trans. Br. Mycol. Soc. **91**:385–388.
Hansen, E.M., and P.B. Hamm. 1988. *Phytophthora* species from remote forests of western North America. Abstract. Phytopathology **78**:1519.
Hansen, E.M., P.B. Hamm, C.G. Shaw III, and P.E. Hennon. 1988. *Phytophthora drechsleri* in remote areas of southeast Alaska. Trans. Br. Mycol. Soc. **91**:379–388.
Harris, A.S. 1970. The loners of Alaska. Am. For. **76**:20–23,55–56.
———. 1971. Alaska-cedar. U.S. Dept. Agric. For. Serv., Am. Woods-FS 224.
———. 1990. *Chamaecyparis nootkatensis* (D. Don) Spach Alaska-cedar. Pages 97–102 *in* R.M. Burns and B.H. Honkala, compilers. Silvics of North America, Vol. 1. U.S. Dept. Agric. For. Serv. Agric. Handb. 654. Washington, DC.

Harris, A.S., and W.A. Farr. 1974. The forest ecosystem of southeast Alaska. 7. Forest ecology and timber management. U.S. Dept. Agric. For. Serv. Pac. Northwest For. and Range Exp. Sta. Portland, OR. Gen. Tech. Rep. PNW-25.

Hennon, P. E. 1986. Pathological and ecological aspects of decline and mortality of *Chamaecyparis nootkatensis* in southeast Alaska. Dissertation. Botany and Plant Pathology, Oregon State Univ., Corvallis.

———. 1990. Fungi on *Chamaecyparis nootkatensis*. Mycologia **82**:59–66.

———. 1992. Survival and growth of planted Alaska-cedar seedlings in southeast Alaska. Tree Planters' Notes **43**:60–66.

———. 1995. *Chamaecyparis nootkatensis*. Pages 1–9 *in*: P. Schutt, H.J. Schuck, G. Aas, and U.M. Lange, eds. Enzyklopadie der Holzgewachse. Ecomed, Landsberg.

Hennon, P.E., and C.G. Shaw III. 1995. Did climatic warming trigger the onset and development of yellow-cedar decline in southeast Alaska? Euro. J. For. Pathol. **24**:399–418.

Hennon, P.E., E.M. Hansen, and C.G. Shaw III. 1990a. Causes of basal scars on *Chamaecyparis nootkatensis* in southeast Alaska. Northw. Sci. **64**:45–54.

———. 1990b. Dynamics of decline and mortality of *Chamaecyparis nootkatensis* in southeast Alaska. Can. J. Bot. **68**: 651–662.

Hennon, P.E., C.G. Shaw III, and E.M. Hansen. 1990c. Dating decline and mortality of *Chamaecyparis nootkatensis* in southeast Alaska. For. Sci. **36**:502–515.

———. 1990d. Symptoms and fungal associations of declining *Chamaecyparis nootkatensis* in southeast Alaska. Plant Dis. **74**:267–273.

Hennon, P.E., G.B. Newcomb, C.G. Shaw III, and E.M. Hansen. 1986. Nematodes associated with dying *Chamaecyparis nootkatensis* in Southeastern Alaska. Plant Dis. **70**:352.

Hill, M.O., and H.G. Gauch. 1980. Detrended correspondence analysis; an improved ordination technique. Vegetatio **42**:47–58.

Hutchinson, O.K., and V.J. LaBau. 1975. The forest ecosystem of southeast Alaska. 9. Timber inventory, harvesting, marketing, and trends. U.S. Dept. Agric. For. Serv. Pac. Northwest For. and Range Exp. Sta. Portland, OR. Gen. Tech. Rep. PNW-34.

Klinger, L.F. 1988. Successional change in vegetation and soils of southeast Alaska. Dissertation. Univ. of Colorado, Boulder.

Owens, J.N., and M. Molder. 1984. The reproductive cycles of western redcedar and yellow-cedar. Victoria, B.C. Min. of For., Forestry Div. Res. Branch. Canada.

Roth, L.R., H.H. Bynum, and E.E. Nelson. 1972. Phytophthora root rot of Port-Orford-cedar. U.S. Dept. Agric. For. Serv. Pest Leafl. 131.

Ruth, R.H., and A.S. Harris. 1979. Management of western hemlock–Sitka spruce forests for timber production. U.S. Dept. Agric. For. Serv. Pac. Northwest For. and Range Exp. Sta. Portland, OR. Gen. Tech. Rep. PNW-88.

Sargent, R.H., and F.H. Moffit. 1929. Aerial photographic surveys in southeastern Alaska. U.S. Geol. Surv. Bull. 797-E: 143–160. U.S. Govt. Printing Office.

Shaw, C.G. III, A. Eglitis, T.H. Laurent, and P.E. Hennon. 1985. Decline and mortality of *Chamaecyparis nootkatensis* in Southeastern Alaska, a problem of long duration but unknown cause. Plant Dis. **69**:13–17.

Sheldon, C. 1912. The wilderness of the North Pacific Coast islands: a hunter's experience while searching for wapiti, bears, and caribou on the larger islands of British Columbia and Alaska. Scribner's Sons, New York.

4

Glen B. Dunsworth

Problems and Research Needs for *Chamaecyparis nootkatensis* Forest Management in Coastal British Columbia, Canada

Chamaecyparis nootkatensis (D. Don) Spach (Alaska yellow-cedar, yellow-cypress) is distributed predominantly in the Pacific region of western North American forests from 61°N to 41°N latitude (Fig. 4.1; see Hennon et al., Chapter 3 for range map; Frenkel 1974, Antos and Zobel 1986, Harris 1990, Klinka 1991). Considered by many to be a high-elevation species, it exists on suboptimal conditions from sea level to 2,000 m in many parts of coastal British Columbia (Harlow and Harrar 1969, Viereck and Little 1972, Harris 1990). Although *C. nootkatensis* will persist on organic soils and ferro-humic podzols, its most rapid growth in coastal British Columbia is on humo-ferric podzols (William J. Beese, personal communication; Klinka 1991). The species requires moist, humid growing conditions and neutral to acidic soils, which predominate north and west on Vancouver Island and on the northwest coast of British Columbia (Krajina 1969, Minore 1979, Harris 1990).

Persistently, poor reproductive success has been a problem for the maintenance and expansion of *C. nootkatensis* populations (Alaback 1991). *Chamaecyparis nootkatensis* seldom occurs as pure stands. However, it is often a scattered, dominant species in a variety of ecosystems in the Mountain Hemlock zone and in the Coastal Western Hemlock biogeoclimatic subzone of coastal British Columbia (Krajina 1969, Klinka 1991; see Green and Klinka 1984 for associated species). Slow advance of this species onto disturbed sites may imply that the current range is an approximation of the extent of recent Pacific Northwest glacial history (Frenkel 1974, Harris 1990, Alaback 1991).

The evolutionary strategy for this species has been persistence through tolerance and longevity (Grimes 1977). Tolerance to, rather than avoidance of, shade, frost, and high soil-moisture conditions has allowed the species to adapt to a variety of suboptimal coastal conditions, from cold climate sites to bogs (Krajina 1969, Levitt 1980,

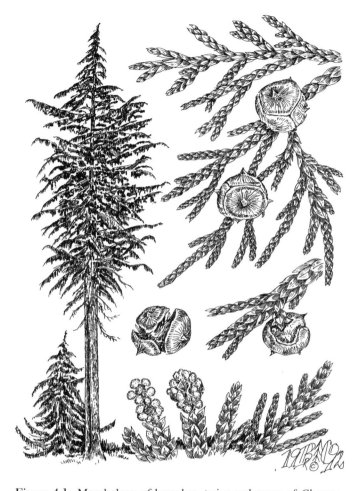

Figure 4.1. Morphology of branches, twigs and cones of *Chamaecyparis nootkatensis*. (Used with permission of B.L. Cunningham, Nacogdoches, Texas.)

Spurr and Barnes 1980). Longevity through allocation of more carbon to defensive chemicals than to sexual reproduction and rapid growth has made this a persistent, late successional species (Bazzaz et al. 1987, Loehle 1988).

Vegetative reproduction may have compensated for poor sexual reproduction and allowed more successful adaptation to moist sites, avalanche disturbance, and volcanic tephra burial (Zobel and Antos 1982, Antos and Zobel 1986, Dunwoodie 1986). However, as with many northwest conifers, this species has considerable genetic variation, much of it within populations and families (Critchfield 1984, Bower et al. 1989, Banerjee and Maze 1991, Russell and Cartwright 1991, Cherry and Lester 1992). Work by Russell et al. (1990) suggests that, based on one- to three-year-old clonal trials, only 5–10% of the genetic variability in growth characteristics can be attributed to different stands (Tables 4.1 and 4.2). This may be a function of longevity and a relatively short postglacial history that did not provide enough generations sufficient time to ecotypically differentiate (Hamrick 1979, Hebda 1983, Critchfield 1984, Dunwoodie 1986). The result has been the evolution of a relatively plastic, shade-tolerant, genetically diverse species persisting successfully as a dominant in old-growth forests on a wide range of coastal British Columbian ecological associations.

Many of the characteristics that have led to a successful evolutionary solution have presented benefits and problems in human use and perpetuation of this species. The current high value of *C. nootkatensis* wood is dependent on natural preservatives (nootkatin) and slow growth characteristics (small heartwood, narrow and consistent ring widths, and small knots), which are artifacts of the tolerance and persistence strategy (Jozsa 1991). However, the reduced emphasis on sexual reproduction may have resulted in small, infrequent cone crops or delayed, poor germination, and limited natural regeneration of this species following clearcut logging (Alaback 1991).

Regeneration Problems

Concern about the maintenance of *C. nootkatensis* as a unique, ecological resource and a high-value fiber source has recently led to increased interest in its ecology and regeneration in British Columbia. Many believe the struggle between these conflicting values will continue well into the 21st century. However, much of the debate has been occurring in the absence of direct evidence about the ecological function of *C. nootkatensis* or the nature of its value.

Natural Regeneration

Persistence and tolerance have allowed *C. nootkatensis* to compete well under suboptimal environmental conditions and become one of the most long-lived species in North America. Most of these suboptimal sites have lower productivity and occur in cold-climate biogeoclimatic zones in coastal British Columbia. This presents a num-

Table 4.1 Percent of Total Variability and Narrow-Sense Heritabilities (H^2) for First-Year Growth Traits of *Chamaecyparis nootkatensis* in a Genetic Architecture Study

Measurement	*Percent Variability*			
	Provenance	Family	Error	H^2
1 Year height	2	24	74	0.73
1 Year diameter	0	11	89	0.33
Root weight	4	11	85	0.34
Root/shoot	11	19	70	0.65

Source: Russell and Cartwright 1991.

Table 4.2 Percent of Total Variability and Heritabilities for Early Height of *C. nootkatensis* in a Clonal Trial

Measurement	Percent Variability				Heritability	
	Provenance	Family	Clone	Error	Broad	Narrow
1 Year height	5	3	53	38	0.84	0.10
3 Year height	0	21	40	39	0.82	0.63

Source: Russell and Cartwright 1991.

ber of site-related forest management problems in addition to the inherent, biological limitations of the species.

Cold-climate sites tend to be inaccessible, have short growing seasons, and be prone to frost and snow-creep problems. These problems have made natural regeneration the most attractive regeneration option following clearcut logging. However, as mentioned above, *C. nootkatensis* has not evolved the physiological mechanisms to compete well in large openings. Many mixed *C. nootkatensis–Tsuga heterophylla* (western hemlock) forests in coastal British Columbia often change to pure hemlock forests after harvesting and natural regeneration (Alaback 1991, Barker 1991). This may be due to poor germination and infrequent cone crops or to early mortality of germlings (Heddon 1986).

The problems of natural seedling establishment can be overcome through artificial regeneration and perhaps through alternative methods of harvesting. Artificial regeneration can also make use of the capability that *C. nootkatensis* has developed over evolutionary time: the capacity for vegetative propagation.

Vegetative Propagation

Chamaecyparis nootkatensis is relatively easy to propagate vegetatively (Karlsson 1974, Karlsson and Russell 1990). However, as in other conifer species, rootability is a function of maturation (chronological, physiological, and ontogenetic age), genetics, and physiology of the donor plant. Recent research into these factors has led to protocols for the production of vegetative propagules known as "stecklings" (Russell and Grossnickle 1990, 1991; Russell et al. 1990). Young, vigorous vegetative material is taken from hedged orchards with a five- to ten-year cycle in order to maximize production. These cuttings are then stimulated to develop into stecklings using rooting hormones (IBA) and containerized nursery culture. Stecklings have become the primary means of artificial regeneration of *C. nootkatensis* in coastal British Columbia (Karlsson and Russell 1990).

Cone and Seed Development

Many of the seed problems in *C. nootkatensis* have also received research interest in recent years, stimulated by seed orchards, which were established as a means of supplying a more consistent seed source and as a vehicle for genetic improvement of high-value traits. The problem of poor seed germination (seed dormancy) has received the greatest effort and has resulted in a three-month, combined warm and cold stratification technique. Current work is directed at understanding the biochemical nature of *C. nootkatensis* seed dormancy in order to reduce the stratification period (Kurz et al. 1989). A potential spinoff benefit of this work may be the development of tissue culture procedures for producing cloned embryos.

Variation in cone-crop size and frequency has been recently addressed through research into the use of the flowering hormone gibberellic acid (GA_3) in seed orchards (Owens and Molder 1977, Bower et al. 1989). *Chamaecyparis nootkatensis* is easily stimulated to flower, but it exhibits considerable clonal variation in the proportions of male and female flowers. Although cone crops can be stimulated, seed development problems such as pollen abortion and nonuniform cone development in low-elevation seed orchards continue to be unresolved.

Pollen abortion has become a problem in low-elevation seed orchards due to *Trisetacus chamaecypari* (cypress bud mite) predation of pollen (Collangelli 1991). The mite girdles and kills the pollen cone during the fall, prior to the winter dormancy period. This has led to 30–60% pollen losses in seed orchards. *Trisetacus chamaecypari* has not been found in wild stands at elevations higher than 200 m on Vancouver Island.

Seed development also appears to be altered under low-elevation growing conditions. In natural stands, cone development occurs over three to four years (Owens and Molder 1975). *Chamaecyparis nootkatensis* requires one more year than most northwest, North American conifers for cone maturation. Branches often have two different cone crops at any one time. Recent evidence from low-elevation seed orchards indicates that cone and seed maturation can occur in one season with greater numbers of filled seeds per cone than in wild stands. This appears to be due to the warmer, drier seasonal conditions and the reproductive plasticity of the species (Collangelli 1991, El Kassaby et al. 1991). One-year cone crops have recently been reported in wild stands in the southern portion of the *C. nootkatensis* range (northern California; John Russell, personal communication; Heddon 1986).

Ecophysiology

Although the development of reliable sources of high-quality propagules is essential to a successful artificial regeneration program for this species, an understanding of its physiological and environmental tolerances is equally important. Recently, Grossnickle (1991) and Grossnickle and Russell (1991) investigated the response of *C. nootkatensis* stecklings to changes in light, evaporative demand, soil moisture, and soil temperature. They found:

- Gas exchange in response to light is similar to most Pacific Northwest species beginning at 5–10% full sunlight and reaching saturation near 30% (Fig. 4.2).
- Stomatal response and photosynthesis are very sensitive to evaporative demand. The species needs low evaporative demand conditions and is one of the most sensitive species in the Pacific Northwest (Fig. 4.3).
- Gas exchange is highly sensitive to low soil temperatures or low moisture (Figs. 4.4 and 4.5).

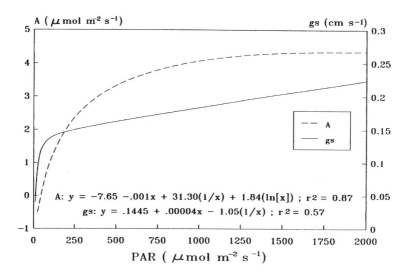

Figure 4.2. CO_2 assimilation rate (A) and stomatal conductance (gs) boundary line analysis response to photosynthetically active radiation (PAR) for *C. nootkatensis* (Grossnickle and Russell 1991).

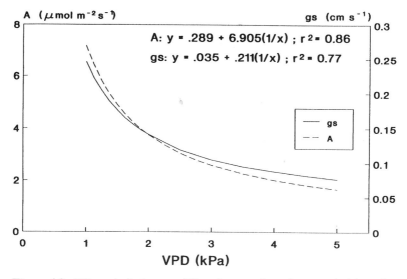

Figure 4.3. CO_2 assimilation rate (A) and stomatal conductance (gs) boundary line analysis response to vapor pressure deficit (VPD, measured in kilopascals) for *C. nootkatensis* (Grossnickle and Russell 1991).

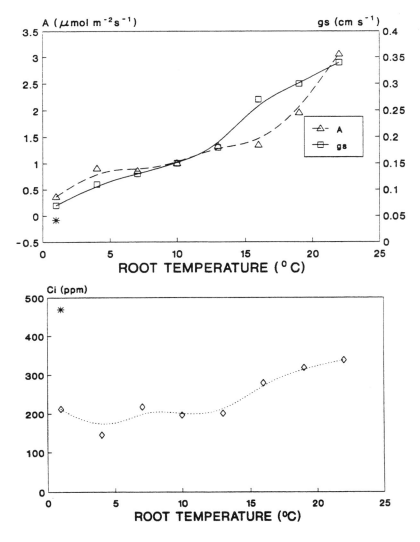

Figure 4.4. CO_2 assimilation rate (A), stomatal conductance (gs), and internal CO_2 concentration (Ci) response to changes in root temperature for *C. nootkatensis*. Mean (n = 20) is shown and the largest SEM (standard error of the mean; ±) during the study was 0.40 for A, 0.002 for gs, and 27.7 for Ci (Grossnickle and Russell 1991). The asterisks indicate mean values for rooted cuttings with A measurements below the compensation point.

Comparisons of hydraulic conductivity in *C. nootkatensis* seedlings and stecklings showed that they were comparable within 21 days from planting (Grossnickle and Russell 1991). Transpirational flux density increased in a curvilinear fashion with new root area. However, seedlings were able to produce nearly double the new root area of stecklings in the 21-day period.

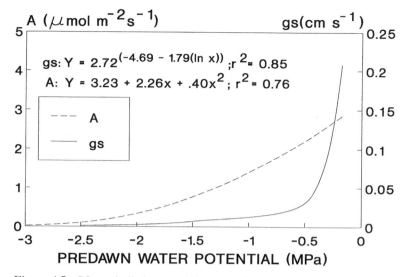

Figure 4.5. CO_2 assimilation rate (A), stomatal conductance (gs) boundary line analysis response to predawn water potential (MPa) for *C. nootkatensis* (Grossnickle and Russell 1991).

These observations confirm that *C. nootkatensis* seedlings and stecklings can tolerate the relatively open-grown conditions of a shelterwood cut or small clearcut if they are associated with high humidity and moist, warm to cool soil conditions. This would also suggest that the lack of seedling regeneration in large clearcuts may be due to mortality associated with extreme and variable moisture, humidity, and temperature conditions during the establishment phase.

Silvicultural Challenges and Research Needs

Recent research on *C. nootkatensis* in coastal British Columbia suggests that the problem of inadequate natural regeneration following clearcut logging can be resolved with artificial regeneration and perhaps alternative harvesting systems. Problems remain in making these solutions reliable and efficient. Top-priority research areas are seed dormancy, pollen and cone development in seed orchards, nursery culture, and physiology of seedlings and stecklings in their natural environments. These areas are receiving the highest priority for current research efforts. Alternative harvesting methods have received little research emphasis.

Looking into the near future, when stands of *C. nootkatensis* can be reliably established, we will still have a very poor understanding of the genetics of *C. nootkatensis* or the ecosystems we are disturbing (Waring and Franklin 1979, Meslo et al. 1981, Maser 1990, Russell and Cartwright 1991). The longer term and currently lower priority questions are those about the nature of value—both from ecological resources and commodities perspectives.

To comprehensively value the ecological resource of the *C. nootkatensis* forests of the Pacific Northwest requires a much greater understanding of ecosystem function and the coevolution of species within those forests (Halpren 1989, Harmon et al. 1990). Thus, a long-term research priority must be forest genetics (genetic architecture) and ecosystem research (see Russell, Chapter 5, on genetics and Hennon et al., Chapter 3, for discussion of ecosystem research).

Chamaecyparis nootkatensis may also play a valuable role in recording changes in past climates. Its sensitivity to humidity and temperature may help forewarn us of future climatic change. A second long-term priority should be dendrochronology and climate change as reflected in *C. nootkatensis* growth patterns.

From a commodity standpoint, high-value characteristics include high relative density; smooth, even grain and color; and high nootkatin content. If these traits retain their value for the next 100 years, the challenge is to determine which silvicultural activities on which sites will enhance those traits in the shortest period of time.

Currently, the cold-climate sites provide poor economics due to slow growth, steep slopes, and long distances from mills. The preliminary ecophysiological evidence suggests that *C. nootkatensis* could be grown on more productive sites at lower elevation in moist, humid climates. This needs to be tested in conjunction with density trials and pruning to establish optimal silvicultural regimes.

Many of the high-value traits can be most easily attained through breeding. The current provenance and progeny testing program of the British Columbia Ministry of Forests will provide an excellent framework for capturing genetic gain (Russell and Cartwright 1991). Traits such as straightness, branch size and angle, and high nootkatin content are likely to be quite simply inherited, and considerable gains may be expected in the first generation. This, in conjunction with the potential for vegetative reproduction from cuttings and embryogenic callus, makes breeding and clonal forestry an attractive option for this species in the future.

A final option from a commodity standpoint may be to maintain *C. nootkatensis* as an ecological resource and to grow its Asian relative, *C. obtusa* (the Hinoki cypress), on the more productive sites at lower elevation (see Yamamoto, Chapter 7, and Zobel, Chapter 2, for further discussion of *C. obtusa*). This would mean moving seed sources 12–15°N. The movement to colder latitudes may be compensated for in part by dropping as much as 1,500 m in elevation. *Chamaecyparis obtusa* requires moist, humid growing conditions and neutral to acidic soils, which predominate on north and west Vancouver Island. Arboretum and horticultural experience supports the feasibility of growing *C. obtusa* on Vancouver Island.

Ultimately, although these forests do have an extensive north–south range in western North America, their global range is small. Thus, they form a globally unique resource; to value them from a commodity perspective alone would be myopic. The long-term nature of the above questions and the state of our current knowledge argue for greater prudence in our harvest of this species.

Conclusions

To provide silvicultural options for forest land managers, reliable stand establishment must be achieved. For *C. nootkatensis,* this means developing high-quality propagules from both seeds and cuttings. Recent research in British Columbia indicates that *C.*

nootkatensis seedlings and stecklings are suitable for regeneration of shelterwoods or small clearcuts. The infrastructure for producing reliable propagules is in place in British Columbia. The current research focus is on making this infrastructure efficient for achieving short-term regeneration objectives.

However, ecosystem function, evolutionary biology, and alternative harvesting research have not been given a high priority by British Columbia's forest research funding agencies. These areas of investigation are important to our understanding of the value of this ecological resource and to human influence through disturbance. We have also only recently recognized the role of *C. nootkatensis* in helping to reconstruct past climates and in providing early warning of regional climatic change.

Reliable stand management of *C. nootkatensis* in British Columbia has been resolved through past research efforts. Forest land managers now have an array of silvicultural options at their disposal. To meet the global challenge of applying these options wisely, a greater understanding is required of *C. nootkatensis* as an ecological resource and of the consequences of human disturbance of its ecosystems.

References

Alaback, P.B. 1991. Yellow-cypress: An overview of biology and evolution. Pages 21–22 *in* Lousier, 1991.

Antos, J.A., and D.B. Zobel. 1986. Habitat relationships of *Chamaecyparis nootkatensis* in southern Washington, Oregon, and California. Can. J. Bot. **64**:1898–1909.

Banerjee, M., and J. Maze. 1991. Within population variation in yellow-cypress: forestry implications. Pages 49–50 *in* Lousier, 1991.

Barker, J.E. 1991. Operational use of yellow-cypress in coastal British Columbia. Pages 44–48 *in* Louiser, 1991.

Bazzaz, F.A., N.R. Chiariello, P.D. Coley, and L.F. Pitelka. 1987. Allocating resources to reproduction and defense. Bioscience **37**:58–67.

Bower, R.C., S.D. Ross, and B.G. Dunsworth. 1989. Effect of GA_3 treatment timing in relationship to natural daylength on flowering and sex expression in *Chamaecyparis nootkatensis*. Can. J. For. Res. **19**:1422–28.

Cherry, M.L., and D.T. Lester. 1992. Genetic variation in *Chamaecyparis nootkatensis* from coastal British Columbia. West. J. App. For. **7**:25–29.

Collangelli, A.M. 1991. Yellow-cypress reproductive biology—pollen, seed, and seed orchards. Pages 29–30 *in* Lousier, 1991.

Critchfield, W.B. 1984. Impact of the Pleistocene on the genetic structure of North American conifers. Pages 70–118 *in* R. Lanner, ed. Proceedings of the Eighth North American Forest Biology Workshop. Utah State Univ., Logan.

Dunwoodie, P.W. 1986. A 6000 year record of forest history on Mount Rainier, Washington. Ecology **67**:58–68.

El Kassaby, Y.A., J. Maze, D.A. Macleod, and M. Banerjee. 1991. Reproductive-cycle plasticity in yellow-cedar *(Chamaecyparis nootkatensis)*. Can. J. For. Res. **21**:1360–64.

Frenkel, R.E. 1974. An isolated occurrence of Alaska-cedar in the Aldrich mountains, central Oregon. NW Sci. **48**:29–37.

Green, R.N., and K. Klinka. 1984. A field guide for site identification and interpretation for the Vancouver forest region. British Columbia Ministry of Forests, Vancouver, BC. Land Management Handbook No. 28.

Grimes, J.P. 1977. Evidence for three primary strategies in plants and its relevance to ecological and evolutionary theory. Amer. Nat. **111**:1169–94.

Grossnickle, S.C. 1991. Physiological and environmental tolerances of yellow-cypress. Pages 31–33 *in* Lousier, 1991.

Grossnickle, S.C., and J. Russell. 1990. Water movement in yellow-cedar seedlings and rooted cuttings: comparison of whole plant and rootsystem pressurization methods. Tree Phys. **6**:57–68.

———. 1991. Gas exchange processes of yellow-cedar *(Chamaecyparis nootkatensis)* in response to environmental variables. Can. J. Bot. **69**:2684–91.

Halpren, C.B. 1989. Early successional patterns of forest species: Interactions of life history traits and disturbance. Ecology **70**:704–720.

Hamrick, J.L. 1979. Genetic variation and longevity. Pages 84–114 *in* O.T. Solberg, S. Jain, G.B. Johnson, and P.H. Raven, eds. Topics in plant population biology. McGraw Hill, New York.

Harlow, W.M., and E.S. Harrar. 1969. Textbook of dendrology. McGraw Hill, Toronto.

Harmon, M.E., W.K. Ferrell, and J.F. Franklin. 1990. Effects on carbon storage of conversion of old-growth forest to young forests. Science **247**:699–701.

Harris, A.S. 1990. *Chamaecyparis nootkatensis* (D. Don) Spach Alaska-cedar. Pages 97–102 *in* R.M. Burns and B.H. Honkala, compilers. Silvics of North America, Vol. 1. U.S. Dept. Agric. For. Serv. Agric. Handb. 654. Washington, DC.

Hebda, R.J. 1983. Late-glacial and post-glacial vegetation history at Bear Cove Bog, northeast Vancouver Island, British Columbia. Can. J. Bot. **61**:3172–92.

Heddon, P. 1986. Silvicultural characteristics of yellow-cedar. MacMillan Bloedel Foresters' Annual Meeting, Nanaimo, B.C., Canada.

Jozsa, L. 1991. Yellow-cypress wood properties. Page 8 *in* Lousier, 1991.

Karlsson, I. 1974. Rooted cuttings of yellow-cedar *(Chamaecyparis nootkatensis)*. British Columbia Ministry of Forests, Research Note No. 66.

Karlsson, I., and J. Russell. 1990. Comparisons of yellow-cypress trees of seedling and rooted cutting origins after 9 and 11 years in the field. Can. J. For. Res. **20**:37–42.

Klinka, K. 1991. Ecology of yellow-cedar sites. Pages 23–28 *in* Lousier, 1991.

Krajina, V.J. 1969. Ecology of forest trees in British Columbia. Vol. 2(1):1–146 *in* V.J. Krajina, and R.C. Brooke, eds. Ecology of Western North America. Dept. Botany, Univ. B.C., Canada.

Kurz, M.L., D.T. Webb, and W.E. Vidaver. 1989. Micropropagation of yellow-cypress *(Chamaecyparis nootkatensis)*. Plant, Cell Tissue and Organ Culture **18**:297–312.

Levitt, J. 1980. Responses of plants to environmental stresses, Volume 1: Chilling, freezing and high temperature stresses. Pages 347–447 *in* T.T. Kozlowski, ed. Physiological Ecology Monographs. Academic Press, New York.

Loehle, C. 1988. Tree life history strategies: The role of defenses. Can. J. For. Res. **18**:209–222.

Lousier, J.D., ed. 1991. Yellow-cypress: Can we grow it? Can we sell it? Proc. Symp. March 26–28, 1990, Richmond, B.C., Canada. Forestry Canada and B.C. Ministry of Forests, FRDA Report 171.

Maser, C. 1990. The future is today: for ecologically sustainable forestry. Trumpeter **7**:74–78.

Meslo, E.C., C. Maser, and J. Verner. 1981. Old-growth forests as wildlife habitat. Trans. North Am. Wildl. Nat. Res. Conf. **46**:329–335.

Minore, D. 1979. Comparative autecological characteristics of northwestern tree species, a literature review. USDA For. Serv., PNW For. Range Expt. Sta., Gen. Tech. Report, Portland, OR.

Owens, J.N., and M. Molder. 1975. Cone initiation and development before dormancy in yellow cedar *(Chamaecyparis nootkatensis).* Can. J. Bot. **53**:186–199.

———. 1977. Cone induction in yellow-cypress *(Chamaecyparis nootkatensis)* by gibberellin A3 and the subsequent development of seeds within the induced cones. Can. J. For. Res. **7**:605–613.

Russell, J., and C. Cartwright. 1991. The genetics of yellow-cypress. Pages 34–35 *in* Lousier, 1991.

Russell, J., and S.C. Grossnickle. 1990a. Rooting of yellow-cypress cuttings, I. Influence of donor plant maturation. Forest Resources Development Agreement Research Memo No. 083, British Columbia Ministry of Forests, Canada.

———. 1990b. Rooting of yellow-cypress cuttings, II. Effect of fertilizer application period on rooting and growth. Forest Resources Development Agreement Research Memo No. 101, British Columbia Ministry of Forests, Canada.

Russell, J.H., S.C. Grossnickle, C. Ferguson, and D.W. Carson. 1990. Yellow-cedar stecklings: nursery production and field performance. Forest Resources Development Agreement Research Report No. 148, B.C. Ministry of Forests, Canada.

Spurr, S.H., and B.V. Barnes. 1980. Forest ecology. Third edition. John Wiley and Sons, New York.

Viereck, L.A., and E.L. Little. 1972. Alaska trees and shrubs. U.S. Dept. Agric. For. Serv., Agric. Handbook No. 410.

Waring, R.H., and J.F. Franklin. 1979. Evergreen coniferous forests of the Pacific Northwest. Science **204**:1380–86.

Zobel, D.B., and J.A. Antos. 1982. Adventitious rooting of eight conifers into a volcanic tephra deposit. Can. J. For. Res. **12**:717–719.

5

John H. Russell

Genecology of *Chamaecyparis nootkatensis*

Chamaecyparis nootkatensis (D. Don) Spach (yellow-cedar) is a long-lived, stress-tolerant species. It has a north–south distribution on the Pacific Northwest coast similar to *Thuja plicata* western (redcedar), *Tsuga heterophylla* (western hemlock), and *Picea sitchensis* (Sitka spruce), occurring over 20° latitude from northern California to southeast Alaska. Within this latitudinal range, however, *C. nootkatensis* occupies a unique geographic distribution. It occurs strictly at high elevations (over 1,200 m) in both the Siskiyou Mountains in northern California and southwest Oregon and on the west side of the Cascade Mountains in Washington and Oregon. It occurs at middle to high elevations in both the Olympic Mountains in Washington and the Coastal Mountains in southern British Columbia. North of approximately 51°N latitude, *C. nootkatensis* occurs from sea level to timberline, where it is restricted to a narrow longitudinal band along the coast. In central Oregon (Frenkel 1974) and southern British Columbia, two isolated stands occur more than 200 km inland from the most easterly coastal populations.

Throughout most of the range of *C. nootkatensis* the climate is very humid, with relatively cool summers and mild winters (Krajina 1969). Winter temperatures rarely go below -20°C. However, within *C. nootkatensis'* natural range, climatic and site conditions vary widely. In British Columbia, *C. nootkatensis* grows on moderately dry to wet soils and on nutrient very-poor to very-rich soils (Klinka 1991). In the southern part of its range (south of Mount Rainier in Washington state), *C. nootkatensis* is found on wet to dry sites and occurs generally in open habitats from bogs to rocky ridges (Antos and Zobel 1986). In the extreme northwest areas of *C. nootkatensis'* distribution (Prince William Sound), temperatures below −20°C are not uncommon (Paul Hennon, personal communication).

Although *C. nootkatensis* has the capacity to survive and grow under a wide range of conditions, it does not occur on many sites it would seem capable of occupying. On disturbed sites, *Abies amabilis* (amabilis fir) colonizes more quickly, and in closed canopy forests it is more shade-tolerant, thus replacing *C. nootkatensis.* On deep, well-drained soils, other species such as *T. plicata, T. heterophylla,* and *P. sitchensis* outgrow *C. nootkatensis.* The main factors that limit the distribution of *C. nootkatensis* within its natural range seem to be its inability to reproduce prolifically and to compete in early growth phases.

Conifer species in the Pacific Northwest region of North America have evolved alternative strategies in adapting to heterogeneous environments (Rehfeldt 1984). *Pseudotsuga menziesii* (Douglas fir), *T. heterophylla,* and *P. sitchensis* are relatively more specialized—i.e., the gene frequencies of a population (a community of inter-breeding individuals) change in response to environmental differences. In contrast, *Pinus monticola* (white pine) and *T. plicata* tend to be more generalized, with little genetic differentiation among populations (Roche 1969, Falkenhagen 1977, Griffin 1977, Kuser and Ching 1980, Rehfeldt 1984, Campbell 1986, Lines 1987, Loopstra and Adams 1989, Ying 1990).

Genetic variation among populations for specialists tends to be adaptive (i.e., correlate with environmental site variables). For example, trees from seed collected in relatively mild and/or wet climates grow faster, grow later into the season, and are more cold-susceptible than populations from colder and/or drier climates (Rehfeldt 1984). Generalists, on the other hand, display more phenotypic plasticity and more genetic variability within populations than among populations (Rehfeldt 1984, Sultan 1987).

Given its wide ecological amplitude, especially if considered in the absence of competition, and its lack of vegetative buds (i.e., indeterminate growth), *C. nootkatensis* should exhibit substantial phenotypic plasticity in fitness traits, typical of a generalist. At the same time, its wide latitudinal distribution and the occurrence of disjunct, isolated populations (resulting in decreased gene flow) suggest the evolution of a specialist mode, represented by the differential changes in gene frequencies between populations in response to environmental selection pressures.

Genecology

In a study involving seedlings originating from a rangewide population collection in which seed from individual trees were kept separate (families within populations), seedling morphological and physiological traits were measured (Russell 1993). Specifically, traits were measured that sampled the developmental sequence of events that influence the adaptation of a population to its environment. Seed collected for this study and subsequent seedlings were kept separate by individual trees, resulting, within each population, in groups of siblings that have one or both parents in common (open-pollinated family within a population). Seedlings from 33 rangewide populations and 171 families were grown for three years in a common garden trial at the Cowichan Lake Research Station, Vancouver Island, British Columbia. Subsets of populations were grown under four greenhouse environments during the second season (see Russell 1993 for details).

Genecological questions that were addressed included: How much of the genetic variability could be attributed to differences among populations as opposed to differences between families within populations? Was any of the genetic variability associated with populations adaptive? Finally, how did these results compare to other Pacific Northwest conifers commonly associated with, or having similar distributions to, *C. nootkatensis?*

The results of this study indicate that *C. nootkatensis* is intermediate between a specialist and a generalist mode. Substantially more genetic variability was attributed to differences from tree to tree within a population than between populations (Fig. 5.1). Rangewide provenance testing with family structure in Pacific Northwest conifers has usually shown that the variation among populations is substantially greater than the variation between families within a population (Namkoong et al. 1972, Namkoong and Conkle 1976, Fashler et al. 1985). This has especially been the case where a species has shown a high degree of specialization in response to heterogeneous environments (Campbell 1979, White et al. 1981, Loopstra and Adams 1989). In contrast, genetic variation among families was substantially larger than between populations for *P. monticola,* which has exhibited a generalist mode of adaptation (Rehfeldt 1979, 1984).

Differences among populations of *C. nootkatensis* seedlings in growth, phenological, and cold-hardiness traits were significant but moderate, when correlated with latitude and elevation of seed origin. Southern and high-elevation populations were taller, had more trees growing into the fall, and were more susceptible to cold damage than northern and low-elevation populations (Figs. 5.2 and 5.3a). Other Pacific North-

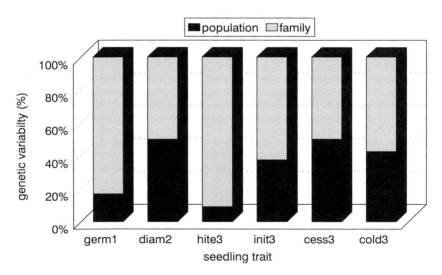

Figure 5.1. Distribution of genetic variability among populations and between families within populations for seedling traits of *C. nootkatensis.* germ1: % germination. diam2: 2-year root collar diameter. hite3: 3-year total shoot height. init3: shoot growth initiation at start of third growing season. cess3: shoot growth cessation at end of third growing season. cold3: cold-injury after third growing season.

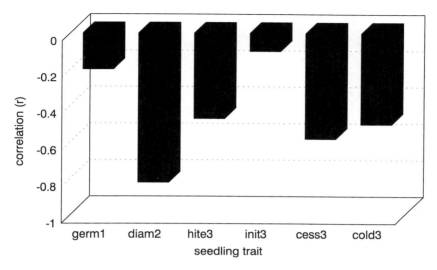

Figure 5.2. Correlation of seed and seedling traits of *C. nootkatensis* with latitude of seed origin. germ1: % germination. diam2: 2-year root collar diameter. hite3: 3-year total shoot height. init3: shoot growth initiation at start of third growing season. cess3: shoot growth cessation at end of third growing season. cold3: cold-injury after third growing season.

west conifers have shown similar, but stronger, trends for increasing growth and cold susceptibility with decreasing latitude. Correlation of growth cessation with latitude (i.e., photoperiodic ecotypes) is also common in conifers with extended latitudinal distributions (Vaartaja 1959, Campbell 1979, Rehfeldt 1979, Kuser and Ching 1980, Lines 1987).

These adaptive trends were less apparent when only comparing populations from the main coastal distribution of *C. nootkatensis* within British Columbia (Fig. 5.3b). This has also been shown for *P. menziesii* and *T. heterophylla* (Ying 1990) and reflects the relatively homogeneous maritime influence on coastal climates.

Environmental factors had a large effect on growth rate, morphology, and physiology of *C. nootkatensis* (Fig. 5.4). A shortened photoperiod and drought stress each induced decreased shoot growth. When either stress was removed, growth of the stressed seedlings exceeded that of unstressed seedlings. These results are similar to those reported in the literature for other *Cupressaceae* (Harry 1987, Krasowski and Owens 1991, Arnott et al. 1993).

Evidence of some degree of specialization among populations in response to drought stress was apparent. When grown under either ideal or stressful moisture levels, seedlings from relatively xeric and interior sites developed a more conservative phenotype than those from relatively mesic and coastal sites: smaller height and diameter growth, decreased lateral branch growth, and more allocation of carbohydrates to roots and less to stem and branches. These characteristics would allow for increased water absorption and retention through the exploitation of a larger volume

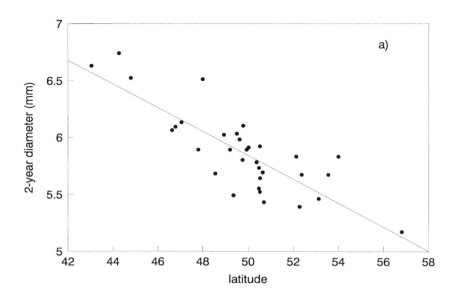

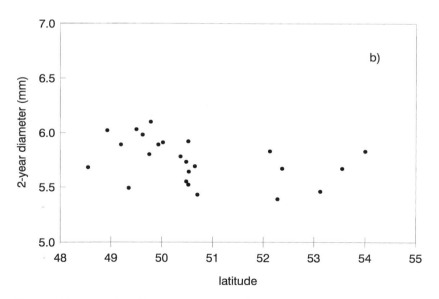

Figure 5.3. Regression of second-year *C. nootkatensis* shoot diameter on latitude of seed origin. a: Rangewide populations. b: Coastal British Columbia populations.

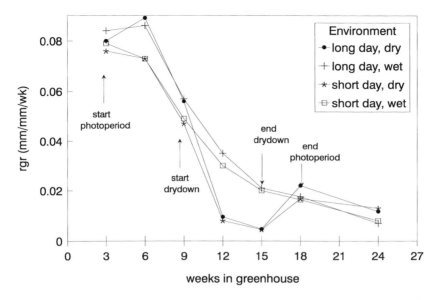

Figure 5.4. Effect of four greenhouse environments on *C. nootkatensis* second-year growth. rgr: relative growth rate (mm of shoot growth/mm of total shoot height/week).

of soil, increased absorption efficiency per unit root area, and decreased transpirational surface (Levitt 1980b).

Seedlings from the more xeric site also had increased photosynthetic efficiency per unit area foliage (Fig. 5.5) and greater stomatal conductance under both well-watered and drought conditions, lower osmotic potential prior to drought, and higher turgor pressure at a given relative water content (% water by weight). These physiological adaptations to drought, as well as the ability to osmotically adjust and increase cell wall inelasticity in response to a drought, would allow for the maintenance of positive turgor and cell growth (Hsiao 1973) and minimize both physical and metabolic damage to cellular processes (Levitt 1980a).

Morphological and physiological adaptations to drought have been reported for other Pacific Northwest xeric-habitat conifers including *P. menziesii* (Ferrell and Woodward 1966, Zavitkovski and Ferrell 1968, Unterscheutz et al. 1974, White 1987, Joly et al. 1989), and *P. contorta* (lodgepole pine; Dykstra 1974).

Summary

The combination of an extensive latitudinal range and geographic isolation, coupled with a wide ecological amplitude and indeterminate growth habit, suggest aspects of both a specialist and generalist adaptive mode for *C. nootkatensis*. This has been supported in the text and is summarized by the following generalizations for seedling

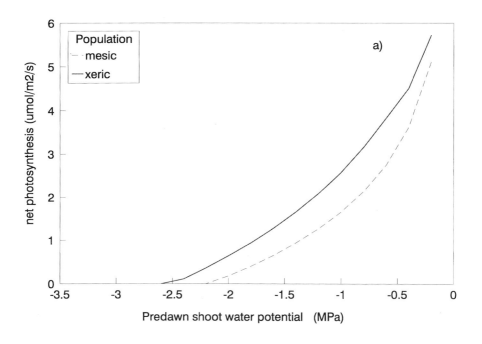

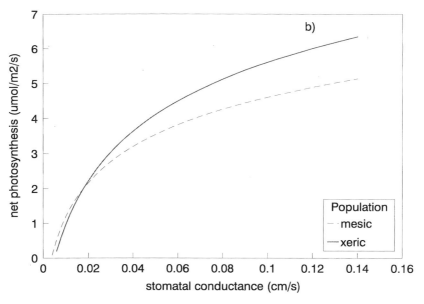

Figure 5.5. a: Net photosynthesis predicted response to predawn shoot water potential for 2-year-old *C. nootkatensis* from two populations. b: Net photosynthesis predicted response to stomatal conductance for 2-year-old *C. nootkatensis* from two populations.

traits that measure the range of the annual developmental sequence of *C. nootkatensis:*

1. There is significant interpopulation and family-within-population genetic variation.
2. Genetic variation among families is greater than interpopulation variability.
3. Genetic variation at the population level is moderately correlated with latitude and elevation of seed origin.
4. Drought-resistant ecotypes were evident from xeric habitats.
5. Seedlings exhibit phenotypic plasticity in response to environmental changes.

The preceding generalities on population effects and associations with geography are more relevant when rangewide populations are measured. If southern populations (i.e., Oregon populations) are excluded, correlations of traits with seed origin are, for the most part, not significant.

Adaptive responses to drought follow a similar pattern. Morphological and physiological responses to drought are only apparent among populations from areas that differ greatly in moisture availability (i.e., coastal, windward versus coast:interior transition).

It seems possible that the extreme southern and inland *C. nootkatensis* populations have responded to environmental selection pressures—most likely aided by reduced gene flow that is due to spatial isolation and poor sexual reproduction—by changes in gene frequency. However, the species has maintained both substantial genetic variation and phenotypic plasticity within populations. *Chamaecyparis nootkatensis* seems to have evolved an intermediate adaptation mode with less genetic differentiation associated with geography than *P. menziesii, P. sitchensis,* and *T. heterophylla,* and more geographic differentiation than *P. monticola* and *T. plicata.*

The genecology of *C. nootkatensis* seems consistent with the natural history and ecology of the species. Reestablishment of coastal British Columbia forests following the retreat of the last glaciation occurred slowly over the last 12,000 years (Wainman and Mathewes 1987, Whitlock 1992). Different species became established at different rates depending on climate and their life histories. *Thuja plicata,* a long-lived, shade-tolerant, climax species, as is *C. nootkatensis,* established itself on north Vancouver Island as recently as 2,000 years ago (Hebda and Mathewes 1984). Thus, for many coastal species, given that they are long-lived and site disturbances such as fire have been minimal, not many generations have passed since recolonization. The moist climate of the coast, with its moderate environmental gradients, results in decreased selection pressures. Thus, the opportunity for the accumulation of adaptive genetic differences among populations has been minimal, and any genetic differences among populations of *C. nootkatensis* on the coast are most likely due to chance movement of seed.

That *C. nootkatensis* is less of a specialist than other Pacific Northwest conifers fits well with its indeterminate growth habit (i.e., non-bud forming). Growth cessation in the late summer is more responsive to temperature than to photoperiod. Shoot growth tends to be slower; however, it occupies more of the total growing season, stopping and resuming growth into the fall depending on seasonal temperatures (Russell 1993). Thus, the species is more in tune with local temperature fluctuations through plastic responses, as opposed to changes in gene frequencies resulting in photoperiodic ecotypes.

Management Implications

Current seed transfer guidelines governing the movement of *C. nootkatensis* seed for reforestation in British Columbia recognize two broad seed zones, both long, narrow north–south zones following the coast (Coastal Seed Transfer Guidelines 1990). The division between the two delineates the coastal forests from the coastal:interior forest transition zone. The transition zone tends to be colder and drier, with wider seasonal temperature and moisture fluctuations.

Within each seed zone, *C. nootkatensis* seed can be transferred from its original locale, $+/-3°$ latitude in the coastal zone and $+/-2°$ latitude in the transition zone, and $+/-400$ m elevation in the coastal and $+/-300$ m in the transition.

Restricting *C. nootkatensis* transfer between the two zones is supported by the results of the study reported here. Seed from the more xeric site located in the transition zone was more adapted to sites with high evaporative demand brought about by less rainfall and higher summer temperatures. However, results from the study indicate that transfer of seed within zones could be wider. Long-term population tests have just been established in a number of wild sites throughout the range of *C. nootkatensis* in British Columbia. Results from these tests, indicating wide adaptability, are necessary before considering any changes in current seed transfer guidelines.

References

Antos, J.A., and D.B. Zobel. 1986. Habitat relationships of *Chamaecyparis nootkatensis* in southern Washington, Oregon, and California. Can. J. Bot. **64**:1898–1909.

Arnott, J.T., S.C. Grossnickle, P. Puttonen, A.K. Mitchell, and R.S. Folk. 1993. Influence of nursery culture on growth, cold-hardiness, and drought resistance of yellow cypress. Can. J. For. Res. **23**:2537–2547.

Campbell, R.K. 1979. Genecology of Douglas-fir in a watershed in the Oregon Cascades. Ecology **60**:1036–1050.

———. 1986. Mapped genetic variation of Douglas-fir to guide seed transfer in southwest Oregon. Silvae Genet. **35**(2–3):85–96.

Coastal seed transfer guidelines. 1990. British Columbia Ministry of Forests.

Dykstra, G.F. 1974. Drought resistance of lodgepole pine seedlings in relation to provenance and tree water potential. British Columbia For. Serv. Res. Note No. **62**.

Falkenhagen, E.R. 1977. Genetic variation in 38 populations of Sitka spruce. Silvae Genet. **26**:67–75.

Fashler, A.,Y.A. El Kassaby, and O. Sziklai. 1985. Interprovenance variability in an IUFRO Douglas-fir provenance-progeny trial. Pages 187–204 *in* W. Ruetz and J. Nather, eds. Proceedings IUFRO Working Party S.2.02.05. Vienna.

Ferrell, W. K., and S. Woodward. 1966. Effect of seed origin on drought resistance of Douglas-fir. Ecology **47**:499–503.

Frenkel, R.E. 1974. An isolated occurrence of Alaska-cedar (*Chamaecyparis nootkatensis* [D.Don] Spach) on the Aldrich Mountains, Central Oregon. Northw. Sci. **48**(1):29–37.

Griffin, A.R. 1977. Geographic variation in Douglas-fir from the Coastal ranges of California: II. Predictive value of a regression model for seedling growth variation. Silvae Genet. **26**(5–6):158–163.

Harry, D.E. 1987. Shoot elongation and growth plasticity in incense-cedar. Can. J. For. Res. **17**:484–489.

Hebda, R.J., and R.W. Mathewes. 1984. Holocene history of cedar and native Indian cultures of the North America Pacific coast. Science **225**:711–713.

Hsiao, T.C. 1973. Plant response to water stress. Ann. Rev. Plant Physiol. **24**:519–570.

Joly, R.J., W.T. Adams, and S.G. Stafford. 1989. Phenological and morphological responses of mesic and dry site sources of *P. menziesii* to water deficit. For. Sci. **35**:987–1005.

Klinka, K. 1991. Ecology of yellow-cedar sites. Pages 23–28 in J.D. Lousier, ed. Proceedings yellow cypress symposium: can we grow it? can we sell it? British Columbia For. Serv. For. Resources Dev. Agreement Res. Report No. 171.

Krajina, V.J. 1969. Ecology of forest trees in British Columbia. Ecol. West. N. Amer. **2**(1):1–146.

Krasowski, M.J., and J.N. Owens. 1991. Growth and morphology of western redcedar seedlings as affected by photoperiod and moisture stress. Can J. For. Res. **21**:340–352.

Kuser, J.E., and K.K. Ching. 1980. Provenance variation in phenology and cold hardiness of western hemlock seedlings. For. Sci. **26**(3):463–470.

Levitt, J. 1980a. Response of plants to environmental stresses. Vol. I. Chilling, freezing, and high temperature stresses. Academic Press, New York.

———. 1980b. Response of plants to environmental stresses. Vol. II. Water, radiation, salt, and other stresses. Academic Press, New York.

Lines, R. 1987. Seed origin variation in Sitka spruce. *In* D.M. Henderson and R. Faulkner, eds. Proc. Royal Soc. Edinburgh **93B**:25–39.

Loopstra, C.A., and W. Th. Adams. 1989. Patterns of variation in first year seedling traits within and among Douglas-fir breeding zones in southwest Oregon. Silvae Genet. **38**(5–6):235–243.

Namkoong, G., and M.T. Conkle. 1976. Time trends in genetic control of height growth in ponderosa pine. For. Sci. **22**:2–12.

Namkoong, G., R.A. Usanis, and R.R. Silen. 1972. Age-related variation in genetic control of height growth in Douglas-fir. Theor. Appl. Genet. **42**:151–159.

Rehfeldt, J. 1979. Ecotypic differentiation in populations of *Pinus monticola* in northern Idaho—myth or reality? Am. Nat. **114**(5):627–636.

———. 1984. Microevolution of conifers in the northern Rocky Mountains: A view from common gardens. Pages 32–46 in R.M. Lanner, ed. Proceedings 8th North American forest biology workshop, Logan, UT.

Roche, L. 1969. A genecological study of the genus *Picea* in British Columbia. New Phytol. **68**:505–554.

Russell, J.H. 1993. Genetic architecture, genecology and phenotypic plasticity in seed and seedling traits of yellow-cedar (*Chamaecyparis nootkatensis* [D.Don] Spach). Dissertation. Univ. British Columbia.

Sultan, S.E. 1987. Evolutionary implications of phenotypic plasticity in plants. Evol. Biol. **21**:127–178.

Unterscheutz, P., W.F. Ruetz, R.R. Geppert, and W.K. Ferrell. 1974. The effect of age, preconditioning, and water stress on the transpiration rates of Douglas-fir (*Pseudostuga menziesii*) seedlings of several ecotypes. Physiol. Plant. **32**:214–221.

Vaartaja, O. 1959. Evidence of photoperiodic ecotypes in trees. Ecol. Monogr. **29**:91–111.

Wainman, N. and Mathewes, R.W. 1987. Forest history of the last 12,000 years based on plant macrofossil analysis of sediment from Marion Lake, southwestern British Columbia. Can. J. Bot. **65**:2179–2187.

White, T.L. 1987. Drought tolerance of southwestern Oregon Douglas-fir. For. Sci. **33**:283–293.

White, T.L., D.P. Lavender, K.K. Ching, and P. Hinz. 1981. First-year growth of southwestern Oregon Douglas-fir in three test environments. Silvae Genet. **30**:173–178.

Whitlock, C. 1992. Vegetational and climatic history of the Pacific Northwest during the last 20,000 years: implications for understanding present-day biodiversity. Northw. Env. J. **8**:5–28.

Ying, C.C. 1990. Adaptive variation in Douglas-fir, Sitka spruce, and true fir: A summary of provenance research in coastal British Columbia. *In* Proceedings joint meeting of Western Forest Genetics Association and IUFRO working parties S2.02, 05, 06, 12, and 14. Olympia, WA.

Zavitkovski, J., and W.K. Ferrell. 1968. Effect of drought upon rates of photosynthesis, respiration and transpiration of seedlings of two ecotypes of Douglas-fir. Bot. Gaz. **129**:346–350.

6

Mel Greenup

Managing *Chamaecyparis lawsoniana* (Port-Orford-Cedar) to Control the Root Disease Caused by *Phytophthora lateralis* in the Pacific Northwest, USA

Chamaecyparis lawsoniana (A. Murr.) Parl. (Port-Orford-cedar) ranges along the Pacific Coast from Coos Bay, Oregon to Eureka, California, and eastward to the vicinity of Grants Pass, Oregon, and Redding, California (Figs. 6.1 and 6.2). The species is highly adaptable and thrives on diverse soil types and sites in coastal and inland regions of the Coast Ranges and Klamath Mountains of both southwest Oregon and northwest California. Within these forest regions, *C. lawsoniana* grows extensively on state, county, and private lands as well as on federal lands managed by the Bureau of Land Management and the Siskiyou National Forest of Oregon and Six Rivers, Klamath, and Shasta-Trinity National Forests of California.

Chamaecyparis lawsoniana commands premium values for its unique wood qualities and special uses. Highest prices are paid for saw logs, leaving no doubt about the species' economic importance (Dan Brattain, personal communication). Logs of young second-growth trees delivered to the mill are valued at $600 to $1500 per thousand board feet ($212 to $530 per 10 m^3), depending on log diameter (1993 prices). Export prices paid by Japan for the logs of old-growth trees also reflect log diameter but, within diameter classes, values increase with the straightness of grain and decreasing number and size of knots. Timber industry reports give a general range of values for classes of logs measured at the small end (Table 6.1; metric volume prices are approximate).

The root disease caused by *Phytophthora lateralis* was first reported by a private horticultural nursery in the Seattle, Washington area in the early 1920s. The disease moved south in landscape plantings and from there into forests on the hooves of grazing mammals and the wheels and undercarriages of vehicles. Disease movement is primarily along roads and streams. Isozyme analysis of the pathogen indicates

Figure 6.1. Thirty-year old *C. lawsoniana,* Siskiyou National Forest, Oregon.

no genetic variation in its current range (Everett Hansen, personal communication),
reinforcing the belief that this fungus was introduced.

The disease moves from tree to tree through root grafts and spreads to new locations by spores in moist soil on vehicles or on the feet of mammals, and also by
spores that swim in overland flows of water (Roth et al. 1987). Tree infection occurs
at the root tip and, over time, the fungus colonizes the root system and kills the tree.

Figure 6.2. Old-growth *C. lawsoniana,* Siskiyou
National Forest, Oregon.

Table 6.1 Port-Orford-Cedar Log Export Values

Log Diameter		Price (1993 U.S. dollars)	
inches	cm	per 1000 bd ft	per 10 m^3
8–11	20–28	$1500+	$530+
12–17	30–43	2000–3000	707–1060
18–29	46–74	3500–4500	1237–1590
30+	76+	6500–8500	2297–3004

Source: Dan Brattain, personal communication.

The effect is similar to what would occur if the tree were girdled mechanically. Rate of mortality depends on tree size. From the time of infection, seedlings will die in a few weeks, saplings in a few months, and large trees in one or more years (personal observation).

The Port-Orford-Cedar Action Plan

In 1985, the Western Natural Resources Law Clinic at the University of Oregon formally notified the Forest Service of its belief that *C. lawsoniana* was not being adequately protected and maintained within its native environment. The Clinic requested that the Forest Service establish a committee with the authority to promote comprehensive management practices to control the root disease throughout the range of *C. lawsoniana*. The Forest Service met this concern by involving preservationists, environmentalists, scientists, private industry people, and Forest Service personnel in a consensus process that produced a list of action items. The Interregional Port-Orford-Cedar Action Plan (U.S. Forest Service 1988) was approved by the regional foresters of the Pacific Northwest and Pacific Southwest regions. The plan entails research, administrative studies, mapping and monitoring of *C. lawsoniana* and the disease, creation and implementation of control strategies, and methods to validate and update those strategies. Funds for implementing the plan are provided by the Forest Pest Management and Timber Management programs of the Pacific Northwest and Pacific Southwest regions.

Program Planning

Planning is done by the Port-Orford-Cedar coordinating group, led by a program manager. Program support, consultation, and oversight are provided by the Forest Supervisor of the Siskiyou National Forest and the program leader and program director of Forest Pest Management for the Pacific Southwest and Pacific Northwest regions, respectively. The group's technical team presently consists of three plant pathologists, two silviculturists, an area ecologist, and the regional geneticist with the Pacific Northwest region; the Pacific yew/Port-Orford-Cedar coordinator and a silviculturist with the Bureau of Land Management; a research geneticist with the Pacific Southwest Research Station; a research plant pathologist and a research forester with the Pacific Northwest Research Station; and a resource assistant with the Smith River National Recreation Area.

Planning sessions are scheduled as needed to develop and review annual and multi-year programs and budgets, to assign specific tasks, and to deal with new issues as they arise. The guide is the Action Plan, and components of the plan are discussed at each meeting. Research progress and management techniques are reviewed continually so that the group can adjust priorities for research needs and technology development. Disagreement is allowed and heard, and discussion is always animated and vigorous. Team members know and understand the objectives and share their opinions and information as progress is made toward agreement on the programs and budgets.

Program Review

Regional office review is ongoing because region personnel participate directly as members of the Coordinating Group. Periodically, the program manager's performance is formally reviewed by the group leaders for Forest Pest Management, Pacific Northwest and Pacific Southwest regions, and by the forest supervisor, Siskiyou National Forest. The Pacific Northwest region evaluated the Port-Orford-Cedar Program as a part of an implementation review of the Siskiyou National Forest Plan.

Action Plan Inputs

Every activity that has the potential for affecting *C. lawsoniana* is assessed by a formal analysis that is developed through the consensus process. Analysis is facilitated by the coordinating group and by direction from the regional foresters. The program manager and group members elicit the ideas, advice, and collaboration of national forest and ranger district people by means of program presentations. Personnel of other forest land ownerships and management units are informed and encouraged to cooperate and participate by means of periodic status reports and ongoing consultations. Whenever new issues arise and affect program operations, the program manager alerts the forest supervisors and regional forester staffs.

Control Strategies and Monitoring

Disease-control strategies are devised through a formal analysis process of each project. Monitoring for success of these strategies is spelled out in this analysis. Monitoring can be either project-specific or broad enough to assess overall movement of the root disease. Mapping of *C. lawsoniana* and the disease in National Forests was completed in 1993. The maps have been digitized into a Geographic Information System (GIS). As the accuracy of mapping improves, so does our ability to track and control the disease.

Our overall goal is to reduce the spread of the disease, and mapping and monitoring were the foremost concerns identified in the Action Plan. The specific objectives are to

1. define and map the boundaries of infection;
2. determine the rates of spread of the disease; and
3. evaluate effects of the mitigation measures.

A plan to accomplish these three objectives was developed and reviewed. To implement the plan and obtain field cooperators, the work accomplished and the efforts needed were presented and discussed at a workshop in 1992. The workshop was attended by representatives of the National Forests and Ranger Districts managing land in the natural range of *C. lawsoniana*.

Control Strategies Implemented

Eight strategies aimed at controlling spread of the root disease have been implemented on the ground:

1. Conducting operations in summertime to reduce chances for spore movement.
2. Cleaning vehicles and equipment before entering an uninfested area and before leaving an infested area to remove soil that may spread the root pathogen.
3. Constructing berms on roadsides to control vehicle splash and surface runoff and reduce chances of infecting roadside trees.
4. Removing *C. lawsoniana* from roadsides to prevent infection along roads and reduce roads as pathways for infestation.
5. Closing roads that run in and through infested areas.
6. Maintaining roads in ways that keep infested soil inside infested areas.
7. Using logging systems that reduce the need for and extent of new forest access roads.
8. Planning future road locations to reduce the chances of introducing the fungus in disease-free areas.

Program Research Needs

Research and administrative study needs initially were identified by the consensus process and formalized in the Action Plan. The coordinating group subsequently prioritized each of the Action Items on research and information needs. The coordinating group can change the list of needs and corresponding priorities if a new need is identified.

Program Accomplishments

Screening for resistance was done through a contract with the Oregon State University (OSU) Department of Botany and Plant Pathology. Some 202 individual trees, tested by branch inoculations with the fungus, were found to have some indications of resistance to the disease. Validation of branch test results is being done by outplanting seedlings and rooted branch cuttings in infested areas. A breeding program to increase resistance levels has been incorporated in the Siskiyou National Forest Tree Improvement Plan.

Analyses of conditions that promote spread of the disease are under way. Methods for detecting *P. lateralis* spores are being investigated. Enzyme-linked immunosorbent assay (ELISA) shows promise but needs work to improve sensitivity (Everett Hansen, personal communication). A study of the longevity of resting spores at the Cedar Rustic Campground has been completed by pathologist J. Kliejunas, Pacific Southwest region (Kliejunas 1992). Resting spores in this environment—cold-moist in winter, warm-moist in spring, and hot-dry in summer and fall—survived four years. An informal study begun in Coos County Forest by pathologist E. Hansen of OSU in 1986 indicates that *P. lateralis* can survive in soil without a host for a much longer time (personal communication). The environment there is warm-moist most of the year and especially favorable for survival of the fungus. Taken together, the results suggest that the longevity of resting spores varies with the site environment. A spore longevity of five years or more is assumed for the purpose of evaluating control strategies.

A soil assay for detecting *P. lateralis* spores in the absence of dying trees is being developed by pathologist P. Tsao (1993) at University of California, Riverside, but the task is complex and may take several years to complete.

Isozyme analyses of rangewide genetic variation in *C. lawsoniana* were completed by geneticist C. Millar and co-workers at the Pacific Southwest Research Station (Millar et al. in prep.). There were relatively strong north–south and east–west multilocus clines (much stronger than in *P. menziesii* [Douglas fir] and *P. ponderosa* [ponderosa pine] populations in the same region), and differences among river drainages contributed greatly to population diversity. Foliage samples were assayed for 10 trees in each of 40 stands chosen to represent upper, middle, and lower portions of the major river drainages inhabited by the species. A study on the correlation of genetic (isozyme) and ecological diversity was completed by Millar and area ecologist T. Atzet of the Pacific Northwest Region (Millar et al. 1991). Correlations between allozyme frequencies and soil types were not strong (an R^2 of 0.3), but were large considering the short geographic distances among population pairs.

A common-garden study of rangewide genetic variation in the adaptive traits of growth and survival has been planned by plant physiologist J. Jenkinson of the Pacific Southwest Research Station (Jenkinson et al. 1992). Seed collections were begun in 1992. Ten or more trees will be collected in each of 40 stands, to sample *C. lawsoniana* on lowland and upland sites and on diverse soil types in the watersheds of coastal and inland regions in southwest Oregon and northwest California. Forest Service cooperators include personnel of the Dorena Tree Improvement Center in Cottage Grove, Oregon; the Siskiyou National Forest, Oregon; the Northern Zone Tree Improvement Group in Yreka, California; the Institute of Forest Genetics at Placerville, California; and the Humboldt Nursery in McKinleyville, California.

A New Issue Confronted

The discovery of *P. lateralis* on *Taxus brevifolia* (Pacific yew) in extreme northwest California in 1990 on the Gasquet Ranger District of the Six Rivers National Forest

(DeNitto and Kliejunas 1991) showed that *C. lawsoniana* is not the only forest tree susceptible to the disease. A forest survey designed to locate infected yew was conducted in autumn 1991 (DeNitto and Greenup 1992). The fungus was found on 17 yew trees in 11 separate localities on the Six Rivers and Siskiyou National Forests, in stands where yew intermingles with infected *C. lawsoniana*. The trees are growing along roads and slow-flowing streams, in environments that enhance increased spore loads and infection by the fungus (Murray and Hansen, in progress). Continued survey and study by M. Murray and Everett Hansen of OSU found more than 100 additional Pacific yew infected or dead.

Mapping and monitoring of infected yew trees within the native range of *C. lawsoniana* is ongoing. Any yew that is suspected of being infected is mapped, and occurrence of the disease is verified by a pathologist. Findings are expected to improve our understanding of the disease in *C. lawsoniana*. Wherever yew is found, whether mixed with *C. lawsoniana* or in a control-project area, disease-control strategies will be developed to protect the yew using the same standards and procedures that are used to protect *C. lawsoniana*. In most areas, the same strategies should work effectively for either species.

Studies to determine the relation between Pacific yew, the root disease, and *C. lawsoniana* are under way. A study of the susceptibility of yew to the fungus and a search for natural resistance to the disease are also under way. That yew may be more variable than *C. lawsoniana* in its susceptibility to infection is suggested by field observations and comparative evaluations of rooted yew cuttings with *C. lawsoniana* seedlings (DeNitto and Kliejunas 1991).

Future Program Success

Approaches taken in the Port-Orford-Cedar Program implement the intents of the National Forest Management Act and New Perspectives Forestry. These policies require that forest management maintain species diversity, and that goal is emphasized in the forest plans. The chief's staff, Washington Office, is kept informed of progress on the Action Plan. Support from the Pacific Southwest and Pacific Northwest regional offices in Portland, Oregon, and San Francisco has contributed markedly to the attainment of Action Plan goals. Such support, together with some modest but reliable funding, remains critical for the continued future success of the program.

Support for and allocation of research funds is essential to certain aspects of the Action Plan. Identification of genetic resistance mechanisms, for example, requires basic research and cannot be achieved without a long-term commitment of funds. The Pacific Northwest and Pacific Southwest regions are concerned that a pest of limited native range may not compete with pests of national concern for forest pest management technology and development funds. Program funding is requested through normal budgeting channels, and program success depends on the extent to which the requests are honored.

In May 1995, the Forest Service determined that most of the work and study items were complete or in progress. Also, the five National Forests containing *C. lawsoniana* have final Forest Plans that identify the need to protect the species and provide direction for its management. The Port-Orford-Cedar coordinating group will continue to integrate activities on federal lands throughout the species' range. Their overall objective is to assure the coordinated management of *C. lawsoniana* and the root disease that afflicts it.

References

DeNitto, G., and M. Greenup. 1992. Pacific yew root disease survey. U.S. Dept. Agric. For. Serv., Redding, CA.

DeNitto, G.A., and J.T. Kliejunas. 1991. First report of *Phytophthora lateralis* on Pacific yew. Plant Disease **75**:968.

Jenkinson, J.L., R. Stutts, and J.A. Nelson. 1992. Geographic patterns of genetic variation in Port-Orford-cedar: common garden tests in coastal and inland regions of Oregon and California. U.S. Dept. Agric. For. Serv., Pacific Southwest Forest and Range Experiment Station, Albany, OR.

Kliejunas, J.T. 1992. Soil monitoring for *Phytophthora lateralis* at Cedar Rustic Campground, Gasquet Ranger District, Six Rivers National Forest. U.S. Dept. Agric. For. Serv., Pacific Southwest Region, Redding, CA. Technical Report No. R92-3.

Millar, C.I., D. Delany, and R. Westfall. In preparation. Genetic diversity in Port-Orford-cedar, rangewide study. U.S. Dept. Agric. For. Serv., Pacific Southwest Forest and Range and Experiment Station, Redding, CA.

Millar, C.I., D. Delany, R. Westfall, and T. Atzet. 1991. Ecological factors as indicators of genetic diversity of Port-Orford-cedar. U.S. Dept. Agric. For. Serv., Pacific Southwest Forest and Range and Experiment Station, Redding, CA.

Murray, M., and E. Hansen. *Work in progress.* Oregon State University, Corvallis.

Roth, L.F., R.D. Harvey Jr., and J.T. Kliejunas. 1987. Port-Orford-cedar root disease. U.S. Dept. Agric. For. Serv., Pacific Northwest Region. Forest Pest Management Report No. R6-FPM-PR-010-91.

Tsao, P.H. 1993. Detection of Port-Orford-cedar pathogen in soil. Univ. of California-Riverside. Project report to the U.S. Dept. Agric. For. Serv.

U.S. Forest Service. 1988. Interregional Port-Orford-cedar action plan. U.S. Dept. Agric. For. Serv., Regions 5 and 6.

7

Shin-Ichi Yamamoto

Regeneration Ecology of *Chamaecyparis obtusa* and *Chamaecyparis pisifera* (Hinoki and Sawara Cypress), Japan

There are two species of *Chamaecyparis* (Cupressaceae) in Japan: *Chamaecyparis obtusa* (Sieb. et Zucc.) Endlicher (Hinoki cypress) and *Chamaecyparis pisifera* (Sieb. et Zucc.) Endlicher (Sawara cypress) (Figs. 7.1–3 and Table 7.1). Geographical distribution of these tall evergreen conifers nearly overlaps, although *C. pisifera* occurs infrequently in southwest Japan and is absent from Shikoku Island (the fourth of the main Japanese Islands). The distributional range of *C. pisifera* extends slightly northward of that of *C. obtusa; C. obtusa* (Fig. 7.4) grows from 30°15′–37°10′N and *C. pisifera* (Fig. 7.5) from 32°48′–39°32′N (Hayashi 1960). In the Kiso district of central Japan, both species are abundant. The altitudinal range of *C. obtusa* is 80–2,200 m (rarely to above 2,500 m; Horikawa 1972) and that of *C. pisifera* is 280–2,590 m above sea level (Asakawa et al. 1981). At present, *C. pisifera* has a limited distribution, mainly in the central part of Japan. It was widely and abundantly found in the Pliocene era and has been gradually decreasing up to the present (Miki 1958). The reverse tendency is found in the past and present distribution of *C. obtusa* (Miki 1958).

Natural *C. obtusa* stands have suffered from timber harvesting dating back to ancient times (600s A.D.). The timber was used for building temples or Shinto shrines (Sato 1973). Today, the remaining natural stands more or less reflect the influence of past human cutting. *Chamaecyparis obtusa* is a very important species for reforestation in Japan. Therefore, research on this species has been mainly confined to traits related to reforestation. Ecological traits, especially in relation to the regeneration of natural stands, have remained relatively unstudied.

There are no clear historical records of timber harvesting in natural stands of *C. pisifera,* although its timber has been used for making bath and rice tubs (Iwata and Kusaka 1954, Hayashi 1960). Research on *C. pisifera* and its forest is very limited,

Figure 7.1. Shape of adult trees of *C. obtusa* (center) and *C. pisifera* (left).

except for dendrological (Iwata and Kusaka 1954, Hayashi 1960) and phytosociological (Maeda 1951, Maeda and Yoshioka 1952, Tanimoto et al. 1984, Miyawaki and Okuda 1990) descriptions. Natural stands of *C. pisifera* have mainly been converted to plantations of *Cryptomeria japonica* (Japanese red cedar); *C. pisifera* has rarely been used for reforestation (Iwata and Kusaka 1954).

Site Differences between *C. obtusa* and *C. pisifera* within Areas of Sympatry

Within areas of sympatry of *C. obtusa* and *C. pisifera,* the two species occur on different sites (Maeda 1951, Iwata and Kusaka 1954, Hayashi 1960, Yatoh 1964, Yamanaka 1979). *Chamaecyparis obtusa* is generally common on xeric sites such as middle to upper slopes and ridges, where extensive pure stands occasionally develop. Sometimes, podzolic soils and serpentinite characterize these sites. In some areas of the Kiso district, a dense cover of *Sasa* species (dwarf bamboo) develops in the understory of *C. obtusa* stands (Akai 1972).

On the other hand, mesic or wet conditions characterize the habitat of *C. pisifera.* Stands often develop in rocky depressions, on lower slopes, and near mountain

Figure 7.2. Interior views of (a) *C. obtusa* and (b) *C. pisifera* stands.

streams. *Chamaecyparis pisifera* does not develop extensive pure stands and has a spotty distribution.

Regeneration Characteristics

Seed production by *C. obtusa* and *C. pisifera* begins between 20 and 25 years of age (Asakawa et al. 1981). Seed production of *C. obtusa* generally fluctuates greatly from year to year (Hasegawa 1943, Sakaguchi 1952, Ozawa 1962, Sato 1973, Asakawa et al. 1981); however, there are very few estimates of seed production in natural stands

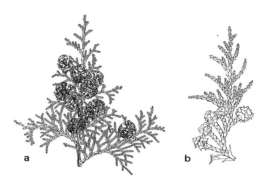

Figure 7.3. Branchlets with mature fruit. a: *C. obtusa* (× 5/7). b: *C. pisifera* (× 3/2) (from Yatoh 1964).

(Table 7.2). *Chamaecyparis pisifera* seed production also fluctuates annually (Y. Sato, unpublished observations). There have been no published studies on seed production in natural stands of *C. pisifera*.

Chamaecyparis obtusa and *C. pisifera* produce smaller seeds than most Japanese conifers; *C. pisifera* seeds (0.64–1.15 g per 1,000 seeds) are smaller than those of *C. obtusa* (1.3–3.2 g per 1,000 seeds; Asakawa et al. 1981). The small winged seeds of both species are widely disseminated by wind. Most *C. obtusa* seeds in natural stands fall from November to the following February (Sakurai 1984). Viable seeds of both species can germinate immediately under favorable conditions (20–30°C; Asakawa et al. 1981). Their seedlings emerge from May to August in *Chamaecyparis* stands in the Kiso district of central Japan. Both species seedlings have two cotyledons, but *C. pisifera* has more primary leaves than *C. obtusa* (Table 7.1). Therefore, emergence of mature (scale) leaves is faster in *C. obtusa* than in *C. pisifera* under the same environmental conditions (Sato 1973).

Chamaecyparis obtusa reproduces only sexually. Therefore, recruitment from seedlings is important for *C. obtusa* regeneration. Mortality of viable *C. obtusa* seeds from seedfall to germination is generally high on forest floors with a thick litter layer. Forest floor disturbance is important in decreasing the mortality of *C. obtusa* seeds (Yamamoto and Tsutsumi 1984). Exposed mineral soil beneath a gap created by a

Table 7.1 Some Morphological Characteristics of *C. obtusa* and *C. pisifera*

Species	No. of 4-Leaved Joints on the Seedling Shoot	Growth Form of Adult Trees	Length of Shoot Internode
C. obtusa	2–5	round, without basal branching	alternately long and short
C. pisifera	15–20	pyramidal, with basal branching	equal

Source: Miki 1958.

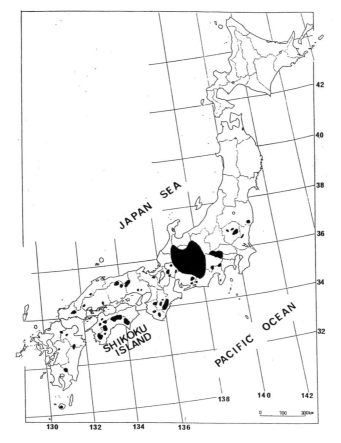

Figure 7.4. Geographical distribution of *C. obtusa* in Japan (redrawn from Hayashi 1960).

tree-fall supports the best seedling emergence for *C. obtusa* (Yamamoto 1988); seedling mortality is lower under a gap than a closed canopy. Beneath a closed canopy, the mortality is lower on elevated surfaces such as fallen decaying logs or rotten stumps (Table 7.3). *Chamaecyparis obtusa* saplings, however, rarely occur under a closed canopy and can usually be found only beneath a gap. Canopy trees of *C. obtusa* with roots raised above the ground, which might have established on stumps, can be found in some districts.

Little is known about the reproduction of *C. pisifera*. Vegetative reproduction in *C. pisifera* is by layering. Layered plants have branches, stems, or whole plants bent to the ground; parts that are buried in litter produce roots (Fig. 7.6). The layered plants generally have a cushion-like form. Many new plants are produced by this process, depending on the number of branches.

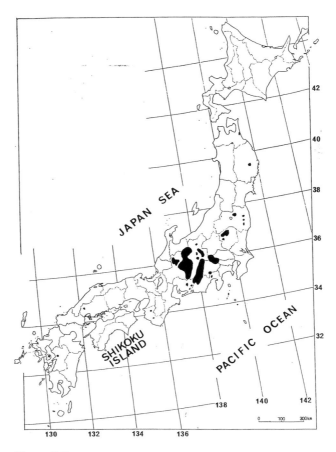

Figure 7.5. Geographical distribution of *C. pisifera* in Japan (redrawn from Hayashi 1960).

Table 7.2 Seed Production in Natural Stands of *C. obtusa* and *C. pisifera*

Species	Number of seeds (m^{-2}) Abundant Year	Poor Year	Stand Age (yrs)	Area	References
C. obtusa	7,155[a]	125[a]	85–108	Kyoto Pref.	Saito et al. (1979)
	5,156	371	60[b]	Kochi Pref.	Sakurai (1984)
	4,479	181	300	Nagano Pref.	Y. Sato (unpublished observations)
C. pisifera	4,324	197	300	Nagano Pref.	Y. Sato (unpublished observations)

[a]The number of seeds was calculated from total seed weight per m^2 using 0.002 g as individual seed weight.
[b]Secondary forest.

Table 7.3 Mean Number (m^{-2}) and Survivorship of *C. obtusa* Seedlings on Different Microenvironments in Old-Growth *C. obtusa* Stands of the Kiso District, Central Japan

Substrate	Year of Emergence	1987 (Autumn)	1989 Summer	1989 Autumn	Survival rate (%)
Beneath a closed canopy					
On litter	1986	6.00		1.75	29.17
	1989		8.63	3.13	36.23
On stumps	1986	5.20		1.80	34.69
	1989		17.39	8.59	49.39
On fallen logs	1986	3.72		2.39	64.15
	1989		5.33	2.95	55.26
Beneath a gap					
On litter	1986	19.25		14.25	74.03
	1989		44.50	21.75	48.88
On exposed	1986	32.59		25.45	78.09
mineral soil	1989		72	38.75	53.82

Source: Yamamoto, unpublished observations.

Although layering is common in *C. pisifera* stands in central Japan's Kiso district (personal observation), there are few reports on this form of vegetative reproduction. Hayashi (1960) briefly describes layering of *C. pisifera* at higher altitudes. Recently, Imai and Ohsawa (1988) reported layering of *C. pisifera* and its role in maintaining the species' population.

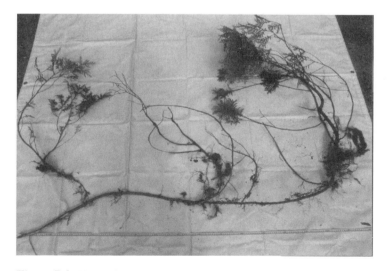

Figure 7.6. Vegetative reproduction of *C. pisifera:* rooted branch.

Table 7.4 Mean Number (m^{-2}) and Survival Rate of New Germinants and Layered or Established Plants of *C. pisifera* on Different Substrata in a *C. pisifera* Stand of the Kiso District, Central Japan

Substrate	Year of Emergence	1987 (Summer)	1988 (Summer)	1989 Summer	1989 Autumn	Survival Rate (%)
New germinants						
On litter	1987	0				
	1988		0			
	1989			3.38	0	0
On stumps	1987	4.41			1.39	31.52
	1988		0			
	1989			12.06	3.94	32.67
On fallen logs	1987	4.21			0.70	16.63
	1988		0			
	1989			2.06	0.78	37.86
Layered or established plants (<1.3 m tall)						
On litter	pre-1985	4.13			4.13	100
On stumps	pre-1985	5.80			3.94	67.93
On fallen logs	pre-1985	1.25			1.12	88.92

Source: Yamamoto, unpublished observations.

Recruitment from seedlings in *C. pisifera* is generally ineffective (Table 7.4). Underneath a closed canopy in natural stands, no new germinants were found in the year following seed production; new germinants may die immediately after emergence on substrata with a thick litter layer. New germinants can survive longer on elevated surfaces such as fallen or decaying logs or rotten stumps, although they suffer high mortality. Once established, *C. pisifera* can tolerate shade (Hayashi 1960) and survive on any substratum (personal observation). Layered plants under a closed canopy grow and reach the canopy layer when a gap is created above them. *Chamaecyparis pisifera* seems to colonize best in stony sites under a gap (Iwata and Kusaka 1954, Imai and Ohsawa 1988). *Chamaecyparis pisifera* cuttings root readily and can resprout from stumps, in seedling form, after cutting or browsing by animals.

Conclusion

Regeneration characteristics of *C. obtusa* and *C. pisifera* are markedly different (Table 7.5). Vegetative reproduction by layering of *C. pisifera* is an effective mechanism for its population maintenance or for regeneration in habitats where seedling recruitment is usually ineffective. Wet conditions characteristic of its habitat may allow reproduction by layering of *C. pisifera* to be effective.

Today, the remaining natural stands of *C. pisifera* are very few compared with those of *C. obtusa,* and they are decreasing further due to cutting. More autecological research needs to be done on *C. pisifera* to conserve its forests.

Table 7.5 Summary of Regeneration Characteristics of *C. obtusa* and *C. pisifera*

Species	Seed Weight[a] (g per 1,000 seeds) and Seed Production	Vegetative Reproduction	Advanced Growth beneath a Closed Canopy	Most Favorable Microenvironments for Seedling Recruitment
C. obtusa	1.3–3.2, fluctuates annually	none, only sexual	generally rare	on mineral soil beneath a gap
C. pisifera	0.64–1.15, fluctuates annually	by layering	usually abundant, but distributed patchily (mainly layered plants)	on stony sites beneath a gap

[a]Asakawa et al. 1981.

Acknowledgments   Financial support was provided by Grants-in-Aid for Encouragement of Young Scientists (62760123) and for Scientific Research (63560151 and 04660168) from the Ministry of Education, Science, and Culture. I thank Yasuhumi Sato for providing his unpublished data and Donald B. Zobel for his advice on improving this chapter.

References

Akai, T. 1972. Studies on natural regeneration. II. The regeneration of *Chamaecyparis obtusa* at Miure experimental forest in Kiso district. Bull. Kyoto Univ. For. **44**:68–87. (In Japanese, with English summary.)

Asakawa, S., M. Katsuta, and T. Yokoyama, eds. 1981. Seeds of woody plants in Japan, Gymnospermae. Japan Forest Tree Breeding Ass'n, Tokyo. (In Japanese.)

Hasegawa, K. 1943. Experimental studies on the viability of forest tree seeds. Teisitsu Rinyakyoku Tokyo Ringyo Sikenjou **4(3)**:1–355. (In Japanese.)

Hayashi, Y. 1960. Taxonomical and phytogeographical study of Japanese conifers. Norin Shuppan, Tokyo. (In Japanese.)

Horikawa, Y. 1972. Atlas of the Japanese flora: an introduction to plant sociology of East Asia. Gakken Co., Tokyo.

Imai, A., and M. Ohsawa. 1988. Distribution characteristics and the population maintenance of *Chamaecyparis pisifera*. Trans. 35th Mtg. Jpn. Ecol. Soc.:112. (In Japanese.)

Iwata, T., and M. Kusaka. 1954. Conifers in Japan, illustrated. Sangyo Tosho, Tokyo. (In Japanese.)

Maeda, T. 1951. Sociological study of *Chamaecyparis obtusa* forest and its Japan-sea elements, with plates I–III. Enshurin (Tokyo Univ.) **8**:21–44. (In Japanese, with English summary.)

Maeda, T., and J. Yoshioka. 1952. Studies on the vegetation of Chichibu Mountain forest. II. The plant communities of the temperate mountain zones, with plates II–III. Bull. Tokyo Univ. For.. **42**:129–150. (In Japanese, with English summary.)

Miki, S. 1958. Gymnosperms in Japan, with special reference to the remains, with plates I–III. J. Inst. Polytech., Osaka City Univ., Ser. D, **9**:125–150.

Miyawaki, A., and S. Okuda, eds. 1990. Vegetation of Japan, illustrated. Shibundo, Tokyo. (In Japanese.)

Ozawa, J. 1962. Seeds of conifers. Chikyu Shuppan, Tokyo. (In Japanese.)

Saito, H., H. Matsushita, and M. Takeoka. 1979. Net production rate in natural forest of *Chamaecyparis obtusa* S. et Z. on poor site near Kyoto city. Scientific Reports of the Kyoto Prefectural Univ., Agriculture **31**:59–69. (In Japanese, with English summary.)

Sakaguchi, M. 1952. Silviculture of *Chamaecyparis obtusa.* Youkendo, Tokyo. (In Japanese.)

Sakurai, S. 1984. Seed production and survival of regenerated current year seedlings in *Chamaecyparis obtusa* stands. Bull. Forestry and Forest Prod. Res. Inst. **331**:167–180. (In Japanese.)

Sato, K. 1973. *Chamaecyparis obtusa* in Japan. Zenkoku Ringyou Kairyou Hukyuu Kyoukai, Tokyo. (In Japanese.)

Tanimoto, T., Y. Kiyono, K. Morisada, K. Kidai, and T. Aizawa. 1984. The relation between topography and structure of natural *Chamaecyparis* stands in the Kiso district. Trans. 95th Mtg. Jap. For. Soc.:**83**. (In Japanese.)

Yamamoto, S. 1988. Seedling recruitment of *Chamaecyparis obtusa* and *Sciadopitys verticillata* in different microenvironments in an old-growth *Sciadopitys verticillata* forest. Bot. Mag. Tokyo **101**:61–71.

Yamamoto, S., and T. Tsutusumi. 1984. The population dynamics of naturally regenerated Hinoki seedlings in artificial Hinoki stands. III. Dynamics of the seed population on the forest floor. J. Jap. For. Soc. **66**:483–490. (In Japanese, with English summary.)

Yamanaka, T. 1979. Forest vegetation in Japan. Tsukizi Syokan, Tokyo. (In Japanese.)

Yatoh, K. 1964. Dendrology, conifers. Asakura Shoten, Tokyo. (In Japanese.)

8

Raymond M. Sheffield, Thomas W. Birch,
William H. McWilliams, & John B. Tansey

Chamaecyparis thyoides (Atlantic White Cedar) in the United States

Extent and Characterization Using Broad-Scale Inventory Data

In the United States, *Chamaecyparis thyoides* (L.) BSP. (Atlantic white cedar) occurs within 250 km of the Atlantic Ocean and Gulf of Mexico (see Figs. 8.1, 8.2, and 8.3). It is found from southeastern Maine southward along the coastline to northeastern Florida and along the Gulf coast from the panhandle of Florida westward to Mississippi (Little 1971). Although the species has been studied extensively, the total area occupied and total inventory volume have not been established. Numerous local studies have attempted to quantify the extent, utilization, and recent trends for this species, but no composite picture for the entire range has emerged. This chapter utilizes broad-scale forest inventory data collected by the U.S. Department of Agriculture (USDA) Forest Service to establish baseline totals for the entire Atlantic white cedar resource.

Methods

Survey Procedures

The data used in this assessment were collected by Forest Inventory and Analysis (FIA) Research Work Units at the Northeastern, Southeastern, and Southern Forest Experiment Stations of the USDA Forest Service. These inventories use permanent sample plots systematically distributed across timberland to attain a proportionate sample of all major forest types, sites, and ownerships in a region. The inventories are broad-scale in scope, with each sample plot representing from 1,000 to 3,000 hectares (ha). Thus, the data have limited utility for describing Atlantic white cedar in localities, but they can be used for regional assessments.

   At each sample location, FIA uses a multi-point cluster plot to assure a representa-

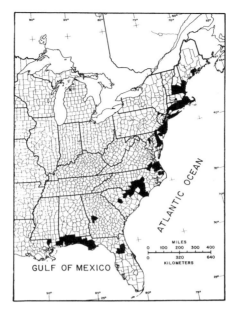

Figure 8.1. Distribution of *C. thyoides,* including field observations, herbarium records, and published sources. Counties in
which *C. thyoides* has been found are inked
in black. (From Laderman 1989.)

tive sample of trees and associated stand conditions. Collectively, the cluster points
sample approximately 0.4 ha of timberland. At each point, trees 12.7 cm dbh (diameter at 1.4 meters above ground level) and larger are selected using a basal area factor
of 8.61 m² per hectare. Trees between 2.5 and 12.7 cm dbh are tallied on circular
fixed-area plots that share common point centers with each variable-radius plot center. Plots established in prior surveys are relocated and remeasured to determine the
elements of change between inventories (removals and mortality). The period between remeasurements varies between 6 and 10 years. Additional details on sampling
procedures ·and data collected by FIA are available (Birdsey and Schreuder 1992,
USDA Forest Service 1967, 1992).

Study Region

The FIA has more than 44,000 sample plots in states where Atlantic white cedar is
known to exist. Each of these samples was examined to determine if white cedar was
present in the tally of trees. At least one stem 2.5 cm dbh or larger was a prerequisite
for inclusion in the study. Atlantic white cedar was not found on any FIA samples in
Maine, New Hampshire, or Rhode Island. The species is known to occur in these

Figure 8.2. *Chamaecyparis thyoides* morphology, life size. a. First-year seedling, juvenile foliage. b. Branchlet with flowers. c. Branchlet with mature fruit. (Adapted from C.S. Sargent 1896.)

states (Little 1971, Laderman et al. 1987, Ward and Clewell 1989, Laderman 1989), but its sparse occurrence combined with the dispersed nature of the FIA plots resulted in the omission of these states from our sample. Atlantic white cedar was found on a total of 169 FIA sample locations; these plots form the basis for the following analysis. All summaries of Atlantic white cedar area, inventory volume, and other resource attributes contained in this chapter are derived from this sample. The inventories were conducted during different years for each state, ranging from 1980 to 1990. We have chosen to display the results for the year 1990.

The states in which Atlantic white cedar was found on FIA samples are grouped into the following regions for presentation purposes:

Region	States
North Atlantic	Connecticut, Delaware, Maryland, Massachusetts, New Jersey, and New York
South Atlantic	North Carolina, South Carolina, and Virginia
Gulf	Alabama, Florida, and Mississippi

Figure 8.3. Old-growth stand of *C. thyoides* in a Massachusetts glacial kettle. (Photograph by A.E. Bye.)

Results

Extent, Distribution, and Characteristics

Atlantic white cedar occurs on a total of 215,000 ha of timberland in the United States. About 27% of this timberland is in the North Atlantic states, 39% is in the South Atlantic states, and 34% is in the Gulf states (Table 8.1). Atlantic white cedar is a minor species in terms of relative abundance; it occupies less than 1% of the total timberland in the states in which it is found on FIA samples. The species is prominent in several localities. The most extensive concentrations occur in New Jersey, principally in the Pinelands region; in the Dismal Swamp area of North Carolina and Virginia; and in Florida, principally in the floodplains of the Appalachicola, Blackwater, and Escambia Rivers. As mentioned previously, white cedar is found in states other than those shown, but the broad-scale nature of the sample did not identify these populations.

About 59,000 ha of Atlantic white cedar timberland, 27% of the total, are under the control of public agencies such as state forests and parks, and national wildlife

Table 8.1 Area of Timberland in the United States with Atlantic White Cedar, 1990[a]

Region and State	All Timberland	Timberland with Atlantic White Cedar
North Atlantic		
Connecticut	719[b]	2
Delaware	157	2
Maryland	996	6
Massachusetts	1,218	3
New Jersey	775	44
New York	6,394	2
Total	10,259	59
South Atlantic		
North Carolina	7,430	77
South Carolina	4,929	1
Virginia	6,247	5
Total	18,606	83
Gulf		
Alabama	8,765	5
Florida	6,167	64
Mississippi	6,747	4
Total	21,679	73
All regions	50,544	215

[a]Data compiled from all Forest Inventory and Analysis sample plots in the eastern United States that indicated the presence of Atlantic white cedar.
[b]In thousands of hectares.

refuges (Fig. 8.4). The forest industry controls another 22%, or 46,500 ha. About half of all timberland with evidence of Atlantic white cedar, 109,000 ha, is owned by other private owners (collectively, all private owners other than forest industry).

Atlantic white cedar is a component of numerous forest cover types; broad groupings of these types are presented in Table 8.2. Approximately 35,000 ha, 16% of the timberland occupied by white cedar, are classified as a pine forest type. The specific pine type varies from *Pinus rigida* (pitch pine) in the North Atlantic states to *P. taeda* (loblolly pine), *P. serotina* (pond pine), and *P. elliottii* (slash pine) in the southern portion of Atlantic white cedar's range. An additional 22,000 ha are classed as pine–hardwood. In these stands, pine species account for 25–50%, with the remainder in hardwoods and softwood species such as *Taxodium* spp. (cypress) and Atlantic white cedar.

The majority of the stands in which Atlantic white cedar occurs are classified as hardwood forest types, including stands dominated by white cedar or cypress. About 14,000 ha are grouped into an upland hardwood category. An assortment of mixed hardwoods is included in this group, mostly species such as *Liriodendron tulipifera* (yellow-poplar), *Liquidambar styraciflua* (sweetgum), and *Acer rubrum* (red maple) that occur on mesic sites and tolerate seasonal flooding. Lowland hardwood forest

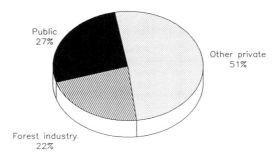

Total area = 215 thousand hectares

Figure 8.4. Distribution of Atlantic white cedar
timberland, by ownership class, 1990.

types are the most common ones associated with the occurrence of Atlantic white
cedar. Approximately 144,000 ha on which white cedar occurs are so classified. Included in this group are those stands in which white cedar dominates, as well as
those in which it occurs as a minor component. Lowland hardwood stands are concentrated in the South Atlantic and Gulf states.

As indicated above, Atlantic white cedar may account for a small portion of total
tree stocking, or it may occur in nearly pure stands (Fig. 8.5). On 21,900 ha, it
accounts for 75% or more of the total. These are essentially the "pure" stands as
defined by Eyre (1980). On another 22,300 ha, Atlantic white cedar composes between 50 and 75% of the stand, bringing the total area on which white cedar accounts

Table 8.2 Area of Timberland in the United States Supporting Atlantic White Cedar, by
Forest Type and Region, 1990[a]

| Forest type | All regions | Region[b] | | |
		North Atlantic	South Atlantic	Gulf
Pine	35.1[c]	10.3	9.2	15.6
Pine–hardwood	21.8	4.2	6.3	11.3
Upland hardwood	14.1	11.8	1.1	1.2
Lowland hardwood	144.1	33.2	65.9	45.0
All types	215.1	59.5	82.5	73.1

[a]Data compiled from all Forest Inventory and Analysis sample plots in the eastern United States that indicated the
presence of Atlantic white cedar.
[b]See Table 8.1 for definitions of regions.
[c]In thousands of hectares.

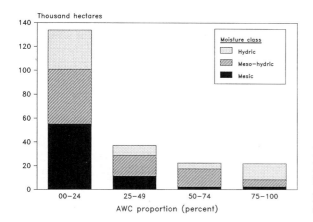

Figure 8.5. Distribution of Atlantic white cedar timberland, by Atlantic white cedar stocking proportion and soil moisture class, 1990.

for the majority of the trees to 44,200 ha, or 1/5 of the total area with Atlantic white cedar occurrence. The bulk of the area is characterized by relatively low Atlantic white cedar stocking relative to that of other species. Nearly 2/3 of the total has white cedar proportions of less than 25%.

Atlantic white cedar is found as a minor component in stands more frequently on mesic, well-drained sites. When it occurs on hydric sites, it commonly constitutes a greater proportion of the total stand. More than 41% of the stands in the 0–24% stocking class are mesic in nature, such as well-drained, but moist, flatwoods. Another 34% are intermediate between mesic and hydric, or mesohydric. These sites are normally poorly drained, and the soil is wet much of the time. On the other end of the white cedar stocking continuum, about 61% of the "pure" white cedar stands are on hydric sites. These areas are very poorly drained, with a water table at or near the soil surface for much of the year. Another 28% of the stands in this category are mesohydric.

Other studies throughout the ecosystem corroborate the findings of Atlantic white cedar habitat associations. Moore and Carter (1987) reported white cedar occurrence in the traditional peaty soils, and also in the floodplains of blackwater rivers, in nonalluvial hardwood–pine forests, in North Carolina bays, and in stream corridors in the Sandhills regions of North Carolina. Other authors that report Atlantic white cedar presence in a diversity of habitats, moisture regimes, and species mixes include Dill et al. (1987) for Delaware, Maryland, and Virginia; Clewell and Ward (1987) and Ward and Clewell (1989) for Florida and the Gulf region; Levy (1987) for the Great Dismal Swamp area of North Carolina; and Laderman et al. (1987) for the northern portion of the ecosystem.

Volume Characteristics

The merchantable volume of Atlantic white cedar in the United States is estimated at 9.8 million m³. This inventory is defined as the cubic volume of the main stem, excluding stump and top, for trees that are 12.7 cm dbh and larger. Height and dbh measurements of individual cedar trees on the 169 FIA plots were used as independent variables in a regression equation to predict tree volumes. Geographically, the white cedar inventory is distributed in much the same pattern as that for all Atlantic white cedar timberland. Inventory concentrations are found in New Jersey, North Carolina, and Florida. These states also have almost all the stands where Atlantic white cedar accounts for a majority of each stand's stocking (greater than 50%).

By ownership, about 39% of the white cedar volume is public timberland, 22% is forest industry, and the remaining 39% is other private ownerships (Fig. 8.6). About 6.0 million m³, or 62% of the total inventory, is contained in stands where Atlantic white cedar accounts for more than 50% of the stocking. These stands are concentrated on timberland owned by public agencies and by owners in the "other private" category. Two-thirds or more of the Atlantic white cedar inventory on these two owner groups is contained in such stands, compared to only one-third on forest industry land. Forest industries have either liquidated much of the Atlantic white cedar inventory on their lands, transferred the lands to other owners, or possibly never owned lands with significant inventories of the species.

In the United States, a common unit of measure for expressing inventory volumes for trees that currently are capable of producing saw logs is board feet. (A board foot is defined as 1 ft x 1 ft x 1 in., but the actual thickness is somewhat less.) Slightly more than one-half the total merchantable cubic volume is in trees that are large enough to qualify as sawtimber trees. The Atlantic white cedar sawtimber inventory totals 850 million board feet (bd ft) in the United States, with the international 1/4-inch log rule used as a standard. (See Husch et al. 1982 or other forest mensuration texts for discussions of log rules.) The board-foot volumes presented can be converted to an equivalent volume in cubic meters by using a factor of 5.3 m³ per thousand board feet. The reader should not convert the total cubic volume in the inventory to board feet, since many trees have cubic volume but not board feet.

Ward (1989) estimated the Atlantic white cedar sawtimber inventory at 170 to 180 million bd ft. The disparity between this and the FIA estimate (850 million bd ft) of sawtimber volume is probably due to differences in estimation methods. Ward's estimate is based on the observations of numerous individuals, primarily in areas where white cedar is concentrated. Observational estimates probably do not adequately account for Atlantic white cedar's occurrence in stands where it is not the predominant species. Figures 8.5 and 8.6 and Table 8.3 indicate that much of the white cedar occurs in stands with a diversity of species. The FIA sample is designed to accurately quantify species occurring in any stand condition. The FIA estimate does have associated sampling errors. The variation measured on FIA samples results in the following confidence intervals for cubic and board-foot inventories at the 95% confidence level:

Million m³: 9.8 ± 1.2
Million bd ft: 850 ± 248

Species Composition

The species composition of the trees tallied on the 169 FIA plots showed that approximately 65 tree species were coexisting with Atlantic white cedar. Herbaceous taxa

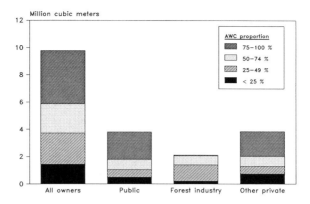

Figure 8.6. Distribution of Atlantic white cedar inventory volume by ownership class and white cedar stocking proportion, 1990.

were not included in this study. The 18 most abundant species coexisting with Atlantic white cedar are shown in Table 8.3 for the white cedar range and by region. Relative occurrence was computed based on number of stems and basal area. Distributions are a composite for all stands within each region.

Atlantic white cedar stands in the North Atlantic states are characterized by a more homogeneous species mix and more concentration of stocking in white cedar compared with the two more southern regions. Clewell and Ward (1987) reported that the number of plant species associated with white cedar in the northern portion of its range was small compared to the southern portion. In the North Atlantic states, the average Atlantic white cedar stand has a makeup of 43% white cedar, 16% *A. rubrum,* and 10% *P. rigida;* seven additional species compose more than 1% of the stocking of these stands. Roman et al. (1987) reported *P. rigida* associations in the New Jersey Pinelands, with understory species including *A. rubrum* and *Nyssa sylvatica* var. *biflora* (blackgum or swamp tupelo).

In the South Atlantic states, *N. sylvatica* var. *biflora* is the dominant component of Atlantic white cedar stands with 25%, followed by white cedar with 19%, and *A. rubrum* with 15%; *P. taeda, Taxodium, Persea borbonia* (redbay), *Magnolia virginiana* (sweetbay), *Nyssa aquatica* (water tupelo), and *P. serotina* also make up sizable portions of these stands. This is consistent with species rankings reported by Levy (1987) for Atlantic white cedar stands in the Great Dismal Swamp of Virginia and North Carolina. In the Gulf states, *N. sylvatica* var. *biflora* is also the dominant species at 25%; white cedar comprises 21%, followed by *P. elliottii* and *M. virginiana* with 14% each. *Sabal* spp. (palms), *Taxodium, Quercus nigra* (water oak), *Q. laurifolia* (laurel oak), and *A. rubrum* also make up significant proportions of Atlantic white cedar stands in this region. These species associations are similar to those reported by Ward and Clewell (1989).

Table 8.3 Relative Species Importance Based on the Occurrence of Trees 2.54 cm dbh and Larger in Stands with Atlantic White Cedar, by Region, 1990[a]

Species	All regions (%)	Region[b] (%)		
		North Atlantic	South Atlantic	Gulf
Acer rubrum (red maple)	10.3	16.2	14.6	2.3
Chamaecyparis thyoides (Atlantic white cedar)	23.3	42.8	18.7	21.3
Fraxinus spp. (ash)	1.1	1.6	1.1	1.0
Liquidambar styraciflua (sweetgum)	1.7	2.2	—	—
Liriodendron tulipifera (yellow-poplar)	1.4	1.8	—	—
Magnolia virginiana (sweetbay)	7.7	2.0	5.0	13.5
Nyssa aquatica (water tupelo)	2.0	—	4.0	—
Nyssa sylvatica var. *biflora* (blackgum, swamp tupelo)	21.6	4.9	25.3	25.0
Persea borbonia (redbay)	2.9	—	6.2	—
Pinus elliottii (slash pine)	5.4	—	—	13.7
Pinus rigida (pitch pine)	1.4	9.8	—	—
Pinus serotina (pond pine)	1.8	—	3.2	—
Pinus taeda (loblolly pine)	4.9	1.6	7.6	—
Quercus laurifolia (laurel oak)	1.5	—	—	2.9
Quercus nigra (water oak)	1.6	—	—	3.1
Sabal spp. (palms)	1.6	—	—	4.6
Taxodium spp. (cypress)	4.5	—	6.4	3.1
Tsuga spp. (hemlock)	0.2	1.7	—	—
Other species	5.1	15.4	7.9	9.5
Total species	100.0	100.0	100.0	100.0

[a]Data compiled from all Forest Inventory and Analysis sample plots in the eastern United States that indicated the presence of Atlantic white cedar.
[b]See Table 8.1 for definitions of regions.

Growth and Removal Estimates

Estimates of net annual growth and annual removals of Atlantic white cedar in the United States are presented in Figure 8.7. Removal of white cedar each year averages 156,000 m^3, including 14.3 million bd ft of sawtimber. This estimate of white cedar removals includes timber cut for a product, removed in clearing land and in silvicultural operations, and merchantable volume that was cut but left during harvesting. It is derived from the remeasurement of trees at all FIA sample plots and depicts the situation for roughly the preceding decade. In comparison, Ward (1989) estimated annual removals of sawtimber at 19 million bd ft. This estimate was derived by canvassing processors of Atlantic white cedar for the year 1985. A comparison of this removal volume with previous periods is not possible, but it is highly likely that the

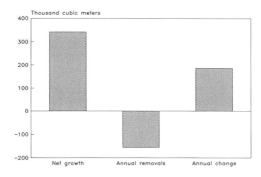

Figure 8.7. Average net annual growth, annual removals, and annual inventory change for Atlantic white cedar in the United States, 1990.

present utilization of the species for timber products is small in a historical perspective (Baines 1989, Ward 1989).

Net growth is defined as the net annual change in volume of Atlantic white cedar (gross growth minus mortality). Estimates of net growth from plot remeasurement were not available for all regions in this study. Therefore, net growth was estimated from measured white cedar growth rates in North Carolina, South Carolina, and Florida. A growth rate was calculated for this three-state area by dividing white cedar net growth by white cedar inventory. This process yielded a growth factor (proportion) that was then multiplied by the total Atlantic white cedar inventory (9.8 million m^3) to arrive at a net growth estimate for the entire resource of 342,000 m^3.

We do not have Atlantic white cedar data across the entire region for consecutive inventories, but we can compare net growth and annual removals of white cedar to gain insight into probable inventory trends. If net annual growth exceeds annual removals, then inventory volumes are increasing. This comparison suggests that the volume of Atlantic white cedar across the eastern United States has increased by some 186,000 m^3 annually during the 1980s. Increases in Atlantic white cedar inventory are probably a recent occurrence. Much of the resource now exists on public land, and environmental regulations have directly or indirectly imposed additional restrictions on harvesting in many localities. As stands dominated by Atlantic white cedar have been cut, the resource exists more and more in mixed stands, thus discouraging the utilization of the species for timber products.

Perhaps the depletion of the Atlantic white cedar resource that has characterized its past has abated. Timber harvesting, altered landscape hydrology, the exclusion of fire, clearing for cropland and other uses, and the lack of planning for the regeneration of the species have all been linked to the long-term reduction in Atlantic white cedar extent (Laderman et al. 1987, Roman et al. 1987, Baines 1989). Regionwide, the trends presented above for merchantable volume suggest that Atlantic white cedar is not in danger of depletion in the near future. Trends in specific localities, however, may be quite different than those presented for the species' entire range.

Summary

Whereas Atlantic white cedar occurs across a substantial portion of the eastern United States, it has limited stocking in many of the stands in which it occurs. There is evidence of the species on almost 215,000 ha, but white cedar accounts for 50% or more of total tree stock on only 44,000 ha. Atlantic white cedar is a more dominant component of the total stand on sites that have hydric soil conditions when compared with more mesic soils. The merchantable volume of Atlantic white cedar trees across the ecosystem is estimated to be 9.8 million m^3. Based on estimates of net annual growth and annual removals, volume of Atlantic white cedar during the past decade has increased by approximately 186,000 m^3 annually.

In Memory   The authors wish to acknowledge the contribution and leadership of the late John B. Tansey in planning and conducting this study. Without his energy and quest for excellence, this study and chapter would never have materialized.

References

Baines, R.A. 1989. Prospects for white cedar: a North Carolina assessment. Forem Mag. **13**:8–11, Duke Univ. Durham, NC.

Birdsey, R.A., and H.T. Schreuder. 1992. An overview of forest inventory and analysis estimation procedures in the eastern United States—with an emphasis on the components of change. U.S. Dept. Agric. For. Serv., Rocky Mtn. For. and Range Exp. Sta. Gen. Tech. Rep. RM-214. Fort Collins.

Clewell, A.F., and D.B. Ward. 1987. White cedar in Florida and along the Northern Gulf Coast. Pages 69–82 *in* A.D. Laderman, ed. Atlantic white cedar wetlands. Westview Press, Boulder.

Dill, N.H., A.O. Tucker, N.E. Seyfried, and R.F. Naczi. 1987. Atlantic white cedar on the Delmarva Peninsula. Pages 44–55 *in* A.D. Laderman, ed. Atlantic white cedar wetlands. Westview Press. Boulder.

Eyre, F.H., ed. 1980. Forest cover types of the United States and Canada. Soc. Am. For., Washington, DC.

Husch, B., C.I. Miller, and T.W. Beers. 1982. Forest mensuration. 3rd ed. John Wiley, New York.

Laderman, A.D. 1989. The ecology of Atlantic white cedar wetlands: a community profile. U.S. Fish Wildl. Serv. Biol. Rep. **85(7.21).**

Laderman, A.D., F.C. Golet, B.A. Sorrie, and H.L. Woolsey. 1987. Atlantic white cedar in the glaciated Northeast. Pages 19–34 *in* A.D. Laderman, ed. Atlantic white cedar wetlands. Westview Press, Boulder.

Levy, G.F. 1987. Atlantic white cedar in the Great Dismal Swamp and the Carolinas. Pages 57–67 *in* A.D. Laderman, ed. Atlantic white cedar wetlands. Westview Press, Boulder.

Little, E.L., Jr. 1971. Atlas of United States trees. Vol. 1. Conifers and important hardwoods. Misc. Publ. 1146. U.S. Dept. Agric., For. Serv., Washington, DC.

Moore, J.H., and J.H. Carter III. 1987. Habitats of white cedar in North Carolina. Pages 177–189 *in* A.D. Laderman, ed. Atlantic white cedar wetlands. Westview Press. Boulder.

Roman, C.T., R.E. Good, and S.B. Little. 1987. Atlantic white cedar swamps of the New Jersey

Pinelands. Pages 35–40 *in* A.D. Laderman, ed. Atlantic white cedar wetlands. Westview Press. Boulder.

Sargent, C.S. 1896. The silva of North America. Vol. X. Houghton Mifflin, Boston.

U.S. Department of Agriculture, Forest Service. 1967. Forest survey handbook. FSH 4813. U.S. Dept. Agric., For. Serv., Washington, DC.

———. 1992. Forest Service resource inventories: an overview. U.S. Dept. Agric., For. Serv., Washington, DC.

Ward, D.B. 1989. Commercial utilization of Atlantic white cedar. Economic Botany **43**:386–415.

Ward, D.B., and A.F. Clewell. 1989. Atlantic white cedar *(Chamaecyparis thyoides)* in the southern states. Fla. Sci. **52**:8–47.

9

Lucinda McWeeney

Reconstruction of the Mashantucket Pequot Cedar Swamp Paleoenvironment Using Plant Macrofossils, New England, USA

Plant macrofossils found in sediment cores taken from wetlands that were once post-glacial lakes and ponds can provide documentation for the changing vegetation patterns and wetland evolution of today's coastal forests in southern New England, USA. Plant macrofossil analyses hold the potential to produce species-level identifications of plants that were growing close to the site of deposition, thereby clarifying local habitat evolution (Watts 1978, Warner and Barnett 1986, McWeeney 1991) and enabling ecologists to trace the entrance of taxa into a locale and follow the fluctuations in population diversity (Jackson 1989). Fluctuations in the vegetation patterns and the water table can be inferred from the preservation of the organic remains or lack thereof, from the autochthonous presence of certain taxa, and changes in the rate of sediment accumulation (Digerfeldt 1986).

Decades of palynological research have resulted in a significant number of pollen reconstructions for northeastern North America (Delcourt and Delcourt 1981, 1987; Gaudreau and Webb 1985). Regional reconstructions of environmental changes since the last deglaciation (Table 9.1), which began approximately 18,000 ^{14}C years before present (BP), are well documented. However, very little plant macrofossil information has been reported to supplement the pollen interpretations for the postglacial period in southern New England (Peteet et al. 1990, Peteet et al. 1993).

Previous palynological investigations of Connecticut wetlands that originated during the late-glacial period between 15,000 to 10,000 years ago produced evidence of a regional plant succession from tundra to *Picea* (spruce) and *Larix* (larch) woodlands followed by conifer–hardwood forests up to modern, mixed *Quercus–Carya* (oak–hickory) forest. One of the earliest chronostratigraphic compilations for the southern New England sequence is shown in Table 9.1.

The zones in Table 9.1 were refined by Margaret Davis (1969) as a result of her

Table 9.1 Pollen Zones for Southern New England

Zone	Southern New England	^{14}C Years Before Present	Dominant Plant Community
	Late Pleistocene		
T1	Older Herb Zone	>14,000	tundra
T2	Pre-Durham Spruce	14,000 to 13,500	spruce, pine, birch
T3	Younger Herb Zone	13,000 to 12,500	park tundra
A	Pre-Boreal	12,500 to 10,000	spruce, fir, pine, oak
A1	Durham Spruce Zone	13,500 to 13,000	spruce rising
A2	Durham Spruce Zone	13,000 to 12,500	1st spruce maximum
A3	Durham Spruce Zone	12,500 to 11,000	pine and spruce
	Younger Dryas		
A4	Durham Spruce Zone	11,000 to 10,000	2nd spruce maximum
	Beginning of Holocene		
B	Boreal	10,000 to 8,500	pine
C1	Atlantic	8,500 to 5,000	oak, hemlock
C2	Sub-Boreal	5,000 to 2,000	oak, hickory
C3	Sub-Atlantic	2,000 to present	oak, chestnut

Sources: Adapted from Deevey 1939, Leopold 1955, Deevey and Flint 1957, Peteet et al. 1993.

work at Roger's Lake in Old Lyme, Connecticut. Based on low pollen deposition rates and the presence of tundra vegetation species, the Tundra Zone was bracketed between 14,000 and 12,000 years ago. The increase in tree pollen 12,000 years ago marked the beginning of the *Picea* spp. (spruce) woodland. According to Davis's interpretation, *Pinus strobus* (white pine), *Tsuga canadensis* (eastern hemlock), *Populus* spp. (poplar), *Quercus* spp., and *Acer* spp. (maple) pollen increased dramatically by 9,000 years ago, indicating forests. Subsequent increases in certain arboreal pollens from *Fagus americana* (American beech), *Carya* spp. (hickory), and *Castanea dentata* (American chestnut) were interpreted as evidence of late Holocene (Table 9.1) migration into the region (Davis 1969).

It is now recognized that low amounts of some pollen types—e.g., 5% *Quercus* spp. pollen—indicate the local presence of these trees (Gaudreau and Webb 1985). Even smaller amounts of heavy pollen such as *Acer* spp., as well as that from non-wind pollinated species such as *C. dentata,* indicate that the plants are close by.

In order to determine the late-Quaternary woodland transitions at Pequot Cedar Swamp in Ledyard, Connecticut, sediment cores were recovered to identify plant macrofossils preserved since the last glaciation of the region, to demonstrate the value of these identifications for interpreting the effects of climate change, and to trace the prehistoric local vegetation patterns that led to the present *Acer rubrum* (red maple) and *Chamaecyparis thyoides* (Atlantic white cedar) swamp. The focus of this chapter will be on the arboreal changes with references to the associated aquatic changes that are presented in more detail elsewhere (McWeeney 1994).

Setting

Pequot Cedar Swamp in Ledyard, Connecticut (41° 24′ N, 71° 58′ W; 37 msl), is 14 km north of the Atlantic coast (Fig. 9.1). Also known as the Great Swamp, Pequot Swamp has been the subject of geological, palynological, and botanical analyses (Shuford 1975, Webb et al. 1990, Thorson and Webb 1991). It lies in a glacially formed basin covering an area of 1.87 km^2 within a watershed of 7.4 km^2. The swamp temporarily stores stormwater runoff from nine minor streams and Main

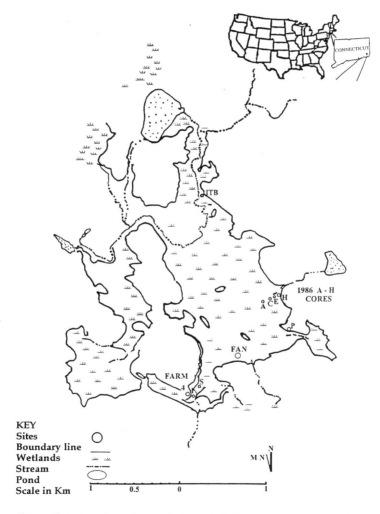

Figure 9.1. Location of cores in Pequot Cedar Swamp, Ledyard, Connecticut. ITB: Indian Town Brook. FAN: Fan Core, 1986. A–H: Transect of cores recovered in 1986.

Brook, which traverses the northern edge of the swamp. The average precipitation of 123.3 cm is evenly distributed throughout the year. Because of this, water table fluctuations, whether seasonal or storm related, are ± 15 cm per year (Thorson and Webb 1991). Modern January temperatures average −2.3°C, with a July mean temperature of 21°C (Shuford 1975). Today's climate extremes are somewhat ameliorated by the proximity to the coast.

The site is within the Oak–Chestnut Forest region. The trees, shrubs, and undergrowth surrounding the pond vary according to aspect, soil, and water conditions. Pequot Swamp is dominated by an *A. rubrum* forest along with numerous *P. strobus* trees. Shrubs include *Alnus* spp. (alders), *Vaccinium corymbosum* (highbush blueberry), *Clethra alnifolia* (sweet pepperbush), *Ilex verticilata* (common winterberry), *Sambucus canadensis* (common elderberry), and *Rhododendron maximum* (great laurel). The herbaceous layer contains *Thelypteris palustris* (marsh fern) in the swamp. *Maianthemum canadensis* (Canada mayflower), *Trientalis borealis* (starflower), *Viola* spp. (violets), and *Lycopodium* spp. (clubmosses) carpet the moist ground, while *Veratrum viride* (false hellebore), *Symplocarpus foetidus* (skunk cabbage), and *Nasturtium officinale* (marsh cress) grow along the streams through the swamp. A mixed deciduous forest with trees in the Erythrobalanus or red oak group (e.g., red, black, and scarlet oaks) and the Leucobalanus or white oak group (e.g., white or swamp oak), *Cornus* spp. (dogwoods), *A. rubrum*, *Carya* spp., *Ostrya virginiana* (hop hornbeam), *Prunus serotina* (black cherry), *Betula populifolia* (grey birch), *Juniperus virginiana* (eastern red cedar), and *Juglans cinerea* (butternut) proliferate in the adjacent uplands surrounding the basin. The well-drained areas support *Vaccinium* spp. (lowbush blueberries), *Rubus* spp. (trailing blackberries), *Fragaria virginana* (wild strawberry), and *Smilax glauca* (greenbrier). Clearings allow Asteraceae (asters), *Solidago* spp. (goldenrods) and other Compositae (composites), *Melilotus* spp. (clovers), *Oenothera biennis* (evening primrose), *Potentilla* spp. (cinquefoils), Gramineae (grasses), and *Toxicodendron radicans* (poison ivy) to develop. The slopes support *F. americana*, *Betula lutea* (yellow birch), and *T. canadensis*. *Dennstaedtia punctilobula* (hayscented fern), *Osmunda cinnamomea* (cinnamon fern), and *Dryopteris novaboracensis* (New York State fern) are found in the upland and clearings (William Niering, personal communication 1993); *Onoclea sensibilis* (sensitive fern) and *Polystichum acrostichoides* (Christmas fern) also grow in the uplands.

Background

The basal lacustrine sediments of the Pequot Cedar Swamp sequence indicate the presence of a small lake, probably formed by a melting ice block following retreat of the glacier approximately 17,000 years ago (Thorson 1993). The water table fluctuated during several periods of climatic change. A drop in the water table during the Early Holocene, 10,000 to 8,000 years BP, allowed the development of a shrub swamp possibly followed by a marsh. The subsequent rise in the water level over the last 4,000 years led to the development of the present wooded swamp (Thorson and Webb 1991, McWeeney 1994).

Methods

The Indian Town Brook (ITB) core and Core C were recovered from the north and east sides of the swamp (Fig. 9.1). Core lithology was described using the U. S. Soil Service terminology (Tables 9.2 and 9.3). Sediment color is based on the Munsell color system. The organic matter or loss-on-ignition was measured according to Dean (1974).

Qualifiers such as "cf." (compare; indicates it closely favors), "type" (looks like a type such as), and "sp. or spp." (indeterminate species) were used depending on the degree of certainty in the identification of a plant remains. The number of seeds, needles, plant fragments, or charcoal is the absolute count per 4 cc sample.

This report follows standard geologic terminology for the late-glacial period that extends from at least 18,000 to 10,000 years ago. The Holocene covers from 10,000 years ago to the present. Radiocarbon dating was achieved using unsorted 5 cm core sediment samples or individual plant remains such as needles and seeds. Dating was by accelerator mass spectrometry (AMS), a method capable of dating samples as small as 25 milligrams. All dates reported in this chapter are in uncalibrated radiocarbon (^{14}C) years before present.

Results

Results from two cores, Core C and the ITB Core (Fig. 9.1), are discussed in this chapter. Radiocarbon dates, plant macrofossil records, and lithostratigraphy show the diverse pattern of prehistoric plant growth, deterioration, preservation, and sedimentation across one wetland basin in southern Connecticut as it evolved into a modern *C. thyoides* swamp forest. The lithology and loss-on-ignition results are given in Tables 9.2 and 9.3. Radiocarbon dates and details are presented in Table 9.4.

Description of Indian Town Brook Core

Nearly 5 m of sediments were recovered from the north end of the Pequot Cedar Swamp basin (Fig. 9.1, Table 9.2). The coring site was selected to obtain material that would have accumulated as the stream entered the pond during the late-glacial and Holocene periods. Numerous changes in the lithology (Thorson 1993) and organic content were apparent (Table 9.2).

Five major depositional units can be distinguished in the Indian Town Brook Core. The basal sandy and silty sediments found from 494 to 172 cm date between 15,000 and 12,000 years ago. An increase in organic matter after 12,000 BP left a fibric peat formation between 172 and 135 cm. The amount of organics increased around 10,000 years ago, decreased suddenly to 11% at 101 cm about 7,000 years ago, and continued to decline until reaching the modern peat surface with 22% organic matter. The absolute chronology established by the radiocarbon dates for the Indian Town Brook Core confirms that the sediments are in stratigraphic order (Table 9.4). The plant macrofossil assemblage is divided to correlate with the recognized late-glacial and Holocene pollen zones for southern New England presented in Table 9.1. The avail-

Table 9.2 Indian Town Brook (ITB) Core Lithostratigraphy[a] and Organic Content

Depth (cm)	Sediment Description	Munsell Color[b]	Loss-on-ignition	
			cm	% Organic
0–20	Surface peat.	10YR 2/1 Black	2	22
20–45	Grading upward from organic stained silty clay to silt.	10YR 3/2 Dark gray brown	10 20–40	7 5
45–52	Ungraded fine sand w/plant fibers and charred streaks.	Dark brown	45 50	6 5
52–97	Bedded silt w/4–5 lamellae of organic staining and light gray silt lenses at 33, 43, 44, and 49 cm. Reversion to pond. 6370 yrs BP @ 91–96 cm[c]	Dark grayish brown	55 61 65 70 96	5 6 6 6 6
97–112	Sapric peat w/ charcoal, grading upward to silt. Mid-Holocene warming. 7440 yrs BP @ 108–111cm.	10YR 2/2 Black to very dark brown	101 106 111	11 26 55
112–135	Woody peat and organic silt. Transition to wooded swamp. 7950 yrs BP @ 127–130 cm.	10YR 2/1 Black	116 121 126 131	51 44 38 35
135–172	Fibrous peat and organic silt. Sharp basal contact at 131 cm for transition to mucky bottom of pond.

8890 yrs BP @ 151–152 cm.

10,050 yrs BP @ 159–160 cm.

11,700 ± 250 yrs BP @ 166–170 cm. | | 136 141 142 143–5 146–8 149–50 151 152 153 154 155 156–7 158 159 160 161–2 163 164–5 166 | 31 26 29 24 22 24 30 27 22 21 25 18 17 21 15 14 10 11 12 |
172–250	Sand w/ graded beds of sandy silt and clayey silt and organic material.	5YR 3/2 Dark olive gray	171 172 182 191	8 1 1 1
172	172–181: massive fine sand.			
181	181–200: sandy silt with sand horizons and organic beds.			
200	200–226: med-fine sand.			
226	226–250: silt interbedded w/			
250	sand.			

(*continued*)

Table 9.2 (*continued*)

Depth (cm)	Sediment Description	Munsell Color[b]	Loss-on-ignition	
			cm	% Organic
250–290	Silt with a few sand beds.			
290–334	Silty sand with organic lenses and occasional granules at 253 and 257 cm.			
334–384	Sandy silts interbedded w/ organic layers.			
to 494	Massive sand with organic lenses overlain by silt and clay. 15,210 yrs BP @ 461–462 cm.			

[a]*Source:* Adapted from Thorson 1993.
[b]*Munsell Soil Color Chart.* Munsell Color Company, Inc., Baltimore, MD, USA. Munsell color charts contain standardized color chips showing hue, value, and chroma divided from zero to ten. The letters designate base color.
[c]@ the sediment depth where material was sampled for a ^{14}C date.

able radiocarbon dates indicate the presence of T1–T3, A1–A4, B, and C chronozones in the ITB Core (see McWeeney 1994 for details).

Description of Core C

Three major depositional units were recorded in the $>$ 3 m Core C (Table 9.3). The basal coarse sands and pebbles were buried by the second unit of fine silty sands. Organic accumulation began nearly immediately in the fine grained material deposited at 305 cm; at least nine subdivisions were identified and described within the peat accumulation.

Two conventional radiocarbon dates were obtained from this core, 8,990 $\pm$ 110 yrs BP and 10,420 $\pm$ 100 yrs BP (Table 9.4). The plant macrofossil assemblage is divided to correlate with four of the pollen zones described for southern New England. The available radiocarbon dates and lithostratigraphy allow for subdivision into the A3, A4, B, and C zones. The results of the macrofossil and invertebrate analyses are described in McWeeney (1994), and the pollen is described in Webb (1990).

Discussion

The sediment cores from Pequot Cedar Swamp provide plant macrofossil evidence for the transition of wetland basin surrounded by arctic/alpine-type plants over 15,000 ^{14}C years ago to the present *A. rubrum–C. thyoides* swamp. The path was not one of direct succession or climax. In many cases, mosaic environments developed in different sections of the basin. By studying more than one core (and in this chapter only 2 of 13 are described) it is possible to trace from the past to the present plant diversity found in one wetland basin.

Table 9.3 Core C Lithostratigraphy[a] and Organic Content[b]

Depth (cm)	Sediment Description	Munsell Color	Loss-on-ignition	
			cm	% Organic
0–10	Surface peat	10YR 2/2	3	88
		Dark brown	5	96
			10	90
10–75	Fibric peat with wood	10YR 2/2	15	93
		Dark brown	30	90
			40	93
			65	93
75–85	Sandy & silty lenses		75	97
85–95	Hemic peat	10YR 2/2	85	91
		Dark brown		
95–100	Sapric peat with charred fibrous material	10YR 1/1 Black	95	90
100–140	Fibric peat	10YR 2/2	105	93
			115	91
			125	93
			130	92
140–147	Disaggregated organic silts			
147–154	Woody fibric peat	10YR 2/2 Dark brown	150	93
155–160	8,990 ± 110 yrs BP (used for date)		160	94
160–185	Woody peat; vertical and horizontal fragments	10YR 2/2 Dark brown	180	95
185–205	Fine-grained silty fibric peat; horizontal layers	10YR 2/2 Dark brown	190	94
			200	96
205–254	Fine-grained silty fibric peat	10 YR 2/2 Dark brown	210	70
			220	59
			230	49
			240	24
			253	25
254–264	10,420 ± 100 yrs BP (used for date)	10 YR 2/2 Dark brown	263	25
254–305	Fine silt and sand with organics	10 YR 2/2 Dark brown	273	17
			283	11
			293	17
			303	17
305–314	Coarse to medium sand with granules	10YR 4/1 Dark gray	310	0.4

[a]*Source:* Adapted from Thorson and Webb 1991.
[b]*Source:* Webb 1994.

Zone T (Tundra)

The ITB core contained the oldest directly dated arctic/alpine-type plant macrofossils for New England, AMS-dated to 15,210 ± 80 yrs BP. High Cyperaceae and low Gramineae pollen counts suggest an area of impeded drainage (Watts 1979). While the pollen typifies an herb-dominated environment, the plant macrofossils document

Table 9.4 Radiocarbon Dates for Pequot Cedar Swamp Cores

Lab Sample Number	Sample Depth (cm)	Material Dated	[13]C Corrected Radiocarbon Age (yrs BP)
Indian Town Brook			
Beta 66346	91–96	peat sediments	6,370 ± 100
Beta 63241	108–111	peat sediments 0.5 g carbon	7,440 ± 120
Beta 64024	127–130	alder twigs and peat 0.4 g carbon	7,950 ± 160
Beta 66347 CAMS9258	151–152	water lily seeds	8,890 ± 60
Beta 66348 CAMS9259	159–160	water lily seeds	10,050 ± 70
Beta 63243	166–170	sediments including the 1st spruce needles 0.2 g carbon	11,700 ± 270
Beta 66349 CAMS9260	461	dwarf willow twigs & driad leaves	15,210 ± 80
Core C[a]			
WIS-1953	154–164	peat sediments	8,990 ± 110
WIS-1954	254–264	peat sediments	10,420 ± 100

[a]*Source:* Webb 1990:26.

an even more diverse community. The only aquatic plant preserved was the free-floating *Batrachium* sect. *Ranunculus trichophyllus* (water crowfoot). The plants that would be found in an arctic/alpine zone but penetrate south of the treeline today (Larsen 1989:114) included *Dryas integrifolia* (mountain avens or driads), *Salix herbacea* (willow), and *Vaccinium* cf. *uliginosum* (bog bilberry) leaves, plus several twigs and mosses. Later, *Betula michauxii* (dwarf birch) joined the prehistoric assemblage; it has an extensive range today throughout the tundra and treeline zones in Canada. The modern treeline/tundra boundary is climatically controlled based on the length of the growing season, temperature, and the average wind direction (Larsen 1989:115). The numerous core sites with tundra macrofossil plant remains strongly indicate that arctic-type vegetation dominated the landscape until nearly 12,000 years ago. The plant remains and the radiocarbon dates from Pequot Cedar Swamp indicate that cold temperatures prevailed in southern New England for several thousand years after the land was ice-free.

A change in the sedimentation rate occurred at Pequot Cedar Swamp just prior to 12,000 years ago. On the north and south margins of the basin, the accumulation rates were rapid, 100 cm/1,000 yrs at ITB based on a 3,000-year average between 15,000 and 12,000 years ago (Table 9.5). Glacial meltwater runoff may have continued to contribute to the basin for a long period. However, dated evidence from the southern end of the basin indicates that nearly 1.5 m accumulated in the last half of the 13th millennium BP. This deposition may reflect the warming period well recognized in Europe as the Allerød Period (Godwin 1975). Although the ice sheet had

Table 9.5 Estimated Sediment Accumulation Rates: Pequot Cedar Swamp

Core Name	Bracketing ^{14}C Dates (yrs BP)		Years (est'd)	Bracketing (cm)		Accumulation Rate (cm/10^3)
Fan	0	5,740 ± 70	6000	0	100	17
	5,740 ± 70	11,260 ± 160	5500	100	170	15
	11,260 ± 160	12,000 ± 60	800	170	183	15
	12,000 ± 60	12,030 ± 90	100	183	233	500
	12,030 ± 90	12,420 ± 110	500	233	322	180
ITB	0	6,370 ± 100	6000	0	100	17
	6,370 ± 100	7,440 ± 120	1000	95	110	15
	7,440 ± 120	7,950 ± 160	500	110	130	40
	7,950 ± 160	8,890 ± 60	1000	130	151	21
	8,890 ± 60	10,050 ± 70	1000	151	159	8
	10,050 ± 70	11,700 ± 270	2000	159	169	5
	11,700 ± 270	15,210 ± 80	3000	169	461	100
Core C[a]	0	8,990 ± 90	9000	0	157	17
	8,990 ± 90	10,420 ± 100	1500	157	258	70

[a]*Source:* Adapted from Webb 1990:27.

moved north of the St. Lawrence River by this time (LaSalle and Chapdelain 1990), the increased meltwater may have generated increased humidity and precipitation in New England. Also, glacially impounded lakes were draining and contributing to fluvial deposition. It is likely that any remaining permafrost in southern New England (Stone and Ashley 1992) was melting during this warming period. The changing sedimentation rate at Pequot Cedar Swamp suggests that major alteration of the local landscape occurred 12,000 years ago.

Zone A3

Around 12,000 years ago, the deposition of sediments switched to fine-grained sand and silt, plus organics. The sediments suggest that deposition took place in a muddy-bottomed pond (Thorson 1993), and the vegetation reflects the pond stage of development (Nichols 1915). Prior to this time, the presence of the free-floating *R. trichophyllus* achene suggests a pond or slow-moving stream environment. However, pond plants such as *Najas flexilis* (naiads), *Potamogeton* spp. (pondweeds), *Cerataphyllum demersum* (coontail), and *Typha* spp. (cattails) entered the macrofossil record for the first time just prior to 12,000 BP, indicating that the basins became impounded. *Najas flexilis* and *Typha latifolia* are rarely found in the northern boreal zone today (Watts and Winter 1966). The modern northern range for these aquatics suggests that the temperature had warmed to conditions found in Nova Scotia.

The AMS date of 12,030 ± 90 yrs BP on the first preserved *Picea* spp. needles and seeds at the south end of Pequot Cedar Swamp marks the beginning of the first trees. At the northern end of the Pequot Cedar Swamp, 5 cm of sediments containing the first *Picea* spp. needles dated to 11,700 ± 270 yrs BP. The first *Picea* spp. needles on Fisher's Island, several kilometers south, off the Connecticut coast, were accompanied by *B. michauxii* and were AMS-dated to 12,455 ± 80 yrs BP. That pair of macrofossils produced a similar date of 12,290 ± 440 yrs BP at Alpine Swamp, New Jersey (Peteet et al. 1993). *Picea* spp. needles from Linsley Pond, Connecticut, dated to 12,590 ± 430 (Peteet et al. 1993). The continuation of *B. michauxii* with the *Picea* spp. could mean that the first spruces looked like those found at the treeline today.

The advent of *Pinus strobus* along with *Picea* spp. and *Abies balsamia* (balsam fir) is also documented by needles found in the same sample that was AMS-dated to 12,030 ± 90 yrs BP. The well-dated presence of *P. strobus* needles is significant as an indicator of warming in southern Connecticut (Peteet et al. 1993). *Pinus strobus* needles were found with the first *Picea* spp. needles in two cores from Pequot Cedar Swamp. A core recovered from Durham Meadows in 1992 also provided *P. strobus* needles from a level AMS-dated on conifer wood to 11,740 ± 85 (AA-10923). While diploxylon (*Pinus banksiana* and *P. resinosa* [Jack pine and red pine]) and haploxylon *(P. strobus)* pollen have been recorded for the late-glacial pollen spectra (Davis 1983, Delcourt and Delcourt 1987, Webb 1988), the *P. strobus* needles show that the five-needle *Pinus* actually grew in southern Connecticut. The minimum average January temperature under which *P. strobus* survives is −6°C, the limiting temperature of the conifer assemblage since *Picea* spp., *A. balsamea,* and *Larix* all tolerate much colder extremes. Moreover, the *Picea* and *Abies* can persist under the warmer summer temperatures found in the *P. strobus* range, where the July average temperature ranges from 17° to 22°C (Burns and Honkala 1990).

Significantly, the *A. balsamea* needles found in the same stratum with *Picea* dating to 12,000 yrs BP document the presence of *A. balsamea* in southern New England 1,000 years earlier than interpreted from the pollen; *Abies* is a low pollen producer and may have been more abundant than *Picea* spp. at times (Davis 1978). *Abies balsamea* prefers cool temperatures and moist soils (Burns and Honkala 1990:26), suggesting that the climate during the 12th millennium BP was cool and wet (Peteet et al. 1993).

Larix laricina (larch) made up the fourth conifer component in the Pequot Cedar Swamp late-glacial woodland; however, it was the last to appear in the A3 zone. Several charred plant macrofossils appeared with the *Larix* in the levels just prior to 11,200 yrs BP at the southern end of the swamp. Wood, stems/fibers, and a burned *Larix* needle were found, followed by charred *Picea* needles that appeared in the last level of this zone. Together, the charred material suggests that boreal fires occurred locally, creating openings available for colonization by new taxa. After boreal conifers colonized the area, the sediment accumulation rate abruptly dropped (Table 9.5, lines 3 and 4). It may be that the vegetation growing in the new woodland anchored more of the sediments on the land.

If it was warm enough for *P. strobus* to survive 12,000 years ago, it was also warm enough for several deciduous trees to coexist under those conditions. Unfortunately, rapid deterioration of deciduous leaves occurs within the first year after abscission and transport to the final resting place (Spicer 1981); therefore, the preservation record is poor, especially in peat (Johnson 1992). The seeds, fruit, and nuts disappear quickly, devoured by numerous predators in the food chain all the way down to fungi and bacteria (Fowells 1965, Watts 1978). For documentation we must turn to the pollen record.

Previously, late-glacial pollen from temperate deciduous trees was regarded as having blown in from outside the area or as being reworked pollen from earlier time periods (Sirkin 1967, Wright 1971, Davis 1969 and 1983). However, with 5% oak pollen, that taxon is local (Gaudreau and Webb 1985). At Pequot Cedar Swamp, oak pollen rose up to 10% between 13,000 and 10,500 yrs BP (Webb 1990); according to the pollen count, the late-glacial peak of 30 grains occurred at 288 cm, around 12,000 yrs BP. The oak trees were clearly growing locally at Pequot Cedar Swamp.

An untraditional use of pollen counts instead of percentages and comparison of the counts between the late-glacial and the late-Holocene periods provided support for a mixed mosaic deciduous and coniferous environment in southern New England 12,000 years ago (McWeeney 1994). *Fagus americana, Fraxinus nigra* (black ash), and *C. dentata* pollen were all present at Pequot during the late-glacial and late-Holocene periods (McWeeney 1994). Comparison of the pollen counts suggests that several temperate deciduous species lived close to Pequot Cedar Swamp 12,000 years ago.

Comparison of the plant macrofossils identified from Pequot Cedar Swamp (Figs. 9.2 and 9.3) with the swamp series described by Nichols (1915) suggests that the pond evolved to the sedge stage during the 11th millennium BP. Along with the mesic herbaceous understory, several Cyperaceae species grew at Pequot Cedar Swamp. Macrofossil evidence of Cyperaceae, probably growing around the pond margin, included seeds from *Sparganium* spp. (bur-reed), *Carex* spp., *Cladium mariscoides* (twig-rush), and *Scirpus* spp. (bulrush). Apparently, the basin rapidly progressed through other stages. The shrub stage could be identified by the presence of *Alnus* spp. plant remains and *Scirpus atrovirens,* a bulrush that thrives in alder thickets. *Decodon verticillatus* (water willow) achenes suggest the evolution was heading toward boggy conditions (Nichols 1915) at the south end of the basin by 11,200 years ago.

Zone A4

The plant macrofossils, lithology, and radiocarbon dates indicate that a significant sedimentation hiatus occurred in Pequot Cedar Swamp at the end of the Pleistocene, approximately 11,000 to 10,000 ^{14}C yrs BP (Table 9.1). The duration of this hiatus corresponds with the Younger Dryas Period (Peteet et al. 1990, Peteet et al. 1993). The Greenland ice core dust record has been interpreted to correlate with evidence supporting the impact of a reversion to cold conditions during the Younger Dryas Period, which lasted approximately 1,000 years in the North Atlantic region (Taylor et al. 1993). The record from Pequot Cedar Swamp supports accumulating evidence for a Younger Dryas impact in southern New England (Peteet et al. 1990, Peteet et al. 1993).

A shift in the sediment and macrofossil accumulation at the north end of the basin suggests that a hiatus occurred during the Younger Dryas Period. Only 10 cm of fibric sediment accumulated over approximately 2,000 years (Table 9.5). The shift in the plant macrofossil assemblage between aquatic and terrestrial dominants emphasizes a discontinuity within this unit. At 171 cm, the boreal conifer components appeared along with Gramineae, *Typha* spp. seeds, *Najas flexilis,* and *Potamogeton* spp. fruit; however, only the aquatics continued up the core to 162 cm. At this level, *Picea* spp., *L. laricina,* and *P. strobus* appeared, briefly followed by *Nymphaea odorata* (water lily) remains that dated to 10,050 yrs BP. There appears to be an absence of a 1,000-year record around 160 cm. At the south end of the basin the *Picea* spp., *A. balsamea,* and *L. laricina* presence ended at 11,200 yrs BP (McWeeney 1994) and that unit formed a sharp contact with hemic peat, also suggesting a hiatus during the Younger Dryas Period.

The lack of sediment accumulation and plant preservation from the Younger Dryas Period suggests that the water table was lower. Decay takes place in the upper peat layers under aerobic conditions (Aaby and Tauber 1975, Aaby 1986), leaving the older strata intact below, as is the case at Pequot Cedar Swamp. A drop in water table and lack of sedimentation may indicate that climatic conditions were not only colder, but also dryer during the 11th millennium BP.

Apparently, to learn what was growing during the A4 zone, we need to examine well-dated deposits from deep, closed-system basins such as kettle holes. At Linsley Pond, Connecticut (Peteet et al. 1993), the pollen influx decreased for all taxa, except *Betula* spp. *(B. papyrifera,* paper birch from the macrofossil evidence) and *Alnus* spp. during the A4 zone. *Picea* spp., *A. balsamea,* and *L. laricina* continued to be represented, while *P. strobus* and *Quercus* spp. decreased in amount. That assemblage was interpreted to indicate a temperature drop of 3–4°C (Peteet et al. 1990, 1993). At Flamingo Pingo, a permafrost formation found in the emptied basin of the postglacial Lake Hitchcock in central Connecticut, it appears from the macrofossil evidence that beginning around 12,000 yrs BP, *Picea* spp. was a major contributor to the environment; that dominance continued with the addition of *L. laricina* and semiaquatic moss sometime prior to 10,175 ± 75 yrs BP (AA-10918), suggesting a deterioration in the environment (McWeeney 1992).

Zone B

Following the Younger Dryas Period, sediment accumulation resumed at Pequot Cedar Swamp. The presence of *Nymphaea odorata* and *Brasenia schreberi* (watershield) fruits on the northern margin and in the center indicates that part of the basin continued as an open pool, with a fluctuating water table.

Although an increase in *Pinus* spp. and *Quercus* spp. pollen traditionally marks the beginning of Zone B, and they were abundant at Pequot Cedar Swamp, it was *Picea* spp. and *L. laricina* needles that continued to be found in the sediments at Pequot. The presence of *Alder* spp. and *Chamaedaphne calyculata* (leatherleaf) suggests that the relict conifers may have been associated with the emergence of a bog (Nichols 1915) on the northern part of the basin. Bog indicators also emerged to the east. Charred *Picea* spp. and *L. laricina* needles represent the last appearance for these boreal tree needles at the swamp. However, the bog affinity for *Picea* spp. and

L. laricina does not rule out the possibility of periodic cold conditions between 10,000 and 9,000 years ago, despite it being the period of maximum solar insolation in the northern hemisphere (COHMAP Members 1988). *Picea* spp. needles were also found in the basal organic sediments from a central Connecticut River Valley meander scar dating to 9,370 ± 100 yrs BP (Beta-52257) (Table 9.4) and *Picea* spp. pollen continued to show up in numerous pollen spectra (Gaudreau and Webb 1985). Evidence from the Greenland ice cores indicates fluctuations of cold temperatures during the 10th millennium BP, although they were never as cold as during the Younger Dryas Period (Taylor et al. 1993).

Fire may have been responsible for killing the *Picea* spp. and *L. laricina,* creating openings for *Betula populifolia* (grey birch), known to pioneer on sterile, dry to wet soil (Fernald 1970) and to colonize after fires (Cwynar 1978). A *B. populifolia* bract made its first appearance at Pequot Cedar Swamp in this zone. Preserved Gramineae (grass) seeds also suggest that openings occurred in the canopy.

Evidence for fluctuating wet and dry episodes came from the ITB Core; chemically altered sediments formed intermittently between 10,050 and 8,890 yrs BP. According to the climate models for the period of maximum solar insolation around 9,000 years ago, the northern hemisphere would have experienced decreased precipitation (COHMAP Members 1988, Webb 1990). The presence of the altered sediments from the cores may be the local expression of that decrease in precipitation causing intermittent lowering of the water table during the 10th millennium BP.

In Core C, a different plant macrofossil record was preserved. *Alnus* spp., *Myrica gale* (sweet gale), *Chamaedaphne calyculatta,* and *Salix* spp. shrubs were joined by *Dryopteris* spp., *Sagittaria* spp. (arrowheads), and abundant moss, particularly *Sphagnum* spp. All these plants suggest that the shoreline was retreating toward the center, shrinking the open pond and indicating a change in water level (Digerfeldt 1986).

Zone C

The most significant plant evidence for the late-Holocene appeared in the cores from the east side of the mire, where woody fibric peat continued to be preserved up to the modern surface. The first temperate deciduous tree macrofossils appeared by 90 cm below the surface (BS), probably about 4,000 years ago. Both the *Acer* spp. and *Ulmus americana* (American elm) were identified from charcoal found at that level, suggesting that fire probably played a role in shaping the plant community. The size of the charred wood indicates that it came from a local fire. Unfortunately, their presence cannot tell us when those taxa began growing in the basin, although the pollen influx suggests they were present throughout the Holocene period.

Apparently, the water level remained consistent enough during the late-Holocene to support the growth of *C. thyoides* trees. *Chamaecyparis thyoides* needles and twigs appeared in the same level as the first *T. canadensis* needles, at 75 cm in Core C. *Tsuga canadensis* and *P. strobus* were found in the most shoreward Core H, as well. The *C. thyoides* macrofossils represent a significant find, since no Cupressaceae (cedar) pollen was reported (Webb et al. 1990). The absence may be due to the difficulty of recognizing Cupressaceae pollen and may not reflect a real absence from the record. The *P. strobus* pollen count decreased dramatically; it reached 18% in recent sediments (Webb 1990), based on as few as 12–14 grains, a low number considering that *P. strobus* still occupies the basin.

Seeds from *Sambucus* spp. (elderberry) and *Rubus* spp. (raspberry/blackberry) shrubs were found 150 m from the edge of the basin, providing evidence of their outward spread. From the identified plant macrofossils, we can tell that the eastern side of the basin supported a bog mat throughout the second half of the Holocene.

Prehistorically, this section of Pequot Cedar Swamp most likely supported the full range of plants typically found with *C. thyoides* (Little 1950, Laderman 1989:36) after 4,000 years ago. *Quercus* spp. continued to be the most prolific pollen producer for the site. *Fagus americana, Carya* spp., *Castanea dentata, Fraxinus* spp. (ash), *Populus tremuloides* (quaking aspen), *Ostrya/Carpinus* spp. (hornbeam/ironwood), and *Platanus occidentalis* (sycamore) all contributed minor amounts of pollen (Webb et al. 1990). Ericaceae (heath/blueberry) pollen increased; probably the species changed from *C. calyculata* to *V. corymbosum* shrubs in light of the seeds, leaves, and wood from the former and the present-day growth of the latter. *Ilex*-type plants were present, probably the *Ilex verticillata* shrub found there today. Herbaceous plants with mesic affinity included Cyperaceae, for which the pollen count rose to 11 grains, the highest number since the 12th millennium BP. Intermittently occurring *Typha* spp. pollen paralleled the representation pattern recorded for the seeds. The taxa identified from pollen expand the environmental picture for the *C. thyoides* swamp, complementing what was learned from macrofossils.

Conclusion

The sediment cores recovered from the Pequot Cedar Swamp document more than 15,000 years of a locally changing wetland environment. The oldest AMS-dated arctic/alpine-type plant macrofossils from southern New England were recovered in the cores from this swamp. Beginning with terrestrial shrub tundra and herbaceous plants that grew around the basin for 3,000 years, the plant macrofossil and pollen identifications make it possible to picture diverse environmental fluctuations that occurred prehistorically. The entrance of *Picea* spp. and *A. balsamea* along with the early advent of *P. strobus* in southern New England occurred 12,000 years ago. A hiatus in sediments and plant remains between 11,200 and 10,000 years ago suggests that the reversion to cold conditions during the Younger Dryas Period left a negative record in the open basin system. During the Holocene, multiple levels of evidence indicate fluctuating water levels and temperatures, especially during the 10th millennium BP and during the mid-Holocene warming period. Aggregated, chemically altered sediments and the outward spread of shrubs indicate that periods of drying altered the sediment accumulation only to be followed by periods of a higher water table sufficient to support diverse plant populations. Charcoal evidence of fires increased during the mid-Holocene, leaving a record of temperate deciduous trees that was otherwise unavailable through anaerobic preservation. Apparently, during the second half of the Holocene, the extreme variations in water availability decreased enough to form an acceptable habitat for *C. thyoides*. Anaerobic preservation left a macrofossil record of a number of constant companions to *C. thyoides* in what today must be classed as a predominantly *A. rubrum* swamp.

This study has demonstrated that more than one sediment core is essential for paleo-reconstruction of a wetland basin. This project used a transect of cores recovered from several parts of the basin. The research spanned several disciplines, including botany, ecology, geology, sedimentology, hydrology, and anthropology. However, more work remains to be done. This significant basin in a coastally restricted forest holds tremendous resources for paleoenvironmental reconstruction.

Acknowledgments The project was funded by the Mashantucket Pequot Indians and a Dissertation Improvement Grant from the National Science Foundation. Thanks to Kevin McBride, Frank Hole, Leo Hickey, Graeme Berlyn, Robert Thorson, Dorothy Peteet, Thomas Webb III, Robert Webb, and Paige Newby. I thank many colleagues for reviewing my paleo-reconstructions, but I accept responsibility for any errors.

References

Aaby, B. 1986. Palaeoecological studies of mires. Pages 145–164 *in* B.E. Berglund, ed. Handbook of Holocene palaeoecology and palaeohydrology. Wiley, New York.

Aaby, B., and H. Tauber. 1975. Rates of peat formation in relation to degree of humification and local environment as shown by studies of a raised bog in Denmark. BOREAS **4**:1–17.

Burns, R.M., and B.H. Honkala. 1990. Silvics of North America. Vol. 1: Conifers. U.S. Dep. Agric. Handb. 654. USDA, Washington, DC.

COHMAP (Cooperative Holocene Mapping Project) Members. 1988. Climatic changes of the last 18,000 years: observations and model simulations. Science **241**:1043–1052.

Cwynar, L.C. 1978. Recent history of fire and vegetation from laminated sediment of Greenleaf Lake, Algonquin Park, Ontario. Can. J. Bot. **56**:10–21.

Davis, M.B. 1969. Climatic changes in southern Connecticut recorded by pollen deposition at Roger's Lake. Ecology **50**:409–422.

———. 1978. Climatic interpretation of pollen in Quaternary sediments. Pages 35–52 *in* D. Walker and J.C. Guppy, eds. Biology and Quaternary Environments. Australian Acad. Sci., Sydney.

———. 1983. Holocene vegetational history of the eastern United States. Pages 166–181 *in* H.E. Wright, ed. Late-Quaternary environments of the United States. Univ. of Minnesota Press, Minneapolis.

Dean, W.E. 1974. Determination of carbonate and organic matter in calcareous sediments and sedimentary rocks by loss on ignition: comparison with other methods. J. Sed. Petrol. **44**:242–248.

Deevey, E.S., Jr. 1939. Studies on Connecticut lake sediments. I. A postglacial climatic chronology for Southern New England. Amer. J. Sci. **237**:691–724.

Deevey, E.S., Jr., and R.F. Flint. 1957. Postglacial Hypsithermal Interval. Science **125**:182–184.

Delcourt, P.A., and H.R. Delcourt. 1981. Vegetation maps for eastern North America: 40,000 yrs B.P. to the present. Pages 123–166 *in* R.C. Romans, ed. Geobotany II. Plenum Press, New York.

Delcourt, P.A., and H.R. Delcourt. 1987. Long-term forest dynamics of the temperate zone. Springer-Verlag, New York.

Digerfeldt, G. 1986. Studies on past lake-level fluctuations. Pages 127–144 *in* B.E. Berglund, ed. Handbook of Holocene palaeoecology and palaeohydrology. Wiley, New York.

Fernald, M.L. 1970. Gray's manual of botany, 8th ed. Van Nostrand, NewYork.

Fowells, H.A. 1965. Silvics of the forests of the United States. U.S. Dep. Agric., Agric. Handb. 271. Washington, DC.

Gaudreau, D., and T. Webb III. 1985. Late-Quaternary pollen stratigraphy and isochron maps for the northeastern United States. Pages 247–280 *in* V.M. Bryant, Jr. and R.G. Holloway, eds. Pollen records of late-Quaternary North American sediments. Amer. Asso. Stratigraphic Palynologists Foundation. Dallas.

Godwin, Sir H. 1975. History of the British flora: a factual basis for phytogeography, 2nd ed. Cambridge Univ. Press, Cambridge, England.

Jackson, S.T. 1989. Post-glacial vegetational changes along an elevational gradient in the Adirondack Mountains (NY): a study of plant macrofossils. Bio. Survey/Mus. Bull. 465. New York State Museum, Albany.

Johnson, L.C. 1992. Species-regulated sphagnum decay in raised bogs. Paper presented at the New York Natural History Conference II. New York State Museum of Science, Albany. Unpublished.

Laderman, A.D. 1989. The ecology of Atlantic white cedar wetlands: a community profile. U.S. Fish Wildl. Serv. Biol. Rep. 85 (7.21).

Larsen, J.A. 1989. The northern forest border in Canada and Alaska. Springer-Verlag, New York.

LaSalle, P., and C. Chapdelaine. 1990. Review of late-glacial and Holocene events in the Champlain and Goldthwait Seas areas and arrival of man in eastern Canada. Pages 1–20 *in* N.R. Lasca and J. Donahue, eds. Archaeological geology of North America. Boulder. Geological Soc. of America. Centennial Special Vol. 4.

Leopold, E.B. 1955. Climate and vegetation changes during an interstadial period in southern New England. Dissertation, Yale Univ., New Haven.

Little, S., Jr. 1950. Ecology and silviculture of white cedar and associated hardwood in southern New Jersey. Yale Univ. Sch. For. Bull. 56, New Haven.

McWeeney, L.J. 1991. Plant macrofossil identification as a method toward archaeo-environmental reconstruction. Bull. Archaeological Soc. Conn. **54**:87–97.

———. 1992. Bog formation processes at the Flamingo Pingo, South Windsor, CT. Prepared for Dr. Jelle DeBoer, Wesleyan Univ. Unpublished.

———. 1994. Archaeological settlement patterns and vegetation dynamics in southern New England in the late Quaternary. Dissertation, Yale Univ., New Haven.

Nichols, G.E. 1915. The vegetation of Connecticut IV. Plant societies in lowlands. Bull. Torrey Bot. Club **42**:168–217.

Peteet, D.M., S.S.Vogel, D.E. Nelson, J.R. Southon, R.J. Nickmann, and L.E. Heusser. 1990. Younger Dryas climatic reversal in northeastern USA? AMS ages for an old problem. Quat. Res. **33**:219–230.

Peteet, D.M., R.A. Daniels, L.E. Heusser, J.S. Vogel, J.R. Southon, and D.E. Nelson. 1993. Late glacial pollen, macrofossils, and fish remains in northeastern USA—the Younger Dryas oscillation. Quat. Sci. Rev. **12**:597–612.

Shuford, W.J. 1975. Vegetation survey of inland wetlands, Town of Ledyard Connecticut. Studies in Human Ecology, No. 6. Connecticut Coll., New London.

Sirken, L.A. 1967. Late-Pleistocene pollen stratigraphy of western Long Island and eastern Staten Island, New York. Pages 249–274 *in* E.J. Cushing and H.E. Wright, Jr., eds. Quat. Paleoecology. Yale Univ. Press, New Haven.

Spicer, R. 1981. The sorting and deposition of allochthonous plant material in a modern environment at Silwood Lake, Silwood Park, Berkshire, England. Geol. Survey Prof. Paper 1143. U.S. Govt. Printing Office, Washington, DC.

Stone, J.R., and G.M. Ashley, with contributions by N.G. Miller, R.M. Thorson, L.J. McWeeney, and H.D. Luce. 1992. Ice-wedge casts, pingo scars, and the drainage of glacial Lake Hitchcock. Pages 305–331 in P. Robinson and J.B. Brady, eds. Guidebook for field trips in the Connecticut Valley region of Massachusetts and adjacent states, Vol. 2. New England Intercollegiate Geological Conf. Amherst.

Taylor, K.C., G.W. Lamorey, G.A. Doyle, R.B. Alley, P.M. Grootes, P.A. Mayewski, J.W.C. White, and L.K. Barlow. 1993. The 'flickering switch' of late Pleistocene climate change. Nature 361:432–436.

Thorson, R.M. 1993. Pieces of the Pequot past. Manuscript prepared for the Mashantucket Pequot museum development project. Unpublished.

Thorson, R.M., and R. Webb. 1991. Postglacial development of the cedar swamp. J. Paleolimnology 6:17–35.

Warner, B.G., and P.J. Barnett. 1986. Transport, sorting and reworking of late Wisconsin in plant macrofossils from Lake Erie, Canada. BOREAS 15:323–329.

Watts, W.A. 1978. Plant macrofossils and Quaternary paleoecology. Pages 53–67 in D. Walker and J.C. Guppy, eds. Biology and Quaternary environments. Australian Acad. of Sci., Sydney.

————. 1979. Late Quaternary vegetation of central Appalachia and the New Jersey coastal plain. Ecol. Monogr. 49:427–469.

Watts, W.A., and T.C. Winter. 1966. Plant macrofossils from Kirchner Marsh, Minnesota—a paleoecological study. Geol. Soc. Am. Bull. 77:1339–1360.

Webb, R.S. 1990. Late-Quaternary water-level fluctuations in the northeastern United States. Dissertation, Brown Univ., Providence.

Webb, R.S., F. Lefkowitz, and T. Webb III. 1990. Pollen and sediment records for Pequot Indian Reservation Cedar Swamp, New London County, Connecticut. Manuscript prepared for the Mashantucket Pequots. Unpublished.

Webb, T., III. 1988. Eastern North America. Pages 385–414 in B. Huntley and T. Webb III, eds. Vegetation history. Kluwe, The Netherlands.

Wright, H.E., Jr. 1971. Late Quaternary vegetational history of North America. Page 426–464 in K. Turekian, ed. The Late-Cenozoic glacial ages. Yale Univ. Press, New Haven.

10

Dwight L. Stoltzfus & Ralph E. Good

Plant Community Structure in *Chamaecyparis thyoides* Swamps in the New Jersey Pinelands Biosphere Reserve, USA

Forested wetlands dominated by *Chamaecyparis thyoides* (L.) BSP. (Atlantic white cedar) occur in the New Jersey Pinelands Biosphere Reserve in the eastern United States Atlantic Coastal Plain. The development of plant community structure in these swamps is influenced by both autogenic and allogenic factors within the physical and biotic environment including water table, disturbance, soil layer, herbivory, and plant competition (Little 1950, Roman et al. 1990). Changes in the physical and biotic environment since human colonization appear to be affecting community structure resulting in the replacement of *C. thyoides*-dominated communities by communities dominated by *Acer rubrum* (red maple) and reductions in cedar swamp area (Roman et al. 1990). Cedar swamps have become increasingly fragmented as habitat patch size decreases, resulting in more isolated units with increased edge exposure to surrounding community types. The importance of physical and biotic environmental factors in determining species composition, dynamics, diversity, and physical structure within the highly disturbed cedar swamps is not completely understood. The effect of increased fragmentation on the community structure in cedar swamps also remains unclear.

The purposes of this study are to evaluate the development of plant community structure in cedar swamps in the New Jersey Pinelands and to determine the extent to which environmental parameters, disturbance, and fragmentation influence community structure.

Study Area

All of the study sites are located in the New Jersey Pinelands within the approximately 400,000 ha Pinelands National Reserve (Fig. 10.1), also designated as a Bio-

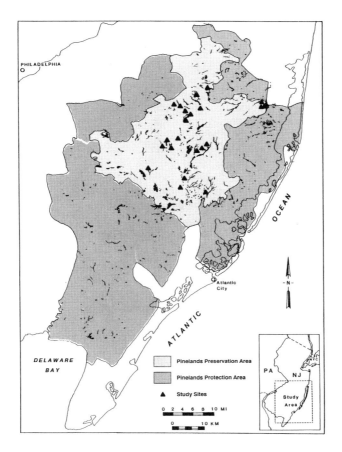

Figure 10.1. Location of cedar swamps included in this study in New Jersey Pinelands. NJ: New Jersey. PA: Pennsylvania. NYC: New York City.

sphere Reserve by the Man and Biosphere Program of the United Nations Educational, Scientific, and Cultural Organization (UNESCO).

Cedar swamps are found along streams and in areas where the water table meets the surface. They are seasonally flooded by acidic, nutrient-poor water with pH between 3 and 5. The continuous flooding and low pH result in low decomposition rates and an accumulation of up to 3 m of organic soils overlying a sandy substrate. A hummock-hollow topography is typical of most swamps, in which hummocks are formed on or around existing trees, old stumps, fallen logs, or uprooted tree roots, and are very seldom completely flooded. Hollows range from *Sphagnum*-covered depressions to flooded areas.

In the New Jersey Pinelands, cedar swamps are interspersed with other wetland types, including hardwood swamps, shrub bogs, and freshwater marshes, and with drier upland areas, resulting in a mosaic of vegetation types (Roman et al. 1990). Flora of the Pinelands have been recorded by Stone (1911), Harshberger (1916), Little (1950), Ehrenfeld and Schneider (1983), and Stoltzfus (1990).

Methods

Study Descriptions

Eighteen cedar swamps dominated by *C. thyoides* were selected for study within the Preservation Area of the Pinelands National Reserve (Fig. 10.1). The swamps range in time since logging from 60 to 130 years and represent an intermediate-growth stage. They were selected according to size and surrounding plant community type in order to study plant community structure and to evaluate the effect of fragmentation (reduced swamp size and increased hardwoods) on community structure. Six sites have been included in each of three size categories: small (0.5–5 ha), medium (5–20 ha), and large (>20 ha). Within each size category, three swamps are surrounded by *Pinus rigida* (pitch pine) lowland or upland, and three are surrounded by hardwood swamps.

Vegetation Sampling

Plant species were sampled in the 18 intermediate-growth cedar swamps in 1985 at 40 random points placed along five transects within each site. Tree diameter measurements and point-to-plant distances obtained with the point-centered quarter method (Cottam and Curtis 1956) were used to calculate density, dominance (basal area), and frequency values. The relative values were determined and averaged to obtain an importance value for each species. Size-class densities of live and dead tree stems were determined from diameter at breast height (dbh) measurement and stem counts within a 5 m^2 circular plot at each point. Shrub and sapling (<2.5 cm dbh and >30 cm high) densities were calculated from stem counts within these same 5 m^2 plots. Seedling (<30 cm high) densities were calculated from seedling counts within a 0.25 m^2 circular plot at each point. Herb species found within these small plots were recorded.

   Stand age was determined by coring *C. thyoides* trees selected at randomly placed points along the well transect within each site. Three trees were cored in each of the following size classes: 2.5–7.4, 7.5–12.4, 12.5–17.4, 17.5–22.4, 22.5–27.4, and 27.5–32.4 (if present) cm dbh. Two cores were taken from each tree at right angles to each other. Width measurements of radial growth for ten-year increments were recorded for each of the two cores and averaged to obtain a representative measurement of each tree.

Data Analysis

Two-way variance analysis (SAS 1982) was used to identify significant differences in relation to size and surrounding community type. Pearson product–moment correlations among vegetation variables and environmental factors were also determined (SAS 1982). Species diversity in each stand was determined by the Shannon-Weaver Index (Ludwig and Reynolds 1988).

Results and Discussion

Species Composition

Tree Layer   Tree species diversity and richness are low, with generally five or fewer tree species (Table 10.1). Community structure within the tree layer typically ranges from pure stands of *C. thyoides* with a closed, dense canopy to a mixture of *C. thyoides* and hardwoods including *A. rubrum, Magnolia virginiana* (sweetbay magnolia), *Nyssa sylvatica* (blackgum), and *Betula populifolia* (gray birch) with a more open canopy. Hardwood swamps and cedar swamps frequently contain mixtures of *A. rubrum* and *C. thyoides* to the point where the two species are codominant.

The tree species sampled in the 18 intermediate-growth swamps include *C. thyoides, A. rubrum, M. virginiana, N. sylvatica, B. populifolia, Sassafras albidum* (sassafras), *P. rigida,* and *Amelanchier canadensis* (shadbush). Comparisons of average importance, density, and basal area values for the tree species illustrate the dominance of *C. thyoides* within these stands (Table 10.2). *Magnolia virginiana, N. sylvatica,* and *B. populifolia* were found primarily in the subcanopy layer in these stands and have low importance, density, and basal area. Species found infrequently in cedar swamps include *P. rigida, S. albidum,* and *A. canadensis.*

Low sapling densities for all tree species present (Table 10.3) reflect the extent of canopy closure within the swamps. Although no direct light measurements were made within the swamps, tree saplings were generally more abundant in swamps where the canopy was partially open. *Chamaecyparis thyoides* saplings were sampled in only two of the 18 intermediate-growth stands and were most abundant where windthrow had resulted in numerous canopy gaps. These results illustrate the low survival of *C.*

Table 10.1 Mean Species Diversity (Shannon-Weaver Index) and Mean Richness Measures for the 18 Intermediate-Growth Cedar Swamps

Layer	Mean
Tree	
Species Diversity	0.69
Species Richness	4.3
Shrub	
Species Diversity	1.95
Species Richness	11.3
Herb	
Species Richness	11.3
Total species richness	22.4

Table 10.2 Frequency (Number of Swamps), Mean Density, Basal Area, and Importance Values for Tree Species in Intermediate-Growth Cedar Swamps

Species	Frequency (no.)	Density (no. ha^{-1})	Basal Area (m^2 ha^{-1})	Importance (%)
Chamaecyparis thyoides	18	2,384	7.59	84.2
Acer rubrum	18	230	0.29	10.7
Nyssa sylvatica	14	42	0.07	2.1
Magnolia virginiana	14	39	0.05	2.0
Sassafras albidum	6	11	0.01	0.5
Pinus rigida	3	3	0.02	0.3
Betula populifolia	4	4	0.01	0.2
Total	18	2,713	8.01	

thyoides except within canopy gaps. *Acer rubrum* saplings were more abundant than *C. thyoides* and were sampled in a greater number of swamps.

Mean seedling densities are high for both *C. thyoides* and *A. rubrum* (Table 10.3). Differences in seed production and germination among the intermediate-growth cedar swamps are evident in the seedling counts, which ranged from 10,000 to 4,360,000 seedlings/ha for *C. thyoides* and from 21,000 to 3,450,000 seedlings/ha for *A. rubrum*. Seedling size for both species is typically less than 5 cm. Low seedling densities for other tree species reflect their low importance values and indicate low recruitment rates for these species.

Table 10.3 Mean Sapling and Seedling Densities for Tree Species in Intermediate-Growth Cedar Swamps

	Total Swamps	Density (no. ha^{-1})
Saplings		
Chamaecyparis thyoides	2	11
Acer rubrum	11	61
Nyssa sylvatica	2	14
Sassafras albidum	2	6
Magnolia virginiana	3	39
Betula populifolia	2	25
Amelanchier canadensis	1	6
Seedlings		
Chamaecyparis thyoides	18	151,889
Acer rubrum	18	135,778
Nyssa sylvatica	15	4,278
Sassafras albidum	1	56
Magnolia virginiana	7	722
Pinus rigida	4	278
Amelanchier canadensis	4	278
Quercus ilicifolia	6	333

Shrub Layer   Shrub species diversity is greater than tree diversity (Table 10.1). The shrub layer is variable in composition and size and ranges from a dense layer up to 4 m high to scattered occasional shrubs. Dominant shrub species include *Vaccinium corymbosum* (highbush blueberry), *Gaylussacia frondosa* (dangleberry), *Clethra alnifolia* (sweet pepperbush), *Leucothoe racemosa* (fetterbush), and *Rhododendron viscosum* (swamp azalea) (Table 10.4). Common, small shrub species with individuals generally less than 30 cm in height include *Rubus hispidus* (dewberry) and *Gaultheria procumbens* (wintergreen) (Table 10.4). Although the same dominant shrub species are present in nearly all swamps, the extent of dominance varies from swamp to swamp.

   Cedar swamps dominated by *C. thyoides* tend to have lower densities of dominant shrub species including *C. alnifolia, R. viscosum,* and *L. racemosa* and higher densities of *G. frondosa* (Stoltzfus 1990). These patterns are reversed in stands with a greater hardwood component. *Vaccinium corymbosum* is evenly spread among all cedar swamps. The extent of canopy closure is an important factor in determining dominant shrub species and the degree of dominance.

Herb Layer   A generally sparse herb layer is present, including a variety of distinctive species, some of which are rare, threatened, or endangered (Fairbrothers 1979).

Table 10.4 Mean Shrub Densities in Intermediate-Growth Cedar Swamps

	Total Swamps	Density (no. ha^{-1})
Large shrubs		
Gaylussacia frondosa	18	13,333
Clethra alnifolia	18	11,360
Vaccinium corymbosum	18	3,544
Leucothoe racemosa	15	3,258
Rhododendron viscosum	17	1,219
Gaylussacia baccata	5	797
Ilex laevigata	13	661
Gaylussacia dumosa	9	539
Kalmia latifolia	9	422
Chamaedaphne calyculata	5	225
Ilex glabra	2	114
Viburnum nudum	5	94
Smilax rotundifolia	6	58
Myrica pensylvanica	2	25
Lindera benzoin	1	25
Rhus vernix	1	8
Viburnum dentatum	1	3
Small shrubs		
Rhus radicans	12	2,722
Rubus hispidus	10	5,889
Parthenocissus quinquefolia	6	833
Gaultheria procumbens	5	3,556
Vaccinium macrocarpon	2	500

Sphagnum moss, which covers portions of the forest floor, includes a number of different species. Diversity and abundance in the herb layer of cedar swamps change as community structure develops. Both increase immediately following moderate disturbance in which the canopy layer is removed, and then decrease as the stand develops. The extent of canopy closure and site hydrologic conditions are primary factors in determining diversity and abundance in cedar swamps where greater moisture and light favor a more developed herb layer.

Herbs found in at least 50% of the swamps include *Trientalis borealis* (starflower), *Carex collinsii* (Collin's sedge), *C. trisperma* (three-seeded sedge), *Drosera rotundifolia* (round-leaved sundew), *Mitchella repens* (partridge berry), *Aralia nudicaulis* (wild sarsaparilla), *Woodwardia virginica* (Virginia chain fern), and *Dryopteris simulata* (Massachusetts fern). *Trientalis borealis, C. collinsii, C. trisperma,* and *D. rotundifolia* were found most frequently. *Sphagnum* cover and herb occurrence tended to be greater in those swamps with a more open canopy.

Stand Structure

Young-growth Stands Rapid shrub and herb layer growth following disturbance is accompanied by the establishment of large numbers of *A. rubrum* and *C. thyoides* seedlings and the regrowth of hardwood species from root systems. Reproduction methods among these species range from seeds only to reestablishment primarily from sprout origin and are important in determining recolonizing abilities. *Chamaecyparis thyoides* regeneration is from seed only and is thus dependent solely on seed sources from surrounding trees and seed stored in the surface peat layer. *Chamaecyparis thyoides* seed production occurs in large quantities. Seed production and the seed bank provide several million seeds/ha.

The abundant seed sources available from surrounding trees and the peat layer generally result in the high densities of cedar seedling establishment reported in this and other studies (Vermeule 1900, Harshberger 1916, Korstian and Brush 1931, Little 1950). Successful *C. thyoides* seedling and sapling establishment is dependent on seed sources from live trees, seed sources in the peat layer, adequate light exposure, microtopography, and proper moisture levels (Little 1950).

Acer rubrum also produces seeds in large quantities (Table 10.3) and maintains its place within the swamps by adaptations for survival in continuously flooded conditions (Day 1987) and its ability to resprout. The other primary species, *N. sylvatica, M. virginiana,* and *B. populifolia,* also have an adaptive advantage in their ability to sprout from roots or stems. Although no *C. thyoides* saplings were sampled in Coopers Branch two years after clearcutting, root sprout growth of *A. rubrum, B. populifolia, M. virginiana, N. sylvatica,* and *S. albidum* resulted in an early presence of saplings up to 5 m in height (Stoltzfus 1990). This rapid establishment of hardwood saplings illustrates the early growth advantage of the hardwood species with the ability to resprout after cutting or fire.

With adequate seed sources in open conditions, dense sapling stands develop. Young regenerating cedar stands may have up to 200,000 *C. thyoides* seedlings/ha and 15,700 stems/ha greater than 30 cm in height (Stoltzfus 1990). Stands composed of *C. thyoides* saplings have up to 20,000 stems/ha above 1.8 m. (Korstian and Brush 1931, Little 1950).

Intermediate-growth Stands High *C. thyoides* densities in young stands result in high mortality levels, as self-thinning begins between 10 and 20 years. A mean dead *C. thyoides* stem density of 8,700 stems per ha in the Coopers Branch fenced plots 28 years after clearcutting indicates that self-thinning has been occurring for several years (Stoltzfus 1990).

An even-aged aspect within developing stands results, as subsequent reproduction and survival are controlled by light availability. All 18 intermediate-growth cedar swamps sampled show even-age characteristics (Fig. 10.2). Fifteen of the 18 swamps are even-aged throughout the entire stand, with age ranges among all of the size classes generally between 15 and 30 years, indicating that recruitment of young individuals within the stand continues for that period of time. The recruitment period within the developing stand is dependent on site conditions, particularly initial *C. thyoides* density, the extent of deer browsing, and moisture levels.

Cedar stands with gaps in the developing canopy layer, which allow increased seedling and sapling survival, have longer recruitment periods than those with a dense vegetative cover, which inhibits seedling and sapling growth. The 20-year recruitment period in Reeds Branch and South Branch Mt. Misery cedar swamps, with relatively closed canopies, is much shorter than the 60-year period in Pole Bridge Branch, where the canopy was more open (Fig. 10.2).

Ages of the trees sampled in Shanes Branch 1 and Middle Branch Mt. Misery cedar swamps (Fig. 10.2) suggest the occurrence of fire or logging. These events cleared entire swamp sections and resulted in the development of a younger even-aged section adjacent to an older even-aged section. In Featherbed Branch, with two distinct age groups, younger trees were scattered among the older trees, suggesting an event, such as storm damage or logging of individual trees, that produced gaps where younger *C. thyoides* trees became established.

Size-class distributions for sampled hardwood species reflect their importance in the mature cedar swamps (Table 10.5). Scattered large *A. rubrum* trees are found along with *C. thyoides* in the canopy layer in several swamps. Greater densities of *A. rubrum* in the smaller size classes show that it is also scattered through the subcanopy layer. The predominance of small *N. sylvatica, M. virginiana,* and *B. populifolia* trees indicates that these species are generally found within the subcanopy layer.

Those individuals that are first to become established within the stand have more rapid growth and remain the dominant individuals throughout stand development (Stoltzfus 1990). Differential growth rates among the different size classes become evident during the first 10 years and continue through at least the first 60 years. Similar growth trends are evident in all size classes, with an initial rapid increase in radial growth followed by a continuing decrease. Growth of *C. thyoides* is moderately rapid, with an annual diameter increase of 2.5–3.8 mm maintained by dominant trees until 100 years. Maximum growth rate occurs at about 50 years (Korstian and Brush 1931). The same study reports an annual height increase of 0.3–0.5 m through the first 50 years, after which the rate decreases to near zero at about 100 years of age.

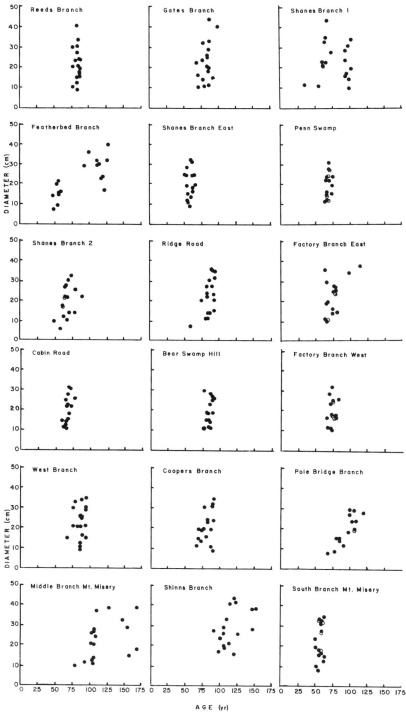

Figure 10.2. Age of *C. thyoides* trees within selected size classes in each of the intermediate-growth cedar swamps. (Open circles are used to note overlapping data points.)

Table 10.5 Size Class (cm dbh) Densities for Tree Species in Intermediate-Growth Cedar Swamps.

Species	Density (no. ha^{-1})					
	2.5–7.4	7.5–12.4	12.5–17.4	17.5–22.4	22.5–27.4	>27.5
Chamaecyparis thyoides	172	394	672	550	303	225
Acer rubrum	117	53	8	19	3	3
Nyssa sylvatica	33	8	11	3	0	0
Sassafras albidum	0	0	3	0	0	0
Magnolia virginiana	39	6	0	0	0	0
Pinus rigida	0	0	0	3	0	0
Betula populifolia	8	8	0	0	0	0

Site quality or productivity estimates have been made based on data collected in cedar stands in North Carolina, Virginia, and New Jersey (Korstian and Brush 1931). A height-age curve based on these data is used to determine a site quality index that represents a measure of height, in feet, of dominant *C. thyoides* individuals after 50 years of growth. The mean site index of 42.5 for this study's stands is in the middle of the 20–70 range of site-index measurements in 47 cedar swamps (Korstian and Brush 1931). The mean from this study supports Little's (1950) suggestion that the majority of southern New Jersey swamps belong in the 40- to 45-foot sites.

Old-growth Stands   Studies of stand structure development through the end of the self-thinning stage and beyond are difficult in New Jersey because of the absence of old-growth stands. We do not know, for example, the extent to which *C. thyoides* dominated the swamps or the proportion of *A. rubrum* abundance in relation to *C. thyoides*. Since the precolonial swamps were not logged as they have been in the last few hundred years, other important questions remain. Were all precolonial cedar swamps even-aged? How extensive were disturbances such as fire, which likely burned at least portions of swamps? It is clear that cedar stands in New Jersey included very large *C. thyoides* trees of much older ages than at present (Little 1950). It is also clear that earlier cedar swamps were much more extensive than they are presently (Stone 1911, Harshberger 1916).

Descriptions of the successional process and interpretations of long-term change have been based on field observations and experiments (Harshberger 1916, Korstian and Brush 1931, Buell and Cain 1943, Little 1950) as well as on greenhouse experiments involving varied light and soil-moisture conditions (Little 1950). The pattern suggested in these earlier studies is that *A. rubrum* eventually replaces *C. thyoides* over time. As large canopy individuals die within even-aged stands, gaps form. *Acer rubrum* individuals invade the gaps and eventually dominate. The mature-stage or old-growth stand thus becomes dominated by *A. rubrum* rather than *C. thyoides*. Wetland forests dominated by *C. thyoides* can be maintained only by continued disturbance whereby the canopy and shrub layers are removed, providing open conditions in which *C. thyoides* can regenerate.

Historically, fire is considered to have been the clearing agent of cedar stands. Cedar stands were periodically burned with a fire frequency sufficient to provide clearing and regrowth before the cedar swamp reached senescence and developed into hardwood swamp (Little 1950, Forman and Boerner 1981). With changing fire frequencies and a general reduction in wildfire intensity resulting from increased development and increased fire control over the last 100 years, the role of fire as a clearing agent has diminished.

Over the last 300 years, logging has replaced fire as the primary clearing agent in the New Jersey Pinelands. With fire's reduced importance at present, logging is considered to be the primary disturbance factor necessary to maintain existing *C. thyoides*-dominated stands. The continuing transition from cedar swamp to hardwood swamp dominated by *A. rubrum* appears to support this pattern of community structure development.

Variability of site conditions and disturbance histories and the differential response by tree and shrub species to these conditions, however, make it difficult to apply this pattern to all swamps. A possible variation is that either *C. thyoides* or *A. rubrum* may become established within gaps and eventually replace the dominant canopy individuals as they die. The development of community structure within the gap would be determined by a combination of factors including *C. thyoides* and *A. rubrum* seed sources, shrub layer thickness, water table, and deer browsing. The presence of both *C. thyoides* and *A. rubrum* saplings within existing gaps in some even-aged *C. thyoides* stands indicates sapling survival with canopy gaps.

Seedling densities for *C. thyoides* and *A. rubrum* do not differ greatly (Table 10.3). Although *A. rubrum* sapling densities are a little higher, the difference does not indicate a great enough difference in shade tolerance to suggest a transition from *C. thyoides* to *A. rubrum* in all mature stands. Many *A. rubrum* saplings were stem sprouts from dead or dying individuals and did not appear very vigorous. Seedling and sapling densities for other hardwood species were low.

The mature stand would thus likely be dominated by *C. thyoides* mixed with *A. rubrum* and other less dominant hardwood species. Mixed-aged stand structures in Featherbed Branch cedar swamp in this study and in Massachusetts and Florida cedar swamps (Laderman 1989) appear to support this pattern of younger *C. thyoides* trees intermixed with older, even-aged trees.

Traditional ideas of succession and community structure development may not be applicable to our understanding of wetland systems (Mitsch and Gosselink 1986). It is appropriate to view cedar swamps as "individual units in which interactions between the disturbance regimes, as described by the intensities, frequencies, and variations in disturbance and the life history, physiology, and behavior of the species influence succession pattern" (Denslow 1985).

Effects of Fragmentation

Few significant differences were detected in species diversity, tree importance, and shrub densities among the different-sized swamps and surrounding community types. The absence of distinct community structure differences resulting from differences in size or surrounding community type may be due, in part, to site variations. Water table and disturbance history variations result in considerable heterogeneity among cedar swamps and may obscure subtle differences. Another reason may be the presence of species with characteristics that enable populations to maintain themselves in patchy environments. Characteristics of such species include high dispersal rates, ability to resprout, high growth rates, early reproduction, and high reproductive rates (Baker and Stebbins 1965, MacArthur and Wilson 1967). These characteristics are associated with dominant trees and shrubs in New Jersey cedar swamps and increase the probability that isolated patches will be maintained.

Fragmentation effects are likely to be greater in the Protection Area than in the more isolated, less disturbed Preservation Area in the center of the New Jersey Pinelands. Increased disturbance by development leads to greater species richness with an increased occurrence of non-Pinelands species, particularly herbaceous species (Schneider and Ehrenfeld 1987). Fragmentation effects on composition and community structure are present in oak–pine forests, a vegetation type located around the perimeter of the New Jersey Pinelands (Gibson et al. 1988). The minimal effect of fragmentation on composition and community structure in cedar swamps in the Preservation Area may reflect this greater isolation.

Although this study has not identified significant differences among existing different-sized swamps, it does not discount the possibility of significant differences over time that can no longer be determined. Significant differences may be identified with additional or continued long-term studies of fragmentation effects in these cedar swamps. The possibility of changing community structure and species relationships in a system where repeated disturbance has altered community structure over an extended period of time needs to be considered in the description of ecosystem structure and function in relation to fragmentation.

Conclusions

Tree diversity is low in cedar swamps, with the canopy layer in some swamps composed almost entirely of *C. thyoides* and, in others, *C. thyoides* along with the primary associated tree species *A. rubrum, M. virginiana, N. sylvatica,* and *B. populifolia.* The shrub layer is more diverse than the tree layer and is dominated by *C. alnifolia, R. viscosum, G. frondosa, V. corymbosum,* and *L. racemosa.* The degree of variability in species composition and distribution within the swamps appears to be determined primarily by disturbance history and hydrology. Differences in burn intensity within the swamp, logging, location of seed sources, effect of deer browsing, and variation in the water table all contribute to differences in the relative abundance of the species present, resulting in increased heterogeneity within the swamps.

Large precolonial cedar swamps have become smaller and more isolated as a result of the cumulative effects of development, harvesting, deer browsing, changes in fire frequency, and hydrologic changes. In order to stop habitat loss and maintain existing cedar swamp area, it is important to provide optimal conditions for cedar regeneration. Factors that encourage *C. thyoides* regeneration include fire occurrence at intervals greater than 75 years, logging at intervals greater than 75 years or logging prevention, seasonal flooding, moderate water table fluctuation, deer browsing pre-

Table 10.6 Physical and Biotic Factors that Influence *C. thyoides* Regeneration

Prevents *C. thyoides* Regeneration	Encourages *C. thyoides* Regeneration
Repetitive severe fires	Fires at intervals > 75 years
Repetitive logging or logging with no slash removal	Logging at intervals > 75 years or logging prevention
Continued flooding	Seasonal flooding
No water table fluctuation or extreme fluctuation	Moderate water table fluctuation
Heavy deer browsing	No deer browsing
Increased *A. rubrum* abundance	Reduced *A. rubrum* abundance
High percent shrub cover	Low percent shrub cover

vention, low *A. rubrum* abundance, and minimal shrub cover (Table 10.6). Where logging continues, selection of stands or stand portions for clearcutting needs to be based on the preceding criteria. These criteria are also applicable in the assessment of potential impacts from development and other disturbances on plant community structure of cedar swamps. Increased protection and management planning in Pinelands wetlands (Roman and Good 1990, Roman et al. 1990) are also necessary in providing for long-term protection of the cedar swamp ecosystem.

Acknowledgments This study was funded in part by the NJ DEP Department of Parks and Forestry (Nos. 85 5573 and 86 4–27343) and NSF grant BSR-8605182 for the Rutgers University Pinelands field station.

In Memory This chapter is in memory of Ralph E. Good, whose influence was critical in developing our present understanding of the Pinelands ecosystem and in initiating continuing research efforts. The continued preservation of this unique ecosystem will always be a tribute to him.

References

Baker, H.G., and G.L. Stebbins. 1965. The genetics of colonizing species. Academic Press, New York.
Buell, M.F., and R.L. Cain. 1943. The successional role of southern white cedar, *Chamaecyparis thyoides,* in southeastern North Carolina. Ecology **24**:85–93.
Cottam, G., and J.T. Curtis. 1956. The use of distance measures in phytosociological sampling. Ecology **37**:451–460.
Day, F.P., Jr. 1987. Effects of flooding and nutrient enrichment on biomass allocation in *Acer rubrum* seedlings. Am. J. Bot. **74**:1541–1554.
Denslow, J.S. 1985. Disturbance-mediated coexistence of species. Pages 307–323 *in* S.T.A. Pickett and P.S. White, eds. The ecology of natural disturbance and patch dynamics. Academic Press, New York.
Ehrenfeld, J.G., and J.P. Schneider. 1983. The sensitivity of cedar swamps to the effects of

non-point source pollution associated with suburbanization in the New Jersey Pine Barrens. Center for Coastal and Environmental Studies, Rutgers Univ., New Brunswick. Unpublished report.

Fairbrothers, D.E. 1979. Endangered, threatened, and rare vascular plants of the Pine Barrens and their biogeography. Pages 229–244 *in* R.T.T. Forman, ed. Pine Barrens: ecosystem and landscape. Academic Press, New York.

Forman, R.T.T., and R. Boerner. 1981. Fire frequency and the Pine Barrens of New Jersey. Bull. Torrey Bot. Club **108**:34–50.

Gibson, D.J., S.L. Collins, and R.E. Good. 1988. Ecosystem fragmentation of oak-pine forest in the New Jersey Pinelands. For. Ecol. and Man. **25**:105–122.

Harshberger, J.W. 1916. The vegetation of the New Jersey Pine Barrens: an ecological investigation. Christopher Sower, Philadelphia.

Korstian, C.F., and W.D. Brush. 1931. Southern white cedar. U.S. Dept. Agric. Tech. Bull. **251**:1–75.

Laderman, A.D. 1989. The ecology of Atlantic white cedar wetlands: a community profile. U.S. Fish Wildl. Serv. Biol. Rep. **85**(7.21).

Little, S. 1950. Ecology and silviculture of white cedar and associated hardwoods in southern New Jersey. Yale Univ. Sch. For. Bull. **56**:1–103.

Ludwig, J.A., and J.F. Reynolds. 1988. Statistical ecology. John Wiley, New York.

MacArthur, R.H., and E.O. Wilson. 1967. The theory of island biogeography. Princeton Univ. Press, Princeton.

Mitsch, W.J., and J.G. Gosselink. 1986. Wetlands. Van Nostrand Reinhold, New York.

Roman, C.T., and R.E. Good. 1990. A regional strategy for wetlands protection. Pages 1–5 *in* D.F. Whigham, R.E. Good, and J. Kvet, eds. Wetland ecology and management: case studies. Kluwer Academic Publishers, The Netherlands.

Roman, C.T., R.E. Good, and S. Little. 1990. Ecology of Atlantic white cedar swamps in the New Jersey Pinelands. Pages 163–173 *in* D.F. Whigham, R.E. Good, and J. Kvet, eds. Wetland ecology and management: case studies. Kluwer Academic Publishers, The Netherlands.

SAS. 1982. SAS user's guide: statistics. SAS Institute, Cary, NC.

Schneider, J.P., and J.G. Ehrenfeld. 1987. Suburban development and cedar swamps: effects on water quality, water quantity and plant community composition. Pages 271–288 *in* A.D. Laderman, ed. Atlantic white cedar wetlands. Westview Press, Boulder.

Stoltzfus, D.L. 1990. Development of community structure in relation to disturbance and ecosystem fragmentation in Atlantic white cedar swamps in the Pinelands National Reserve, New Jersey. Dissertation, Rutgers Univ., New Brunswick.

Stone, W. 1911. The plants of southern New Jersey, with special reference to the flora of the Pine Barrens and geographical distribution of the species. N. J. State Museum Ann. Rept. **1910**:23–828, Trenton.

Vermeule, C.C. 1900. The forests of New Jersey. Pages 13–101 *in* New Jersey Geological Survey, Ann. Rept. State Geol. for 1899. Trenton.

11

Ronald W. Phillips, Joseph H. Hughes, M.A. Buford,
W.E. Gardner, F.M. White, & C.G. Williams

Atlantic White Cedar in North Carolina, USA

A Brief History and Current Regeneration Efforts

Chamaecyparis thyoides (L.) BSP. (Atlantic white cedar), known locally as juniper or white cedar, is an obligate wetland species whose natural botanical range extends along the Atlantic and Gulf Coasts of the United States from southern Maine to northern Florida and through Mississippi. In eastern North Carolina, it has been an economically and ecologically important component of forested swamps, freshwater bogs, pocosins, wet depressions, and streambanks.

Only small remnants are left of the Atlantic white cedar forests that once occupied a significant expanse of North Carolina's eastern coastal plain. White cedar acreage in North Carolina may have declined by as much as 90% in the last two centuries (Frost 1987). From 1880 to 1930 intensive exploitation occurred, with heavy logging in swampy areas where land was considered worthless after the timber had been removed (Lilly 1981). This practice, combined with natural regeneration failure and an absence of artifical regeneration, drainage impacts, fire exclusion, and a lack of competition control has led to the present depleted status of white cedar in North Carolina.

This chapter describes joint research efforts by the U.S. Forest Service, Weyerhaeuser, and North Carolina State University to grow Atlantic white cedar. First, we present a brief history of white cedar use in North Carolina.

History of Exploitation

Historically, the largest concentrations of Atlantic white cedar occurred in North Carolina. The strong, lightweight, easily worked, decay resistant, aromatic wood is still preferred for boat planks, house siding, shingles, tank boards and fencing.

As chronicled by Little (1950), Frost (1987), Baines (1989), and others, there have been three waves of exploitation of the once-vast Atlantic cedar resource in North Carolina: the arrival of the European settlers (1653–1750), the post-Civil War Reconstruction (1865–1910), and a recent period late in the 20th century (1970–1980). There was little impact on local stands in North Carolina during initial European settlement, although northern Atlantic United States stands were heavily logged. The two latter periods proved more destructive to the extensive southern Atlantic white cedar stands.

Up through the 1880s, harvest had been low technology (i.e., by hand). After the Civil War, changes in harvest technology began to drive the history of white cedar stands. In the Reconstruction era, logging occurred with the advent of steam-powered trains and dredges (Earley 1987, Frost 1987). One supplier of North Carolina Atlantic white cedar was a company that bore its founder's name, John L. Roper. The Roper Lumber Company became the largest supplier of North Carolina Atlantic white cedar to the northeastern United States for building materials (Krinbill 1956).

In the 1930s, steam-powered logging trains ran on light railroads through stands of white cedar at 250-yard intervals. Along the railroad, "breadths" of 25 yards (75 meters) were marked off, and survey lines were cut every 125 yards at right angles to the railroad. Individual loggers were assigned to these breadths where each cut, bucked, and moved logs to the railroad, to be stacked for loading (see Figs. 11.1– 11.3).

This method of logging was used until the remaining stands were too small and isolated to be economically harvested with the available technology. A lull in cedar logging ensued from the 1920–30s through the 1970–80s. The majority of cut stands did not regenerate; however, some did return to Atlantic white cedar (John F. Williams, personal communication). By the 1970–80s, these regenerated stands were 70 to 90 years old. But it was not the volume of the stands that allowed the third wave as much as the arrival of new technology: hydraulic equipment and wide, 6-foot (1.8-meter) tracks. This technology allowed the last, or third, wave of cutting.

By this time, the Roper Lumber Company was no longer in business. Other companies, including agribusinesses (Carter 1975) and timber companies, such as the Atlantic Forest Products Company, found that new logging technology provided cost-effective methods to extract Atlantic white cedar. Mechanical feller-bunchers, portable sawmills, and dredges were readily used. At the peak of the third wave, Atlantic Forest Products ran 13 million board feet (bd ft) a year through its saws.

Unfortunately, these swamps and bogs almost never regenerated back to Atlantic white cedar after timber harvest. After each wave of harvesting, fewer stands regenerated. Drainage patterns were altered by clearing for agriculture, blockage, elimination of wildfires, and replacement by competing vegetation. The result has been damaging; white cedar acreage has declined by more than 90% from its original area.

Today it is estimated that only about 170 to 180 million bd ft of usable Atlantic white cedar remain in North Carolina, a little more than a decade's supply at the current rate of 19 million bd ft brought to market annually (see also Sheffield et al., Chapter 8). To compound the problem, there continues to be a strong market for white cedar because it is highly sought for boat building and appearance lumber in housing. As Baines (1989) notes in his survey of forest products companies in North Carolina, the prices are high. The 1991 retail price on finished lumber was at least

Figure 11.1. "A bunch of fine juniper (white cedar) growing on the holdings of the John L. Roper Lumber Co. in the famous Dismal Swamp of Virginia and North Carolina." *American Lumberman* 1907.

$750 to $1,000 per thousand bd ft, if it could be found. By contrast, *Pinus taeda* (loblolly pine), a commercial plantation species, retailed at $450 per thousand bd ft.

In 1989, the Nature Conservancy listed Atlantic white cedar as globally endangered because (1) it exists only in a shadow of its original range, (2) it does not regenerate easily, and (3) it continues to be harvested. Its endangered status reflects concern not only for survival of the species but also for the wetlands and wildlife habitat provided by cedar bogs and for the future supply of white cedar lumber.

Atlantic white cedar is also potentially valuable in wetlands restoration and creation; it is registered as an obligate wetlands species on the federal plant list. An area

A

B

Figure 11.2. A. "Logging road through juniper swamp near Roper, North Carolina." B. "Steam skidder No. 4 on juniper swamp operations near Roper, North Carolina." *American Lumberman* 1907.

A

B

Figure 11.3. A. "One of the many log trains arriving daily at the New Bern Mill of the John L. Roper Lumber Co." B. "Birdseye view of the juniper (white cedar) log pound of the John L. Roper Co.'s Juniper Mill at Roper, North Carolina, showing the remarkably fine character of logs from which the famous 'Roper' shingles and lath are made." *American Lumberman* 1907.

may be designated as a wetland by the presence of Atlantic white cedar and other indicator species, hydric soil type, and hydrological patterns according to federal wetlands delineation guidelines. Under the "no net-loss" wetlands policy of the federal Clean Water Act, wetlands lost through any activity must be restored or new wetlands must be created to take their place. There have been few examples of successful Atlantic cedar stand creation (Laderman 1989, Buford et al. 1990), but more attempts are in progress as planting stock becomes available.

Regeneration Problems

Atlantic white cedar usually fails to regenerate naturally after logging if no measures are taken to control competing vegetation. Most literature indicates that white cedar is intolerant of shade. Korstian and Brush (1931) reported an absence of white cedar seedlings in the understory of closed-canopy stands. Little (1950) found that seedlings growing under closed-canopy conditions survived less than three years. Current Atlantic white cedar restoration work in the Dismal Swamp (in Virginia and North Carolina) imitates the natural disturbances that might have historically regenerated white cedar. This work consists of overstory removal, site preparation, natural regeneration, and, if necessary, seedling release using herbicides or mechanical methods to kill unwanted vegetation (Carter 1987).

In North Carolina, the species occurs in several situations: (1) dense, relatively pure stands with little understory; (2) stands supporting a dense understory of *Gordonia lasianthus* (loblolly bay), *Ilex coriacea* (shining gallberry), *Ilex glabra* (gallberry), *Lyonia lucida* (fetterbush), and *Persia borbonia* (red bay); and (3) in small stands and as scattered individuals in mixed bottomland hardwood forests throughout the Sandhills region. Atlantic white cedar usually occurs on sites characterized by either saturated, acidic, deep organic soils or wet sandy soils. Across its range, white cedar is most frequently found on deep peat deposits underlain by a sandy substrate. Observations by Akerman (1923) in Virginia and Waksman (1942) in New Jersey indicate that white cedar may grow best under those conditions. Korstian and Brush (1931) corroborated that assessment and further concluded that as the percentage of clay and silt increased, the proportion of hardwoods increased until white cedar apparently could no longer compete.

Herbivory by *Odocoileus virginianus* (white-tailed deer), *Sylvilagus* spp. (rabbits), and *Reithrodontomys* spp. (meadow mice) can be a serious problem to successful establishment of white cedar (Little 1950, Laderman 1989). Several plantings in North Carolina have failed because of deer browsing. To reduce browsing, mechanical barriers, such as electric fences and seedling cages, as well as repellents have been used with some success.

In general, white cedar has been considered so difficult to plant that natural regeneration, where possible, has remained the preferred option (Little 1950, Laderman 1989). Therefore, there have been few efforts before 1994 to artificially establish the species. In North Carolina, plantings initiated since the 1960s on the Hofmann Forest (Jones and Onslow Counties) and Bladen Lakes State Forest (Bladen County) have largely been unsuccessful or unimpressive. One planting on the Hofmann Forest was destroyed by wildfire in 1972. A second Hofmann Forest planting had only marginal survival, and only remnant individual trees remain. However, recent plantings in the 1980s by Weyerhaeuser Company and the U.S. Forest Service in eastern North Carolina (described later) reflect the growing interest in the species.

Forestation with *C. thyoides*

The widespread failure of white cedar to regenerate naturally after harvest when its habitat has been altered, combined with human exploitation, has created some urgency for artificial regeneration. However, wetlands legislation, the species' endangered status, and a strong market for sawn timber weigh in favor of its recovery. Successful establishment of white cedar plantations could contribute significantly to the resource from which further work, both research and production, might continue.

Value-added Conservation In the summer of 1989, Duke University and Weyerhaeuser Company commissioned a regional survey (Baines 1989) to determine attitudes and motives for those individuals and organizations participating in the comeback of Atlantic white cedar. The 13 respondents included state agencies (the North Carolina Division of Forest Resources and Natural Heritage Program, East Carolina University, North Carolina State University, and the North Carolina Mariner's Museum), federal agencies (the U.S. Army Corps of Engineers and Marine Corps), industries (employees of Weyerhaeuser Company, Goldsboro Milling Company, TexasGulf Company, and several private forestry consulting firms), and one wildlife magazine author.

The survey revealed diverse motives, but profit was the main reason for interest in white cedar recovery. Atlantic white cedar plantations were seen as an opportunity to generate income in coastal North Carolina counties unlikely to attract much commercial interest otherwise. As Baines (1989) noted, 7 of the 13 cited the profit motive or economic benefit as the reason for their interest in Atlantic white cedar. These respondents discussed the high merchantable value of the timber, increase in jobs, and increased tax base in counties with white cedar habitat.

Based on this survey, conservation of the remnant stands of Atlantic white cedar appears to be strongly motivated by profit or the "value-added" opportunity. Conservation is perceived as a means to bring jobs and a stronger tax base into coastal rural communities that have a preponderance of wetlands or marginal agricultural lands. Indirectly, the profit motive could save an endangered species.

Clearly, Weyerhaeuser has a profit motive, but it has sprung from unlikely sources within the company. The company owns a few isolated groves of white cedar. These are inclusions within pine and hardwood forests that were purchased to create the company landbase during the past 50 years. These cedar groves are sparse remnants of large cedar stands that failed to regenerate at the turn of the century. The species provides only a small portion of the company's total receipts, but it has produced occasional log sales to local markets in the past. No company mills produce Atlantic white cedar products.

Weyerhaeuser's involvement in Atlantic white cedar began as a "grass roots" effort on the part of concerned employees in the late 1970s. Several employees transplanted wild seedlings from company lands to their own gardens and yards. Surprisingly, the seedlings grew extremely well on agricultural soils and in response to cultivation. The best trees were 35–40 feet (10.6–12.1 m) tall after 15 years, far exceeding the normal growth rate in a natural cedar bog. The results were encouraging, so white cedar was studied as an alternative species on the company's fee ownership in coastal North Carolina.

In 1981 and 1982, employees attempted to grow bare-root and containerized nursery seedlings on a limited scale. Seed viability and poor germination were a problem, but the nursery produced a few thousand seedlings. These seedlings, planted as demonstration trials on sites prepared for loblolly pine, grew well using standard loblolly pine silvicultural practices. Nearly all were destroyed by deer browse during the winter after planting. With no practical deer-control method, white cedar pilot plantings were halted.

In early 1988, Weyerhaeuser was asked to participate in the recovery of the white cedar timber supply for boat building as a community goodwill gesture. By this time, the threatened status of white cedar had become a concern not only of those using the lumber but of the wetlands conservation community in general. To Weyerhaeuser's silvicultural research group, it was apparent that the only way to increase the supply of white cedar would be through plantation culture. However, the species was too difficult to regenerate through natural or nursery seeding to support plantation culture.

In the following years, Weyerhaeuser's North Carolina foresters, seedling sales, and research groups directed tree-growing skills developed for pine management to the recovery of white cedar through plantation culture. The remaining natural white cedar groves were set aside for gene conservation.

Rooted cuttings technology has now been successfully developed on a pilot scale for white cedar; up to 28,000 cuttings per year have been produced from shoots taken from young Atlantic white cedar seedlings that serve as motherstock. The seed or seedlings collected for the motherstock are collected on Weyerhaeuser lands or in stands where owners have granted collection privileges. The source of the seed is tracked with regard to tract and county of origin. Large-scale seedling production continues to be difficult and does not have the reliability or flexibility of rooted cuttings.

A market for Atlantic white cedar propagules has developed since this work was begun. In 1994, Weyerhaeuser grew 280,000 rooted cuttings for commercial sale. The planting was up from the previous years' 28,000, in response to high demand for use in wetland mitigation and restoration. The Clinton administration's 1993 wetland policy included a provision for wetland banking. This spurred the demand for cedar to fulfill mitigation cases.

In December 1988, a meeting of North Carolina State University Extension Forestry, North Carolina Division of Forest Resources (NCDFR), and Weyerhaeuser Company was held in northeastern North Carolina to discuss the status of white cedar and to consider initiating demonstration plantations. Research interests included comparison of seedlings versus rooted cuttings. The resultant project, described below, continues as a cooperative effort among these groups.

Forestation with *C. thyoides* Rooted Cuttings and Bare-Root Seedlings

Material and Methods

Site Descriptions Demonstration plantings were installed on a range of sites, from wet mineral to deep peat soils. Previous land use on the sites chosen ranged from cutover woodland to highly productive farmland. The demonstration sites were planted to white cedar in 1989, 1990, and 1991. Most sites with intense competition received directed herbicide treatments for weed control to promote white cedar survival and ensure growth. Repellents and/or physical barriers were used to help prevent deer browsing. A brief description of the demonstration sites follows (Table 11.1).

SITE 1: Hofmann Forest, Onslow County, North Carolina is a pocosin (characterized by organic soil, a high water table, and a predominance of wetland plants, in its

Table 11.1 Site Descriptions, Site Preparation, and Previous Land Use of the Sites Chosen for Atlantic White Cedar Artificial Regeneration

Location	Soil Series	pH	% Organic Matter	Previous Use and Preparation
Hofmann (bedded and non-bedded)	Croatan	3.2	25–55	cutover pocosin pond pine ditched, sheared, windrowed and partially bedded; planted pine 1987
Camp Lejeune	Torhunta	4.0	7–11	ditched, harvested woodland; windrowed and bedded 1984; planted pine 1985; burned 1988
Bertie County	Johnston	4.4	7–12	ditched and cultivated farmland; sod cover; ripped prior to planting
Chowan County	Roanoke	4.4	4–5	harvested woodland; sheared and windrowed prior to planting
Pitt County	Rains	6.9	6–9	harvested forestland; ditched and converted to pasture 1980; bedded prior to planting

natural state) that had been converted from natural vegetation to a loblolly pine plantation. The one-year old planted pines that survived were purposely destroyed following the white cedar planting. In addition to draining, site preparation (for loblolly pine establishment) included pushing and piling logging debris into windrows. The site was "bedded" by the cooperator prior to planting ("beds" are elevated strips of soil pulled up by a tractor and plow during site preparation to improve drainage and weed control). The soil, a Croatan muck, was a deep organic with a low pH value (3.0–3.5). Depth of organic matter exceeded 1.2 m with organic matter content ranging from 25 to >55% by weight.

SITE 2: Hofmann Forest, Onslow County, North Carolina is adjacent to the site described above. With the exception of bedding, site preparation was the same as for Site 1.

SITE 3: Camp Lejeune Marine Base, Onslow County, North Carolina is cutover woodland that was originally planted to loblolly pine in 1985 following ditching and bedding, but the pine trees were largely killed by wildfire in 1988. The soils were either Torhunta or Onslow Series mineral soils with organic matter accumulations of 7–11% by weight in the upper 20.3 cm of the profile.

SITE 4: The North Carolina State University Agricultural Research Station in Bertie County, North Carolina had been ditched and intensively managed for annual agricultural crops. Mineral soils, of the Johnston series, had organic matter accumulations in the upper horizon of 7–12% by weight. The pH values ranged from 4.3–4.5.

SITE 5: The Bunch Tract, Chowan County, North Carolina is cutover woodland. The site was prepared by pushing logging debris into windrows. Soils, of the Roanoke series, were mineral with low pH values and may have been affected by nearby minor agricultural drainage. The seedlings were flat planted (not bedded), and competition was mainly from woody vegetation and sedges. Organic matter ranged from 4–5% by weight in the top 20.3 cm of the soil profile.

SITE 6: Monk Brothers Farm, Pitt County, North Carolina was converted from forest to pasture in about 1980. The soils, of the Rains series, were mineral with abnormally high pH values (6.5–7.0), probably due to intensive agricultural management. The site was bedded immediately prior to planting. Organic matter ranged from 6–9% by weight in the top 20.3 cm of the soil profile.

One-year old seedlings produced by the NCDFR were hand-planted at each site in alternate rows on April 10–13, 1989, using spade-shaped planting bars. Containerized rooted cuttings, supplied by Weyerhaeuser Company, were planted at each site on May 16–18, 1989, on the remaining alternate rows using round planting bars. Tree spacing in these nonreplicated trials ranged from 1.5–1.8 m within rows and 2.4–3.6 m between rows. Distances between rows on bedded sites were predetermined by bed location. Seedling and cutting heights were recorded immediately after planting.

Physical barriers were installed at all sites to test the effects of deer browsing on height growth and survival. Control row plots were not caged for comparison. Polyethylene mesh cages (46 cm) were used as physical barriers to deer browsing, although at Camp Lejeune, 1.2 m tall cages of woven wire were installed and repellents were used. Herbicides were applied on Chowan, Pitt, and Bertie County sites on June 23, July 3, and July 13, respectively. Arsenal (registered trade name for Imazapyr, manufactured by the American Cyanamid Company) herbicide was applied at 295 ml/ha in 1.2 m wide bands over tree rows with a backpack sprayer. The Hofmann Forest site did not develop enough competition the first year to warrant control, and the Camp Lejeune site was not sprayed because of a policy prohibiting herbicide use.

To evaluate each plantation, height and survival were measured after the first, second, and third growing seasons. On each site, one row of seedlings and one row of cuttings adjacent to the seedling row ("measurement rows") were selected for measurement. Measurement row length ranged from 750 m at Hofmann Forest to 1,250 m at Chowan County. The total height of each tree on the measurement row was measured to the nearest centimeter, and percent survival was calculated by dividing the number of living trees in each measured row by the total number of trees planted in the row.

In the second year of the study (1990), a second cycle of planting was accomplished. Seedlings and rooted cuttings were planted in alternate rows April 3–5 and May 1–3, respectively, using the same methods as for the first year. Tree heights were again tallied on measurement rows immediately after planting.

On the Camp Lejeune site, large (1.2 m) wire cages were again installed and chemical retardants applied. The other five sites received no mechanical barriers or chemical retardants due to the absence of any significant predation the first year.

Arsenal, at 295 ml/ha, was tank-mixed with Oust (registered trade name for Sulfometuron methyl, manufactured by the Dupont Company) at 140 g/ha and applied April 18 in Pitt County and April 27 in Chowan County. A later application of Arsenal at 4 ounces per acre was applied on the Bertie County site.

First- and second-year height and survival measurements were taken for the 1990 planting. The methods of measuring were similar to the methods used for the 1989 planting.

Discussion of herbicides used in this project does not imply that such uses are registered or recommended. These applications were experimental in order to ascertain the tolerance of the studied species and to learn how well they respond to chemical release.

Results

1989 Planting   The sites with the most acidic soils (Hofmann Forest) supported the best-performing trees following two growing seasons (Table 11.2). Seedling mean height on the Hofmann bedded site averaged 141.7 cm, while seedling mean height on the nonbedded site averaged 132.3 cm. The next-best site was at Camp Lejeune, where mean seedling height was 87.6 cm after two years. Rooted cutting mean heights on the Hofmann bedded and nonbedded sites averaged 117 and 96.5 cm, respectively. These means are higher than the next best-performing site at Bertie County, where cutting mean height was 68.8 cm.

The Camp Lejeune, Bertie County, and Chowan County sites had similar performance results. Seedling mean height averaged 73.4 cm in Chowan County, 87.6 cm in Camp Lejeune, and 78.2 cm in Bertie County. Cutting mean heights are also comparable on these sites, with averages ranging from 63.7 cm at Camp Lejeune to 68.8 cm at Bertie County. The Pitt County site was characterized by poor performance (seedling mean height 48.5 cm) after two growing seasons.

A relationship between soil pH and tree survival appears similar to that of height performance (Fig. 11.4). Percent survival is highest, exceeding 90%, on sites with the lowest pH. Survival on the Hofmann bedded and non-bedded sites for both seedlings and cuttings ranged from 92–97% percent after two years. Seedling and cutting survival was on the Camp Lejeune site (soil pH 4.0) was 91 and 92%, respectively, after two years. Most mortality on these sites occurred the first year with negligible or no mortality in year two. Seedling and cutting survival was poorer on the Chowan and Bertie County sites compared to the Hofmann and Camp Lejeune sites.

Table 11.2 Atlantic White Cedar Planted 1989. Year 1 and Year 2 Mean Tree Heights by Site for Seedlings and Cuttings Showing the Relationship to Soil pH

| Site | Soil pH | Average Height (cm) | | | |
| | | Year 1 | | Year 2 | |
		Sdlgs	Ctgs	Sdlgs	Ctgs
Hofmann, bedded	3.2	49.7	38.3	141.7	117.0
		n = 60	n = 60	n = 40	n = 40
Hofmann, non-bedded	3.3	56.1	33.2	132.3	96.5
		n = 60	n = 60	n = 40	n = 40
Camp Lejeune	4.0	44.7	33.2	87.6	63.7
		n = 100	n = 100	n = 60	n = 60
Bertie County	4.4	36.5	29.4	78.2	68.8
		n = 80	n = 80	n = 70	n = 70
Chowan County	4.4	29.4	23.6	73.4	64.5
		n = 100	n = 100	n = 50	n = 50
Pitt County	6.9	27.4	19.8	48.5	42.4
		n = 100	n = 100	n = 30	n = 30

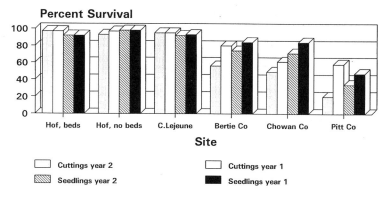

Figure 11.4. Percent survival of Atlantic white cedar seedlings and cuttings after one and two growing seasons, 1989 planting.

Seedling survival after two years on the Chowan County site was 71%, while cutting survival was 52%. Seedling survival on the Bertie County site was 74%, while cutting survival was 55%. Unlike the Hofmann and Camp Lejeune sites, mortality continued following the initial year of establishment. Poorest survival was at the Pitt County site where seedling survival was 34% and cutting survival has 20% after two years.

In general, seedling mean height was slightly better than cuttings (Table 11.2). The largest differences exist on the Hofmann and Camp Lejeune sites. Seedling mean height on the Hofmann bedded site averaged 141.7 cm compared to cutting mean height on the same site of 117 cm. On the Hofmann nonbedded site, seedling mean height averaged 132.3 cm compared to cutting mean height of 96.5 cm. Likewise, seedling mean height on Camp Lejeune was superior to cutting mean height (87.6 cm vs. 63.7 cm, respectively). Mean height differences between seedlings and cuttings were smaller on the remaining sites.

Smaller differences occurred between seedling and cutting survival. The largest difference was on the Chowan County site where seedling survival after two years was somewhat higher compared to cutting survival. Cutting survival was slightly higher than seedling survival in Bertie County.

1990 Planting Preliminary observations after one growing season were made on all six plantings established on the same sites in 1990 (Table 11.3). In many respects, these results reflect trends that are similar to the 1989 results. Again, the Hofmann site showed superior average heights compared to other sites. Camp Lejeune did not outperform other sites with moderately acid soils. This may be attributable to heavier weed competition developing on the Camp Lejeune site in the absence of any chemical control, or it may reflect improved weed control on the other sites. Once again, the Pitt County site performance was worse than all other sites. Survival after one growing season appeared adequate to excellent on all sites except the Pitt County site, where cutting survival (53%) exceeded seedling survival (23%). Seedlings,

Table 11.3 Atlantic White Cedar Planted 1990. Year 1 Mean Tree Heights and Survival by Site for Seedlings and Cuttings

Site	Average Height (cm)		Survival	
	Seedlings	Cuttings	Seedlings	Cuttings
Hofmann, bedded	72.8	55.6	100.0	90.0
Hofmann, non-bedded	55.6	43.4	97.5	100.0
Camp Lejeune	44.1	32.0	90.0	90.0
Bertie County	46.7	35.8	98.5	90.0
Chowan County	48.7	40.1	94.0	82.0
Pitt County	33.7	23.6	23.3	53.3

again, exhibited a small, but significant, advantage over cuttings. Both seedlings and cuttings showed excellent survival and acceptable growth on all sites except Pitt County.

Seedling and cutting survival were both excellent on most sites (Table 11.3). Similar to the 1989 planting, both seedling and cutting survival after two years exceeded 90% on the Hofmann and Camp Lejeune sites, with no additional mortality occurring after the first year. Seedling survival on the Bertie and Chowan County sites averaged 98.5 and 94.0%, respectively. Cutting survival on Bertie County is 90% compared to a lower cutting survival of 82% on Chowan County.

Discussion

Plantability　There is sufficient evidence to conclude that Atlantic white cedar can be successfully planted given the proper site, preparation, and care following establishment. Survival was excellent on those sites to which white cedar is naturally best adapted. No significant technical problems occurred during planting of the seedlings or cuttings. The same planting tools and techniques that are used in planting southern pine were effective, and there is no apparent need to develop other planting techniques. Seedlings planted in 1989 on the Hofmann site showed particular resilience in their ability to survive harsh conditions. On the day of planting, high winds, snow, and cold conditions hampered planting efforts. Many seedlings following planting were barely above water level, and many floated to the top of the water when planted and had to be replanted. One month later, a few seedlings were found lying on the ground with their roots exposed. These seedlings were subsequently replanted. In spite of these poor conditions, seedling survival on the Hofmann site remained well above 90% after three years.

Performance Related to Site Conditions　A correlation seems to exist between tree performance and some soil physical and chemical properties. Trees planted on sites with acidic, organic soil performed better. Naturally occurring white cedar in North Carolina is found predominately on deep organic soils characterized by poor drainage and high acidity. Croatan mucks found on the Hofmann site are characteristic of the soils where white cedar occurs. Tree performance at Hofmann Forest exceeded that at all other sites, which may reflect white cedar's adaptation to highly acid organic soils.

The Camp Lejeune, Chowan County, and Bertie County sites were comparable to each other, but poorer than the Hofmann site. Soils from these sites were largely mineral, but there were organic matter accumulations in the upper horizon. Soils from these sites were moderately acid, but not to the extent of the Hofmann site. Noticeably poor performance on any of these three sites may be the result of heavy woody or herbaceous competition. The Pitt County site offers a good example of white cedar's poor performance on high pH soils. Soils on this site were typically moderately acid; however, as a consequence of agricultural amendments, soil pH had been raised to approximately 7.0. We suspect that high pH values as well as intense weed and grass competition interfered with performance.

Deer Browsing To date, these studies have suffered surprisingly little damage from deer browsing, in spite of large populations of deer in the study areas. Only in isolated cases has any browsing been observed. Most tree shelters were removed during or after the first year, because the seedlings and cuttings were growing out of the tops or sides of the cages. In some cases, the cages were causing physical injury and/or stunted growth. The possibility of deer browsing remains, although this threat is considerably reduced as the trees continue to grow in height. Some of the reasons for the lack of browsing may include (1) other more preferred food nearby, (2) lack of any nearby edge effect (two different vegetative sites adjoining) since most of the plantings are within large cleared sites, and (3) lack of recognition of white cedar by deer, because white cedar may be a new food plant to them.

Conclusions

White cedar's natural regeneration failures and depletion underscore the value of intensive artificial regeneration to establish a base from which continued research can be conducted. This study has demonstrated that white cedar can be successfully established on appropriate sites suitably prepared for planting. Growth and survival are best on sites characterized by deep organic, low pH soils. Weyerhaeuser has found that rooted cuttings are more reliable and easier to grow than bare-root cedar seedlings, yet both appear to have potential for artificial regeneration of white cedar. Severe competition will have to be controlled in order to assure successful establishment. Herbivory problems remain an area of concern, even though no substantial browsing was observed in this test. Although preliminary information from this study points to the potential success of white cedar plantations, additional research is necessary to refine establishment recommendations. Sound, research-based management will be needed in order to restore Atlantic white cedar to its former position of ecological and economic importance.

Acknowledgments The sections on history and value-added conservation were written by Joseph H. Hughes, Claire G. Williams, and Marilyn A. Buford. The section on forestation with

C. thyoides rooted cuttings and bare-root seedlings was written by Ronald W. Phillips, W.E. Gardner, F.M. White, and J.H. Hughes.

References

Akerman, A. 1923. The white cedar of the Dismal Swamp. Va. For. Publ. **30**:1–21.

American Lumberman. April 27, 1907. A trip through the varied and extensive operations of the John L. Roper Lumber Co. in eastern North Carolina and Virginia. Pp. 51–114.

Baines, R.A. 1989. Prospects for white cedar: a North Carolina assessment. Duke University FOREM. Fall 1989, **13**(1):8–11.

Buford, M.A., C.G. Williams, and J.H. Hughes. 1990. Growth and survival of Atlantic white cedar on a South Carolina Coastal Plain site: first-year results. Pages 579–583 *in* S.S. Coleman and D.G. Niery, eds. Proc. 6th Biennial Southern Silvicultural Research Conference, Memphis.

Carter, A.R. 1987. Cedar restoration in the Dismal Swamp of Virginia and North Carolina. Pages 323–325 *in* A.D. Laderman, ed. Atlantic white cedar wetlands. Westview Press, Boulder.

Carter, L.J. 1975. Agriculture: a new frontier in coastal North Carolina. Science **189**:271–275.

Earley, L.S. 1987. Twilight for junipers. Wildlife in North Carolina **51**(12):9–15.

Frost, C.C. 1987. Historical overview of Atlantic white cedar in the Carolinas. Pages 257–264 *in* A.D. Laderman, ed. Atlantic white cedar wetlands. Westview Press, Boulder.

Korstian, C.F., and W.D. Brush. 1931. Southern white cedar. U.S. Dept. Agric. Tech. Bull. 251. 75 pp.

Krinbill, H.R. 1956. Southern white cedar: the forgotten tree. Southern Lumberman, November. Pp. 26, 28, 36, 45.

Laderman, A.D. 1989. The ecology of Atlantic white cedar wetlands: a community profile. Biol. Rep. 85 (7.21). U.S. Fish Wildl. Serv.

Lilly, J.P. 1981. A history of swamp land development in North Carolina. Pages 20–39 *in* C.J. Richardson, ed. Pocosin wetlands. Hutchinson Ross, Stroudsburg, PA.

Little, S. 1950. Ecology and silviculture of white cedar and associated hardwoods in southern New Jersey. Yale Univ. Sch. For. Bull. 56.

Waksman, S.A. 1942. The peats of New Jersey and their utilization. N.J. Dept. of Cons. and Devt. Geol. Series Bull. 55A.

12

Robert T. Eckert

Population Genetic Analysis of *Chamaecyparis thyoides* in New Hampshire and Maine, USA

Chamaecyparis thyoides (Atlantic white cedar) is a widespread but relatively uncommon eastern United States tree species, generally occurring in disjunct palustrine forested wetlands within 200 kilometers (km) of the coast. It ranges from Mississippi to its northernmost limit in Maine and New Hampshire (Baldwin 1963, Laderman et al. 1987). In New England, its occurrence is generally limited to glacial features such as moraine hollows, kettleholes, and old lake beds. New Hampshire has 30 swamp systems containing *C. thyoides,* 15 of which average 0.7 hectares (ha) in extent and 15 average 12 ha (Sperduto and Ritter 1994). Maine has 11 recorded stands (Laderman et al. 1987). In Maine and New Hampshire, the species exists in complex wetlands with areas of almost pure cedar, often containing hundreds of stems, separated by more sparsely distributed cedars interspersed with other tree species. Commonly associated tree species at the northern reach of its range are *Acer rubrum* (red maple), *Betula papyrifera* and *B. alleghaniensis* (paper and yellow birch, respectively), *Picea rubens* and *P. mariana* (red and black spruce, respectively), *Larix laricina* (tamarack), *Tsuga canadensis* (eastern hemlock), *Pinus strobus* (eastern white pine), and *P. rigida* (pitch pine). A recent New Hampshire inventory found 190 vascular plant species associated with *C. thyoides* (Sperduto and Ritter 1994). *Chamaecyparis thyoides* is known to grow very slowly, with eight-inch (20 cm) diameter trees over 200 years old reported by Baldwin (1963). Rate of reproduction and relative reproductive success have been studied only in New Jersey populations (Little 1950), but are described in some detail.

In the northern reaches of its range, the species is highly susceptible to loss because of changes in water level within swamps, often accompanied by lowering of water quality, especially when urbanization is a factor (Laderman 1989). Baldwin repeatedly (1961, 1963, and 1965) comments on the loss of New Hampshire stands known to have existed at earlier dates (Svenson 1929). Commercial harvesting of *C. thyoides* has a significant impact which is greatest in North Carolina, New Jersey, and the western panhandle of Florida. From Maine to Connecticut harvest is less intense (Laderman 1989), with current uses being primarily for posts and shingles. In both New Hampshire and Maine there has been a general decline of *C. thyoides* populations; however, some excellent examples still remain.

Chamaecyparis thyoides is the only member of its genus east of the Rockies and is an uncommon species within New Hampshire, but it is not listed as endangered by the state. Some representative habitat is under protective ownership in New Hampshire by towns and the Society for the Protection of New Hampshire Forests (SPNHF). The species is less abundant in Maine, but the swamps in York County— the Massabesic Experimental Forest and the Saco Heath—are under the jurisdiction of the U.S. Forest Service and The Nature Conservancy (TNC), respectively. TNC also protects portions of cedar swamps at Appleton Bog in Knox County, and the St. Claire Preserve in Waldo County, Maine (B. Vickery, personal communication).

Postglacial migration and establishment patterns of *C. thyoides* are likely to have influenced the genetic structure of populations. Rates of migration are unknown, but dissemination could be expected to be slower than other northeastern conifers since *C. thyoides* is restricted to disjunct swampy or boggy sites that are often isolated. Seed dispersal starts in the fall and may not be complete until the next spring (Laderman 1989), suggesting that transport across snowpack may be important. Belling (1977) discerned from her studies of macrofossils, by which *Chamaecyparis* can be distinguished from *Thuja* and *Juniperus* (Belling 1987), that although *C. thyoides* arrived 4,000 to 6,800 years before present (BP) in southern New England, there appear to have been times when, after its arrival, conditions were not conducive to expansion for periods of up to 1,400 years. Conditions were favorable for cedar in the south-central area of New Hampshire at least 4,000 yrs BP, but similar conditions were not present on the coast of New Hampshire until 380 yrs BP. Belling's evidence suggests that, farther south, *C. thyoides* followed the same pattern of establishment inland at sites in the Appalachian foothills and the adjacent Piedmont Plateau earlier than on the coastal plain.

The present distribution of northern populations in disjunct patches, with a hypothesized colonization history of long periods of establishment before relatively rapid expansion, suggests the potential for long periods of inbreeding followed by shorter periods of gene flow, with the potential for founder effects. The resulting genetic structure of these populations has the potential for being highly complex, and, given the macrofossil evidence for past migration, the genetic patterns may also reflect past migration patterns.

Any adequate conservation efforts for this species must include an evaluation of its genetic structure. The following presents an example of isozyme analysis useful in assessing the genetic variability of protein polymorphisms in *C. thyoides.* It is limited in scope since only a few northern populations of this widespread species were sampled, and the small sample sizes present some limitations in estimates of allele frequencies and detection of rare alleles. It does illustrate the kinds of information obtainable using this approach and should help the reader understand the complexity of population genetic structure, an understanding of which is needed for planning thorough conservation strategies. The work reported here describes genetic variation in populations of *C. thyoides* in New Hampshire and southern Maine.

Materials and Methods

Twenty trees were chosen in each of eight populations of *C. thyoides* (Fig. 12.1). Sample trees were at least 15 meters (m) from each other to increase the chances of including several "neighborhoods" of relatedness. Samples were taken from four populations at the U.S. Forest Service Massabesic Experimental Forest in Alfred, Maine. This cedar swamp complex covers an area of approximately 30 ha, but is made up of much smaller swamps separated variously from one another. Samples were also taken from four populations more widely separated within New Hampshire.

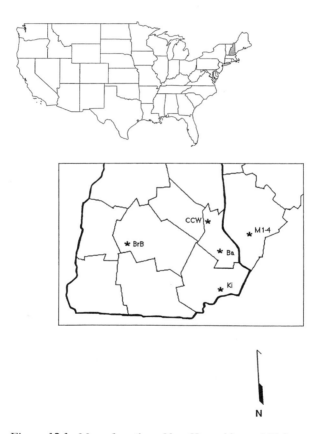

N

Figure 12.1. Map of southern New Hampshire and Maine showing locations of *C. thyoides* stands sampled for genetic analysis. County boundaries are outlined. Ba: Barrington cedar swamp. BrB: Bradford Bog. CCW: Cooper Cedar Woods. Ki: Kingston, Cedar Swamp Pond. M1–4: Massabesic stands.

Two of these are located on SPNHF property: Cooper Cedar Woods in New Durham, New Hampshire, is approximately 7 ha; Cedar Swamp Pond in Kingston, New Hampshire, is the largest stand sampled, at approximately 16 ha. Cedar Swamp Pond is part of a larger complex of five stands separated by less than 0.8 km, totaling 98 ha. The Bradford Bog in Bradford, New Hampshire, of approximately 10 ha, and a cedar swamp in Barrington, New Hampshire, about 6 ha, were also sampled.

Foliage samples for electrophoresis were collected from February to April 1988. Sun-exposed branches, not in contact with branches of neighboring trees, were removed with pole pruners and maintained on ice until extraction of proteins. No more than three hours elapsed between removal of branches from trees and refrigeration in the laboratory.

Foliage samples were extracted (O'Malley et al. 1987) and processed by starch-gel electrophoresis following procedures described by Gagnon et al. (1988). Extraction procedures used the extraction buffer described by Mitton et al. (1979), but modified by the addition of 0.1% bovine serum albumen and 12% glycerol. Samples were electrophoresed 5–6 hours on 12% starch gels in six different systems with pH values ranging from 5.7 to 8.8 to resolve different enzymes. Amperage was maintained at 50 mA, while voltage was maintained at different levels ranging from 50 v to 200 v maxima, depending on which enzyme systems were being analyzed. Variation in isozymes among individual trees was scored for a total of 34 alleles at 16 putative loci (Table 12.1). Designation of loci and alleles remains putative until assessment of haploid megagametophyte tissue from seed is conducted. However, no isozyme variation patterns on gels were used for this study that did not match patterns predicted by known protein structures and genetic models.

Allelic frequencies, heterozygosities, number of alleles per locus, chi-square analyses of frequencies, genetic distance, and inbreeding coefficients were calculated according to methods of Swofford and Selander (1981). Heterozygosity calculations estimate the proportions of homozygotes to heterozygotes, either at a given locus or over all loci for a given individual, and can be summarized for populations. The number of alleles per locus reflects within-population diversity at the locus level. It reflects any history of severe genetic "bottlenecks" caused by drastically reduced breeding population size, which would have resulted in loss or fixation of alleles.

The heterogeneity chi-square analysis presented in this study compares the observed gene frequencies against expected frequencies within loci, across populations. Statistically significant chi-square values indicate differences among populations at individual loci. Significant values suggest that gene frequency differences have arisen as the result of founder effects, isolation, and/or adaptation to specific environmental conditions.

Genetic distances among population samples were calculated as Cavalli-Sforza and Edwards (1967) arc distances. This measure is a transformation of allele frequencies such that populations are represented on a hypersphere in Euclidean space. Arc distances provide the measure of distance among populations. Slatkin (1985) indicates that this measure of genetic distance reflects gene flow and genetic drift, rather than gene flow and mutation rates, as are estimated by some other measures. This seems most appropriate for these measures when used in a conservation context in which populations are widely separated. With reduced gene flow among populations,

TABLE 12.1 Allele Frequencies in Eight New Hampshire and Maine Populations of *Chamaecyparis thyoides*

Locus[b]	M1	M2	M3	M4	CCW	Ki	Ba	BrB
				Population[a]				
ACO								
1	1.000	1.000	0.975	0.925	1.000	0.975	0.975	1.000
2	0.000	0.000	0.025	0.075	0.000	0.025	0.025	0.000
AAT-1								
1	1.000	1.000	1.000	1.000	1.000	1.000	1.000	1.000
AAT-2								
1	1.000	1.000	1.000	1.000	1.000	0.917	1.000	1.000
2	0.000	0.000	0.000	0.000	0.000	0.083	0.000	0.000
ADH								
1	0.875	1.000	0.950	1.000	0.825	1.000	1.000	0.950
2	0.125	0.000	0.050	0.000	0.175	0.000	0.000	0.050
DIA								
1	1.000	1.000	1.000	1.000	1.000	1.000	1.000	1.000
G2D								
1	1.000	1.000	1.000	1.000	1.000	0.950	1.000	1.000
2	0.000	0.000	0.000	0.000	0.000	0.050	0.000	0.000
IDH								
1	1.000	1.000	1.000	1.000	1.000	0.400	1.000	1.000
2	0.000	0.000	0.000	0.000	0.000	0.575	0.000	0.000
3	0.000	0.000	0.000	0.000	0.000	0.025	0.000	0.000
ME								
1	0.500	0.275	0.324	0.525	0.425	0.450	0.289	0.175
2	0.500	0.725	0.676	0.450	0.575	0.350	0.289	0.225
3	0.000	0.000	0.000	0.025	0.000	0.200	0.263	0.550
4	0.000	0.000	0.000	0.000	0.000	0.000	0.158	0.050
PGM								
1	0.300	0.225	0.275	0.350	0.500	1.000	0.342	0.275
2	0.175	0.175	0.150	0.200	0.375	0.000	0.447	0.300
3	0.525	0.600	0.575	0.450	0.125	0.000	0.211	0.425
6PG-1								
1	0.975	1.000	1.000	1.000	1.000	0.875	0.500	0.500
2	0.025	0.000	0.000	0.000	0.000	0.125	0.500	0.500
6PG-2								
1	0.632	0.500	0.825	0.750	0.775	1.000	0.975	0.875
2	0.368	0.500	0.175	0.250	0.225	0.000	0.025	0.125
PGI								
1	1.000	1.000	1.000	1.000	1.000	1.000	1.000	1.000
SKD								
1	1.000	0.850	0.868	0.763	0.575	1.000	0.600	0.778
2	0.000	0.075	0.000	0.000	0.000	0.000	0.000	0.000
3	0.000	0.075	0.132	0.132	0.350	0.000	0.200	0.194
4	0.000	0.000	0.000	0.105	0.075	0.000	0.200	0.028
SOD 1	1.000	1.000	1.000	1.000	1.000	1.000	1.000	1.000
UGPP-1								
1	0.950	1.000	1.000	1.000	0.925	1.000	1.000	1.000
2	0.025	0.000	0.000	0.000	0.000	0.000	0.000	0.000
3	0.025	0.000	0.000	0.000	0.075	0.000	0.000	0.000
UGPP-2								
1	1.000	1.000	1.000	1.000	1.000	1.000	1.000	1.000

[a] Populations M1–M3 are adjacent to each other in the USFS Massabesic Experimental Forest, Alfred, ME. M4 is part of the same complex, but is located approximately 0.3 km away. Populations CCW and Ki are located near Rochester and Kingston, NH, respectively. Populations Ba and BrB are located in Barrington and Bradford, NH, respectively. All populations, except M1–M4, are located at least 55 km apart.

[b] Loci are defined in the first paragraph of "Results."

the genetic distance between a pair of populations presumably increases. Arc distance values greater than 0.6 indicate relatively little gene flow, while values near zero indicate high levels of gene flow. The upper limit value for completely isolated populations with no gene flow would be 1.0.

Fixation indices F_{ST} and F_{IS} are another way of assessing genetic differentiation. F_{ST} measures the correlation between random gametes within subpopulations relative to gametes of the total population (Wright 1978). They are calculated for individual loci in the present study. Low F_{ST} values (i.e., low correlation among gametes sampled from within the subpopulation), indicate low levels of inbreeding in subpopulations relative to the total population. Low F_{ST} values therefore suggest high levels of gene flow. F_{IS} is the correlation between uniting gametes within populations (Wright 1978) and is equivalent to the mean deviation of genotypic proportions from Hardy-Weinberg expectations (Linhart et al. 1981). Positive F_{IS} values are associated with deficiencies of heterozygotes and suggest inbreeding; negative values indicate excesses of heterozygotes. For any one population

$$F_{IS} = 1 - H_o/H_e$$

where H_o denotes the observed heterozygosity and H_e denotes the number of heterozygotes expected if random mating is occurring in the population.

Multilocus relationships among stands can be summarized as canonical scores calculated from gene frequency data. Multilocus scores were calculated using the Candisc procedure (SAS Institute, Inc. 1990). Prior to canonical analysis, allelic data for individual trees were transformed to genotype scores (Smouse et al. 1982) to ensure multivariate normality. Calculation procedures follow those described by Westfall and Conkle (1992). The utility of plots of these scores is to illustrate genetic similarities and differences among stands.

Results and Discussion

Examination of allele frequencies detected for these population samples of *C. thyoides* reveal substantial differences among stands for certain loci (Table 12.1). Of the 16 loci examined, five were monomorphic: the first locus of aspartate amino transferase (AAT-1), diaphorase (DIA), phosphoglucose isomerase (PGI), superoxide dismutase (SOD), and the second locus of uridine diphosphoglucose pyrophosphorylase (UGPP-2). Six loci contained two alleles per locus: aconitase (ACO), AAT-2, alcohol dehydrogenase (ADH), glycerate-2-dehydrogenase (G2D), and the two loci of 6-phosphogluconic dehydrogenase (6PG-1 and 6PG-2). Three loci contained three alleles each: isocitric dehydrogenase (IDH), phosphoglucomutase (PGM), and UGPP-1; and two contained four alleles each: malic enzyme (ME) and shikimate dehydrogenase (SKD). Four of the polymorphic loci, ACO, AAT-2, G2D, and UGPP-1, exhibit one very frequent allele ($\geqslant 0.90$) with one or two less frequent ($\leqslant 0.10$) alleles. The loci ADH, IDH, ME, PGM, 6PG-1, 6PG-2, and SKD, however, exhibit greater diversity, and the distribution of alleles in these loci varies drastically according to population.

All polymorphic loci except ACO and G2D exhibit statistically significant ($P \pm 0.05$) chi-square values across population samples (Table 12.2), suggesting large ge-

Table 12.2 Contingency Chi-Square Analysis for Variable Loci among Eight *C. thyoides* Stands in Maine and New Hampshire

Locus[a]	Number of Alleles	F_{st}	Chi-square	P[b]
ACO	2	0.032	10.09	0.18
AAT	2	0.074	23.32	0.00
ADH	2	0.082	21.34	0.00
G2D	2	0.044	13.09	0.07
IDH	3	0.543	181.62	0.00
ME	4	0.127	142.88	0.00
PGM	3	0.180	106.22	0.00
6PG 1	2	0.356	114.05	0.00
6PG 2	2	0.150	47.79	0.00
SKD	4	0.119	84.53	0.00
UGPP	3	0.047	23.25	0.05

[a]Loci are defined in the first paragraph under "Results."
[b]This value is the probability of obtaining the indicated chi-square value by chance. Nine of 11 variable loci are statistically significant at the 5% level in heterogeneity among the eight stands. This is evidence for a large amount of genetic variation in allele frequencies among northern populations of *C. thyoides*.

netic differences among them for these loci. The three Massabesic stands (M1–M3), which are separated by approximately 20 m, are similar in allele frequencies to one another (Table 12.1). Positive F_{IS} values for M1–M3 indicate greater homozygosity than expected for randomly mating populations and suggest some inbreeding (Table 12.3). Significantly higher ($P \leq 0.05$) levels of homozygosity (data not shown) existed for alleles of the ME locus in M1 and M2, and for ME and ADH in stand M3. The M4 stand, 300 m distant, had a small negative F_{IS}, which, when coupled with lack of significant homozygosity (data not shown), indicates that allele frequencies are consistent with the random mating model for this stand, in contrast with M1–M3. M4 also has an observed heterozygosity level of 14.2%, in contrast with values less than 10.0% for M1–M3 (Table 12.3). The M4 stand may be part of a different mating

Table 12.3 Alleles per Locus (All/loc), Observed Heterozygosity (H_o), F-Statistics (F_{IS}), and Genetic Distances[a] among Eight *C. thyoides* Stands[b] Sampled for Population Genetic Differences

	All/loc	H_o	F_{IS}	M1	M2	M3	M4	CCW	Ki	Ba	BrB
M1	1.5	9.7	0.22								
M2	1.4	9.4	0.15	0.106							
M3	1.4	9.0	0.16	0.093	0.086						
M4	1.5	14.2	−0.08	0.122	0.108	0.083					
CCW	1.5	14.1	0.08	0.143	0.154	0.120	0.122				
Ki	1.5	8.5	0.21	0.266	0.289	0.263	0.255	0.273			
Ba	1.6	14.6	0.10	0.222	0.230	0.206	0.177	0.201	0.260		
BrB	1.6	15.3	0.02	0.201	0.215	0.196	0.186	0.209	0.267	0.096	

[a]Cavalli-Sforza and Edwards (1967) arc distance.
[b]See Table 12.1 (footnote a) and Figure 12.1 for location of stands.

system than M1–M3, because alleles at all loci are at frequencies expected for random mating (data not shown), suggesting that separation of 300 m may be great enough to place M4 under different mating conditions than M1-M3. For management purposes, however, it must be recognized that viable pollen may be carried more than 1 km (Sarvas 1967, Wang et al. 1969); thus, 300 m is not a distance to be recommended for isolation purposes.

Genetic distances (Table 12.3) among the Massabesic stands are low in comparison with most other genetic distances calculated for the other four cedar stands. Of the Massabesic stands, M4 is the most genetically distant. There are no *C. thyoides* in the 300 m distance separating stands M1–M3 from M4. Viable pollen can be dispersed for kilometers, but most probably falls to the ground within the first 100 m from the point of release (Sarvas 1967, Wang et al. 1969). Inspection of the allele frequency data and genetic distances suggests that the four Massabesic populations can be treated as one population for statistical analysis, as their frequencies are similar. The fact that the allele frequencies of population M4 are the result of random mating, and its H_o is 4.8% greater than the three nearby populations, suggests that different conditions for mating may exist there. The Cooper Cedar Woods (CCW) population, 103 km distant, is similar in allele frequencies to the four Massabesic populations and is a low genetic distance from them. The large chi-square values for allele frequency differences among stands (Table 12.3) are due primarily to differences between this group of five populations and the Kingston (Ki), Barrington (Ba), and Bradford Bog (BrB) populations. Genetic distances (Table 12.3) also reflect this pattern.

The Cedar Swamp Pond population in Kingston is especially notable in that although it is the largest investigated (16 ha) and is within 0.5 km of neighboring cedar swamps, it is missing alleles found in the other populations at intermediate frequencies for the loci PGM, 6PG-2, and SKD. It also has alleles the others do not (at the loci AAT-2, G2D, IDH), with one IDH allele at an intermediate frequency of 0.57. This distribution of alleles in Cedar Swamp Pond is unexpected and justifies additional genetic studies of the Kingston cedar swamps, especially since Cedar Swamp Pond is so close to other cedar swamps within this complex. Cedar Swamp Pond is the most genetically distant of the cedar swamps investigated (Table 12.3), with distances of ≥ 0.25 indicating very great differentiation from the others. Gene flow is possible from the two neighboring swamps and could explain the presence of alleles not found in the other stands studied, but the genetic composition of those stands is unknown. Selection can be invoked to explain both the presence of unique alleles and absence of other alleles. Since Cedar Swamp Pond is the only cedar stand sampled that is technically part of a kettlehole bog, its environment may be different from the other swamps sampled. Bogs are known to contain mineral-depauperate soil (Johnson 1985), and conditions may be variable enough to result in some genetic differentiation within Cedar Swamp Pond not found in the others. Obviously, more study is needed.

The number of alleles per locus for *C. thyoides* averaged over all stands is 1.5, ranging from 1.4 to 1.6 (Table 12.3). This value compares with 1.9 for *C. lawsoniana* (Port Orford cedar; Millar and Marshall 1991), 1.6 for *P. rubens,* 3.0 for *P. mariana,* 2.8 for *P. glauca* (white spruce) (Eckert 1989), and 2.1 for *P. strobus* (Ryu and Eckert

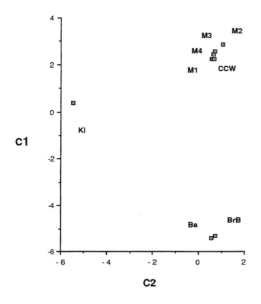

Figure 12.2. Plot of multilocus means for eight stands of *C. thyoides*. C1 and C2 are statistically significant canonical vectors. Differences in locations of stand means are due to differences in allele frequencies. Abbreviations are as in Figure 12.1.

IDH, PGM, and the 6PG-1 and -2 loci, indicating low differentiation among populations for seven of the loci. Values for the IDH and 6PG-1 loci indicate strongest differentiation among the stands at these loci, due to fixation of the most common IDH allele in all but Cedar Swamp Pond, and fixation of the most common 6PG-1 allele in the Cooper Cedar Woods stand and Massabesic stands 2, 3, and 4. These are in strong contrast to Cedar Swamp Pond, the Barrington cedar swamp, and Bradford Bog, where allele frequencies for 6PG-1 are at intermediate values.

Macrofossil evidence (Belling 1977) indicates that the range of this species was once more extensive than it is now. Belling's work suggests that migration into the northeastern United States took place after the recession of the last glacier, during a period of climate warmer than that of today. Her evidence suggests that the Piedmont Plateau was the most favorable route for the northward movement of the species. However, Belling also suggests that migration may have taken place along the now submerged coastal plain, moving inland as the plain was submerged by the rising sealevel. Laderman (1989) summarized evidence for the death of cedar forests that is due to incursion of seawater into coastal freshwater wetlands. Remains of *C. thyoides* stems may be seen at low tide, below the level of saltmarsh peat, on the coasts of New Hampshire, Massachusetts, New Jersey, Virginia, and elsewhere (Bartlett 1909; Heusser 1949, 1963; Belling 1977), and may be found buried deep in offshore marine sediments (Redfield and Rubin 1962). These now-submerged cedar forests may have been a major source of genetic variation for some relict stands in New England.

The cedar stands sampled in this study do not appear to be genetically depauperate, nor do their variation levels suggest they are the result of founder events. Rather, the genetic and fossil evidence suggests that they are relict populations from a time when the species was more widespread. The presence of unique alleles in the Cedar Swamp Pond population suggests that (1) this population is retaining unique alleles in response to low-level selection pressures, and/or (2) it is large enough to retain unique alleles, and/or (3) it is undergoing gene flow from nearby populations containing similar allelic frequencies. If the stands in this area are undergoing gene migration, and there is a large breeding population (N_e), then it is possible that a large portion of the gene pool may be preserved from an earlier time when the species was more widespread in south-central New Hampshire.

Acknowledgments The author acknowledges the valuable cooperation of Dr. Peter Garrett, U.S. Forest Service, Durham, NH; Bruce Hoveland of the Society for the Protection of New Hampshire Forests, Concord, NH; Barbara Vickery, The Nature Conservancy, Topsham, ME; Deborah Dunlop, Department of Plant Biology, University of New Hampshire, Durham, NH; and the faithful field and lab crew of the UNH Forest Biology Group: Karin Gagnon, the late Susan Urban, Andrea Hazard, Tom Hulleberg, Karen Stapelfeldt, and Melinda and Eric Ferland.

Thanks to R. Westfall, P. Garrett, C. Millar, and Y.T. Kiang for helpful early criticism of this manuscript, and K. Stapelfeldt for copyediting. The comments of two anonymous reviewers are appreciated.

This research was supported by the New Hampshire Agricultural Experiment Station and Northeastern Forest Genetics Regional Cooperative Project NE-27, scientific contribution number 1712 from the New Hampshire Agricultural Experiment Station.

References

Baldwin, H. 1961. Further notes on *Chamaecyparis thyoides* in New Hampshire. Rhodora **63**: 281–285.

———. 1963. Outposts of Atlantic white cedar. Society for the Protection of New Hampshire Forests, Forest Notes **77**:8–9.

———. 1965. Additional notes on *Chamaecyparis thyoides* in New Hampshire. Rhodora **67**: 409–411.

Bartlett, H. 1909. The submarine *Chamaecyparis* bog at Woods Hole, Massachusetts. Rhodora **11**:221–235.

Belling, A. 1977. Postglacial migration of *Chamaecyparis thyoides* (L. B.S.P.) (southern white cedar) in the northeastern United States. Dissertation. University Microfilms International, Ann Arbor.

———. 1987. A comparison of morphological characteristics of *Chamaecyparis thyoides, Thuja occidentalis,* and *Juniperus virginiana.* Pages 19–34 *in* A. Laderman, ed. Atlantic white cedar wetlands, Westview Press, Boulder.

Brym, P. 1990. Genetic variability and population structure of eastern white pine in a two hectare stand, Durham, New Hampshire. Ph.D. Dissertation, Univ. of New Hampshire, Durham.

Cavalli-Sforza, L.L., and A.W. Edwards. 1967. Phylogenetic analysis: models and estimation procedures. Am. J. Human Genetics **19**:233–257.

Eckert, R.T. 1989. Genetic variation in red spruce and its relation to forest decline in the northeastern United states. Pages 319–324 *in* J.B. Bucher and I. Bucher-Wallin, eds. Air Pollution and Forest Decline. Proc. 14th Int. Meeting for Specialists in Air Pollution Effects on Forest Ecosystems, IUFRO P2.05, Interlaken, Switzerland, Oct. 2–8, 1988.

Eckert, R., and D. O'Malley. 1988. Population genetic variation in New England red spruce. Pages 333–337 *in* G.D. Hertel, ed. The effects of atmospheric pollution on spruce and fir forests in the eastern United States and the Federal Republic of Germany. U.S. Dept. Agric. Forest Serv. Gen. Tech. Rep. **NE-120**, Northeastern For. Exp. Sta., Radnor, PA.

Fowler, D., and R. Morris. 1977. Genetic diversity in red pine: evidence for low genetic heterozygosity. Can. J. For. Res. **7**:343–347.

Gagnon, K., D. O'Malley, and R. Eckert. 1988. Starch gel electrophoresis of spruce foliage. Dept. For. Resources Tech. Note **17**, Univ. of New Hampshire, Durham.

Harlow, W.M., E.S. Harrar, J.W. Hardin, and F.M. White. 1991. Textbook of dendrology. McGraw-Hill, New York.

Heusser, C.J. 1949. History of an estuarine bog at Secaucus, New Jersey. Bull. Torrey Bot. Club **76**:385–406.

———. 1963. Pollen diagrams from three former bogs in the Hackensack tidal marsh, northeastern New Jersey. Bull. Torrey Bot. Club **90**:16–28.

Johnson, C.W. 1985. Bogs of the northeast. University Press of New England, Hanover, NH.

Laderman, A.D. 1989. The ecology of the Atlantic white cedar wetlands: a community profile. U.S. Fish Wildl. Serv. Biol. Rep. **85** (7.21).

Laderman, A.D., F. Golet, B. Sorrie, and H. Woosley. 1987. Atlantic white cedar in the glaciated Northeast. Pages 19–34 *in* A. Laderman, editor. Atlantic white cedar wetlands. Westview Press, Boulder, CO.

Lesica, P., and F.W. Allendorf. 1992. Are small populations of plants worth preserving? Conservation Biology **6**:135–139.

Linhart, Y.B., J.B. Mitton, K.B. Sturgeon, and M.L. Davis. 1981. Genetic variation in space and time in a population of ponderosa pine. Heredity **46**:407–426.

Little, S. 1950. Ecology and silviculture of white cedar and associated hardwoods in southern New Jersey. Yale Univ. Sch. For. Bull. **56**.

Millar, C.I., and K.A. Marshall. 1991. Genetic conservation of a single species: Implications from allozyme variation in Port-Orford-cedar *(Chamaecyparis lawsoniana)*. For. Sci. **37**:1060–1077.

Mitton, J.B., Y.B. Linhart, K.B. Sturgeon, and J.L. Hamrick. 1979. Allozyme polymorphisms detected in mature needle tissue of ponderosa pine. J. Heredity **70**:86–89.

Neale, D. 1978. Allozyme studies in balsam fir. M.S. Thesis. Univ. of New Hampshire, Durham.

O'Malley, D., K. Gagnon, J. Dolan, and R. Eckert. 1987. Preparation of spruce foliage samples for starch gel electrophoresis. Dept. For. Resources Tech. Note **15**, University of New Hampshire, Durham.

Redfield, A.C., and M. Rubin. 1962. The age of a salt marsh estuary. Science **147**:50–55.

Ryu, J., and R.T. Eckert. 1983. Foliar isozyme variation in twenty-seven provenances of *Pinus strobus* L.: genetic diversity and population structure. Pages 249–261 *in* R.T. Eckert, ed. Proc. 28th Northeastern For. Tree Improvement Conf., July 1982, Durham, NH.

Sarvas, R. 1967. Pollen dispersal within and between populations. Pages 332–345 *in* XIV IUFRO Cong., Sect 22, Munich.

SAS Institute. 1990. SAS/STAT User's Guide. Volume 1, Version 6, 4th Ed. Inc., Cary, NC.

Slatkin, M. 1985. Gene flow in natural populations. Annu. Rev. of Ecol. Syst. **16**:393–430.

Smouse, P., R. Spielman, and M. Park. 1982. Multiple-locus allocation of individuals to groups as a function of the genetic variation within and differences among human populations. Am. Nat. **119**:445–463.

Sperduto, D.D., and N. Ritter. 1994. Atlantic white cedar wetlands of New Hampshire. New Hampshire Natural Heritage Program, Dept. of Nat. Res. and Econ. Dev., P.O. Box 1856, Concord.

Svenson, H. 1929. *Chamaecyparis thyoides* in New Hampshire. Rhodora **31**:96–98.

Swofford, D., and D. Selander. 1981. BIOSYS-1. A computer program for the analysis of allelic variation in genetics. Release 1. Dept. Genetics and Development, Univ. of Illinois at Urbana-Champaign, Urbana.

Tigerstedt, P.M.A. 1973. Studies on isozyme variation in marginal and central populations of *Picea abies*. Hereditas **75**:47–59.

Wang, C.H., T.O. Perry, and A.G. Johnson. 1969. Pollen dispersal of slash pine *(Pinus elliottii)* with special reference to seed orchard management. Silvae Genetica **9**:78–86.

Westfall, R.D., and M.T. Conkle. 1992. Allozyme markers in breeding zone designation. New Forests **6**:279–309.

Wright, S. 1978. Evolution and the genetics of populations. Variability within and among natural populations, Vol. #4. Univ. of Chicago Press, Chicago.

Part II

SYSTEMS
WITH
DIVERSE
DOMINANTS

Forests Not Dominated by
Chamaecyparis *Species*

13

Hiroko Fujita

Characteristics of the Soil and Water Table in an *Alnus japonica* (Japanese Alder) Swamp

Alnus japonica (Thunb.) Steud. (Japanese alder) is the most common tree composing swamp forests in the cool-temperate zone of Japan (Fig. 13.1). It is further distributed in Korea, Manchuria, Ussuri, and Taiwan (Fig. 13.2; Murai 1962, Kurata 1968). It can grow in a wide range of habitats, even under mesic conditions (Fujita and Kikuchi 1986, Fujita 1987). It is common in northern and central Japan, and is widely distributed from lowlands to the montane zone. Its typical habitat is a water-saturated condition, with poorly drained soils and a high water table or flood waters at least part of the year (Yoshioka 1974, Suzuki 1975, Fujita and Kikuchi 1984). It is not strongly influenced by coastal conditions.

However, all wet sites are not occupied by *A. japonica* swamps. In the same wet conditions, there are various hydrophilous communities, such as willow forests, reed swamps, and many kinds of grass and sedge communities. This chapter focuses on the height and fluctuation of the water table and physicochemical features of soils as very important site factors that differentiate *A. japonica* swamp sites from those of other communities.

Types of Sites Occupied by the *A. japonica* Swamp

Japan is an archipelago that extends from north to south over 2,000 km. It consists of narrow and mountainous islands that receive abundant rainfall. The rivers are short, flow rapidly, and make up alluvial plains along the coast. There are numerous lakes and small ponds in volcanic terrain, as well as small depressions that are damp due to impermeable layers or tephra (volcanic material such as ash and lapilli—small stony or glassy lava fragments) in the soil. *Alnus japonica* establishes at these water-saturated sites.

Figure 13.1. Leaves, catkins, and cones of *Alnus japonica.*

According to Sasaki (1978), *A. japonica* forests are classified into eight communities, three alliances, and one order in the phytosociological method. Under this classification, which considers site and biological factors, such as landforms, sediment materials, growth of *A. japonica,* and species composition of stands, it is possible to divide *A. japonica* swamps into two types of sites. Type 1, peatland, is further divided into bog (Type 1-A) and fen (Type 1-B). Type 2 is non-peatland, such as an alluvial plain or valley bottom.

Type 1-A, bog type, is found in an area of high-moor or transitional-moor (Fig. 13.3). *Alnus japonica* here is always poor in growth; the mean annual diameter increment is 0.79 mm/year (Shinsho 1982), and the trees show characteristic twisted-tree form with many small coppices from the base (Shinsho 1978, 1982). At the herbaceous layer, *Sphagnum* species are the main dominant and *Moliniopsis japonica* (purple moor grass), *Vaccinium oxycoccus* (bog cranberry), *Myrica gale* var. *tomentosa* (myrica), *Carex limosa* (limosa sedge), *C. middendorffii,* and *Ledum palustre* var. *nipponicum* often dominate (see the Index for Japanese common names).

Type 1-B is a fen type (Fig. 13.3). *Alnus japonica* growth is moderate and the trees usually form pure stands. *Phragmites australis* (common reed) is the most common dominant species at the herbaceous layer. Many other species, such as *Calamagrostis langsdorffii, Carex pseudocuraica,* and *Equisetum fluviatile* (horsetail), sometimes dominate.

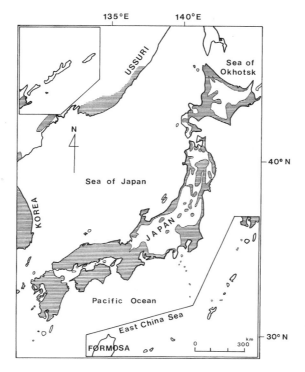

Figure 13.2. Distribution of *A. japonica*. The insets at top left and bottom right show continuations of the island chain to the northeast and southwest, respectively. Hatched area shows range. (Adapted from Murai 1962.)

Type 2 stands were once widely established on alluvial plains or valley bottoms in hilly land (Fig. 13.3). Most of these stands have been lost due to drainage and reclamation for paddy fields (cultivated for rice) or development.

Alnus japonica growth in Type 2 stands is very good. The mean annual diameter increment is 3.25 mm/year (Fujita, unpublished observations). Tree height is usually 10–15 m and sometimes exceeds 20 m. Species diversity of the forest floor is high. Many kinds of forbs, grasses, climbing plants, and ferns can be seen. There are some types of stands that are different in species composition, such as Rhododendro–Alnetum and Cirsio–Alnetum (Asano et al. 1969), *A. japonica–Carex biwensis* community, *A. japonica–Sasa senanensis* community and *A. japonica–Miscanthus sinensis* community (Makita et al. 1976 and 1979), *A. japonica–Irex crenata* var. *paludosa* stand, *A. japonica–Polygonum thunbergii* stand, and *A. japonica–Filipendula kamtschatica* stand (Fujita and Kikuchi 1986). *Ulmus davidiana* (Japanese elm) or *Fraxinus mandshurica* var. *japonica* (Japanese ash) stands often neighbor the *A. japonica* stands. According to Tatewaki et al. (1967), *Alnus* occurs at extremely wet sites, *Ulmus* at relatively mesic ones, and *Fraxinus* dominates at sites intermediate in water condition.

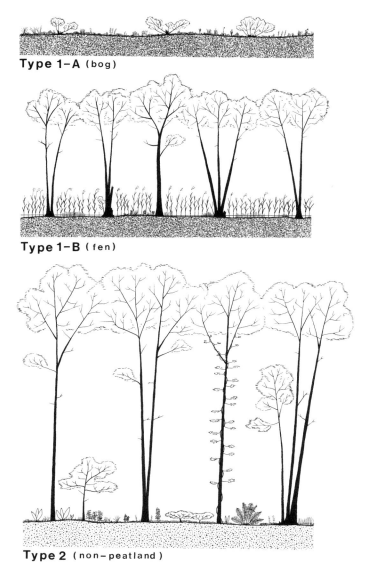

Type 1-A (bog)

Type 1-B (fen)

Type 2 (non-peatland)

Figure 13.3. Sketch of the three *A. japonica* site types.

Methods

Study areas were situated in the Tohoku District, in the northern part of Honshu Island in Japan (140°–141°30′E and 37°20′–41°20′N, 3–570 m above sea level). To check the water table, 52 stands were chosen that include all the *A. japonica* forest types (bog, fen, and alluvial plain). Thirty-three selected points within the 52 stands were prepared for soil surveys. The soil surveys were carried out in the summers of 1981–1984. A pit was dug at each point, 1x1 m^2 wide and to the depth of the water table or, at most, 1 m. The soil profile was checked on several ordinal points, such as depth and thickness of horizons, color, texture, organic matter content, gravel features, existence of mottling, compactness, root distribution, and 2–2′–dipyridyl reaction.

A small tributary basin named Tashiro was selected for the study area to measure seasonal changes of the water table. The study area is on the bottom of a small tributary valley that dissects a hilly area near the northwestern part of Miyagi Prefecture, Japan (140°15′E and 38°44′N, 470 m above sea level). The tributary often flooded near the center of the study area, and the area was divided at that point into upper and lower parts. A thick gravel bed underlies the fine surface on the upper part of the study area; *Ulmus* stands are prominent there, although *A. japonica* stands are found occasionally in the wet places. The lower part consists mainly of clay and silt with a thin gravel bed, except for the streambanks, where coarse and fine sand are deposited. This lower part is usually occupied by *A. japonica* stands, although streambanks are sometimes occupied by *Ulmus* stands.

Plastic pipes (2.8 cm inside diameter) with holes at their lowermost part were driven perpendicularly into the ground. At the beginning of the study (in spring) they were driven 80 cm; in winter, 140 cm. Distance from the ground surface to standing water in the pipe was measured as the level of the water table. Measurements were conducted approximately every two weeks from May 9 to November 21, 1982, and monthly from December 9, 1983, to May 20, 1984.

Features of the Soil and Water Table

Type 1, Peatland

Figure 13.4 shows the frequency distribution of the water table heights that were checked at the soil survey pits. The water table in peatland exists from 10 cm above to 15 cm below the ground surface. Peatland always has high groundwater and is often flooded. There are no obvious differences in water-table height between *A. japonica* Type 1-A (bog) stands and other bog stands dominated by *L. palustre* var. *nipponicum* and *C. middendorffii* or between *A. japonica* Type 1-B (fen) stands and other fen stands dominated by *P. australis* (Umeda 1985).

Figure 13.5 shows the soil profiles of *A. japonica* sites of the peatland type (Type 1). The soil type of Type 1-A (bog) sites is classified as High-moor Peat. The peat is formed from *Sphagnum* species or *Moliniopsis japonica*. It is important to note that muck layers containing a little clay or silt material are always found within 1 m depth at these sites. In summer the water table height ranges from 10 to 20 cm below the ground surface. However, the soil is almost anaerobic, except for the thin upper layer. (Anaerobic conditions were judged by using the Dipyridyl reaction method.)

On the other hand, the soil type of a Type 1-B (fen) site is Low-moor Peat. The peat is very loose and composed of about 80–90% water in volume and dead root

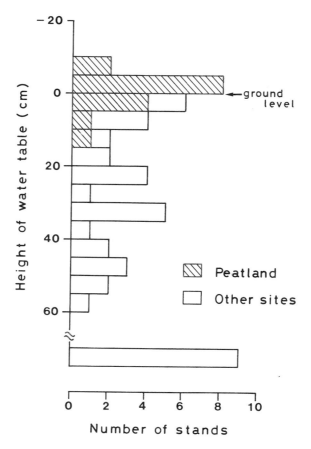

Figure 13.4. The frequency distribution of the average height of the water table of 52 A. *japonica* stands. The − 20 indicates 20 cm above ground. Water tables lower than 60 cm could not be checked; such stands are represented in the lowest bar.

systems of *P. australis* and *A. japonica*. The soil is usually anaerobic. It is also an extreme soil feature, similar to Type 1-A, that muck or mineral soil layers are always found in the soil profile. This fact suggests that *A. japonica* sites in peatland are characterized by the mineral soils supplied from rivers or slopes surrounding the peatland.

Type 2, Non-Peatland

Figure 13.6 shows the soil profile of the Type 2 *A. japonica* sites. The soil is an alluvial type composed not of peat but of mineral material. The soil horizon is com-

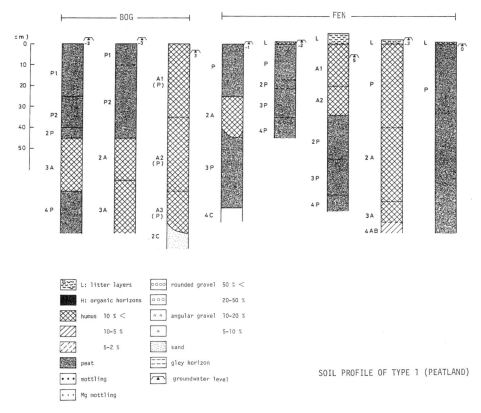

Figure 13.5. Soil profiles of the *A. japonica* Type 1 (peat soil) sites. The soil survey was carried out in the Miyagi and Aomiri Prefectures.

plex. Many layers consist of coarse deposit material brought by a destructive flood or landslide. The A layers contain much organic matter, and their color is brownish black. These A layers tend to be anaerobic; other layers are usually aerobic and are better drained than peatland soil (Type 1).

The soils are eutrophic, similar to the soils of *Ulmus* or *Fraxinus* stands (unpublished observations). Nutrient content of the soil is not the essential factor differentiating Type 2 forests from the others.

Figure 13.7 shows the seasonal change of the water table and the daily amount of rainfall at the Tashiro basin (Fujita and Kikuchi 1984). The figure shows nine selected points of 18 that were examined in 1982 and all six stands examined in 1983–84. The change of the groundwater level depends greatly on rainfall and excess water from melting snow in late April. At the wettest sites, where the water table sometimes reaches the ground surface, the water table fluctuates within as narrow a range as 10 cm or less to a depth of 30 cm below the ground surface. At the most mesic site,

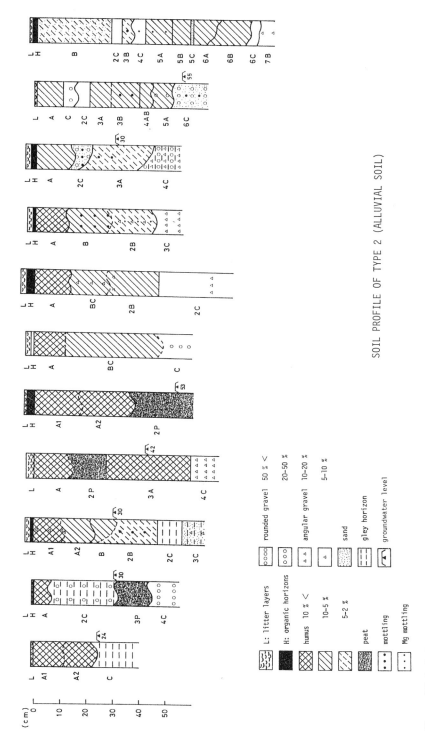

SOIL PROFILE OF TYPE 2 (ALLUVIAL SOIL)

Figure 13.6. Soil profiles of the *A. japonica* Type 2 (alluvial soil) sites. The soil survey was carried out in the Miyagi and Aomori Prefectures.

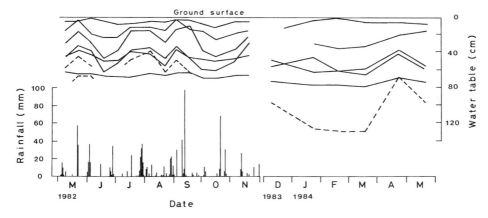

Figure 13.7. Seasonal change of the water table and the daily amount of rainfall at the Tashiro basin. Solid lines: *A. japonica* stands. Broken lines: *Ulmus* stands. (Adapted from Fujita and Kikuchi 1984.)

the level of the water table is very stable, with its highest level at 70 cm below the surface. In contrast, *Ulmus* stands occupy well-drained sites where the water table, always lower than 40 cm, fluctuates widely. It rises to about 40 cm at a time of heavy rainfall, but the duration of these high levels is shorter than at *A. japonica* stands.

Figure 13.8 shows the position of 18 stands plotted against the fluctuation range and highest water table at Tashiro (Fujita and Kikuchi 1984). The line shows the apparent limit of the *A. japonica* stands' distribution. (The limit for *Ulmus* stands could not be determined in this case because only two stands were examined.) The depth of the water table and its fluctuation seem to be the primary factors controlling the formation of these swamp forests. The height and fluctuation patterns of the water table are closely related to physical soil properties and landform.

Conclusion

In addition to wet conditions, the height and fluctuation of the water table and the physicochemical features of the soil are important site factors for *A. japonica* swamps, as summarized in Table 13.1.

The physicochemical features of soil related to the supply of available nutrients are important in determining the border between *A. japonica* swamps and other hydrophilous communities under water-saturated conditions. However, the height and fluctuation pattern of the water table are important in differentiating between *A. japonica* swamp stands and more mesic *Ulmus* or *Fraxinus* forests.

Table 13.1 Summarized Features of Alder Forests and Their Habitats

Type	Soil Type		Nutrient Condition	Water Table		Growth of Alder	Height of Alder (m)	Diversity of Forest Floor
				Range[a] (cm)	Fluctuation			
Type 1, peatland 1-A (bog)	organic soil	High-moor peat	mesotrophic	−20–30	small	poor (e.g., dwarfing, stools)	0.5–2	low to intermediate
Type 1, peatland 1-B (fen)	organic soil	Low-moor peat	eutrophic	−20–30	small	good	3–10	low to intermediate
Type 2, non-peatland	mineral soil	Alluvial	eutrophic	0–80	large	good	10–20	high

[a]minus sign means above ground surface.

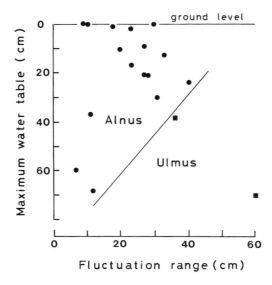

Figure 13.8. Relation of forest type to the fluctuation range and maximum water table. Solid circles: *A. japonica* stands. Solid squares: *U. davidiana* stands. (Adapted from Fujita and Kikuchi 1984.)

References

Asano, K., I. Hayashi, K. Hirabayashi, S. Ito, K. Nakayama, T. Shimizu, and K. Tsuchida. 1969. Vegetation of Sugadaira. I. Plant community. Sugadaira kenkyu houkoku 3:1128. (In Japanese.)

Fujita, H. 1987. Differentiation of some *Alnus japonica* forests based on species composition and their soil condition. Ecol. Rev. **21**:77–85.

Fujita, H., and T. Kikuchi. 1984. Water table of alder and neighboring elm stands in a small tributary basin. Jap. J. Ecol. **34**:473–475.

———. 1986. Differences in soil condition of alder and neighboring elm stands in a small tributary basin. Jap. J. Ecol. **35**:565–573.

Kurata, S. 1968. Illustrated important forest trees of Japan. Vol. 2. Chikyusha Co., Tokyo.

Makita, H., T. Kikuchi, O. Miura, and K. Sugawara. 1976. A geobotanical study of alder forest and elm forest in a small tributary basin. Ann. Tohoku Geogr. Ass. **28**:83–93. (In Japanese, with English summary.)

Makita, H., T. Miyagi, O. Miura, and T. Kikuchi. 1979. A study of an alder forest and an elm forest with special reference to their geomorphological conditions in a small tributary basin. Pages 237–244 *in* A. Miyawaki and S. Okuda, eds. The Yokohama Phytosociological Society, Yokohama.

Murai, S. 1962. Phytotaxonomical and geobotanical studies on genus *Alnus* in Japan. I. Comparative studies on tree species. Government Forest Exp. Sta. Bull. (Japan) **141**:141–166. (In Japanese, with English summary.)

Sasaki, Y. 1978. Vegetation of Ise Bay district. Pages 15–95 *in* Report of Studies on Green Planning of Ise Bay District. Transport Ministry of Japan, Tokyo. (In Japanese.)

Shinsho, H. 1978. Notes on the alder swamp forest *Alnus japonica* Sieb. et. Zucc. var. *arguta* in the Kushiro Moor, Eastern Hokkaido. I. Memoirs of the Kushiro Municipal Museum **5**:31–44. (In Japanese.)

———. 1982. Notes on the alder thickets *Alnus japonica* Steud. in the Kushiro Moor, Eastern Hokkaido. II. Memoirs of the Kushiro Municipal Museum **9**:27–36. (In Japanese.)

Suzuki, Y. 1975. Alder forest in Chiba Prefecture: their habitat. Pages 103–114 *in* Biological Society of Chiba Prefecture, ed. Flora and Vegetation of Chiba Prefecture. Inoue Book Co., Tokyo. (In Japanese.)

Tatewaki, M., M. Tohyama, and T. Igarashi. 1967. The forest vegetation on the lake-side of Abashiri, Prov. Kitami, Hokkaido, Japan. Memoirs of the Faculty of Agriculture Hokkaido Univ. **6**:286–324. (In Japanese with English summary.)

Umeda, Y. 1985. Studies on hydrological characteristics of peatland. Report of a Grant-in-Aid for Scientific Research B (No. 58460221) from the Ministry of Education, Science, and Culture, Japan. (In Japanese.)

Yoshioka, K. 1974. Aquatic and wetland vegetation. Pages 211–236 *in* M. Numata, ed. The flora and vegetation of Japan. Kodansha, Tokyo.

Appendix 13.1 Flora in *A. Japonica* Forests

	Common Name	
Scientific Name	Japanese	English
Trees		
Alnus japonica	hannoki	Japanese alder
Fraxinus mandshurica var. *japonica*	yachidamo	Japanese ash
Ulmus davidiana	harunire	Japanese elm
Shrubs		
Ilex crenata var. *paludosa*	haiinutusge	(holly)
Ledum palustre var. *nipponicum*	isotsutsuji	Labrador tea
Myrica gale var. *tomentosa*	yachiyanagi	myrica
Vaccinium oxycoccus	tsurukokemomo	bog cranberry
Herbs		
Calamagrostis langsdorffii	iwanogariyasu	reed bentgrass
Carex biwensis	matsubasuge	sedge
Carex limosa	yachisuge	limosa sedge
Carex middendorffii	horomuisuge	sedge
Carex pseudocuraica	tsurusuge	sedge
Equisetum fluviatile	mizutokusa	horsetail
Filipendula kamtschatica	onishimotsuke	filipendula
Miscanthus sinensis	susuki	Chinese silvergrass
Moliniopsis japonica	numagaya	purple moor grass
Phragmites australis	yosi	common reed
Polygonum thunbergii	mizosoba	knotweed
Sasa senanensis	kumaizasa	bamboo
Sphagnum	mizugoke	sphagnum moss

14

Peter A. Khomentovsky

Pinus pumila (Siberian Dwarf Pine) on the Kamchatka Peninsula, Northeast Asia

Ecology of Seed Production

This chapter describes some of the ecological features and reproductive characteristics of *Pinus pumila* (Pall.) Regel (Siberian dwarf pine) in typical habitats in one of its easternmost locations, the Kamchatka Peninsula.

Some years ago Khomentovsky and Khomentovskaya (1990) observed that seed parameters and cone crops were not strongly correlated with elevation or position in the landscape, but appeared to be closely related to microclimatic factors. In a preliminary characterization of the impact of conophagous insects feeding on *P. pumila* Khomentovsky and Efremova (1991) found that the only two known insect species inhabiting *P. pumila* cones, *Cecidomyia pumila* (Diptera, Cecidomyidae) and *Eupithecia abietaria* (Lepidoptera, Geometridae), have an almost negligible effect on crop quality, not injuring seeds themselves.

These observations stimulated us to proceed with a more detailed investigation of environmental factors related to seed production variability. A portion of that study is presented here.

General Characteristics of the Species

Pinus pumila (also called Japanese mountain stone pine, dwarf cedar, creeping cedar, mountain pine, dwarf stone pine, and subalpine stone pine) belongs to the subgenus Strobus (Haploxylon), section Cembra, row Pumilae (Krylov et al. 1983). The North American *P. albicaulis* (whitebark pine) is the only other species in this row, ecologically equivalent to *P. pumila,* capable of growth both in the upright and dwarf form.

Pinus pumila is a creeping, dwarf evergreen tree (Fig. 14.1) from 0.5–6 m in height in summer (its "seasonal height"), up to 30 cm in diameter at its base, and up to 20–22 m in skeleton-branch ("trunk-branch") length. The crown has a normal

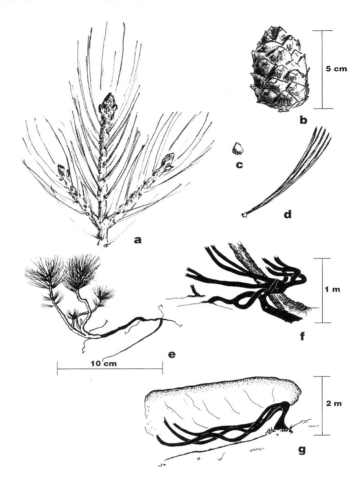

Figure 14.1. *Pinus pumila* morphology. a: One-year shoots (portion of needles removed for better view). b: Mature cone. c: Mature
seed. d: Cluster of needles. e. Young plant at upper tree limit (1,400
msl, 25 yrs old). Marks (/) show places where roots were cut. f. One
variant of the multi-trunk form, showing how *P. pumila* shares sites
with *B. ermanii* in a sparsely forested slope (Bolgit field base). Ap-
prox. 820 msl; 250–300 yrs old. g: Hanging form of crown on
smooth 10° slope at approximately 800–850 msl.

shape in summer and bows down close to the ground in winter; low temperatures,
not the pressure of the snow pushing down the trunk-branches, switch on this "lay-
ing" mechanism (Grossett 1959).

The tree has a short trunk, very often not more than 30–50 cm. Skeleton branches
(Fig. 14.1e, f, h) give the tree a multi-trunk, shrub-like appearance. The basic form
of the crown varies with location. Under upright trees it is asymmetric; on the slopes
of narrow valleys it adopts a hanging form (Fig. 14.1g); while on the slopes of wide

Figure 14.1 (*continued*). h: Cup-like shape of *P. pumila* crown, slightly asymmetrical due to neighboring larch.

valleys in highlands it has a creeping form, and on watershed plateaus, it becomes prostrate. Where there is no shelter above—on tundra, seashore, and wide river valleys in lowlands—the form is cuplike (Fig. 14.1h). Cones develop over two years. Female buds of the first year (immature, or yr-1 cones) become the following year's mature (yr-2) cones.

Pinus pumila can live on very poor soils, absorbing nutrients mainly with the help of mycorrhiza. It starts to produce adventitious roots in its first decade and grows with the help of new ones for centuries, while slowly decaying at the base. Where it is still possible to count tree rings, the wood can reach 350–400 years of age. Starting at the age of 30–50 yrs, the tree produces wingless seeds almost yearly (depending on site conditions), with a slightly noticeable two-year cycle.

Its range (Fig. 14.2) extends from the utmost northeastern limit of the Eurasian forest-tundra zone at Chukotka Peninsula (65°N, 178°E) south-southwest along the Pacific coast (including the Commander and Kuril Islands) to Japan's Hokkaido and Honshu mountains (35°N). It reaches west along the Korean mountains and to the northwest via the Hingan mountain range to Lake Baikal.

Pinus pumila on the Kamchatka Peninsula

Pure stands of *P. pumila* occupy about 42% of the forested area in the Kamchatka district (Fig. 14.3). About 50% is shared with *Betula ermanii* (stone birch) (Atlas of the USSR Forests 1973). In Russia, relatively pure *P. pumila* stands occupy about 12% of the forested area of the whole Far-Eastern Economic Region (A.S. Sheingauz, personal communication) and mixed stands extend further.

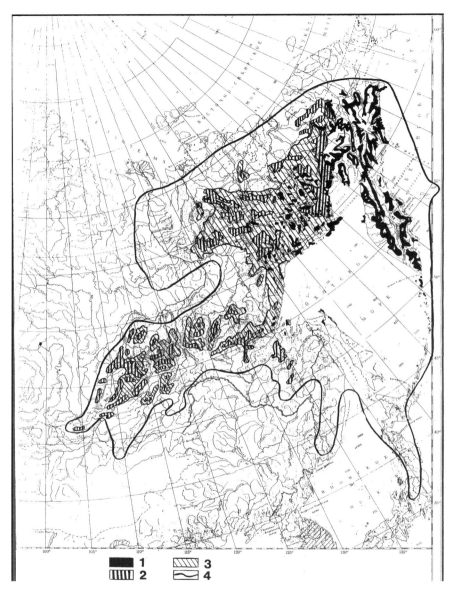

Figure 14.2. Range of *P. pumila*. 1: (Solid black) Areas of absolute dominance. 2: (Heavy vertical stripe) Growing in favorable places among mountain tundra, not dominating. 3: (Finer slanting stripe) As a lower canopy in sparse larch forests. 4: (Wavy line) Area border. (From Tikhomirov 1949, Sochava and Lukicheva 1953.)

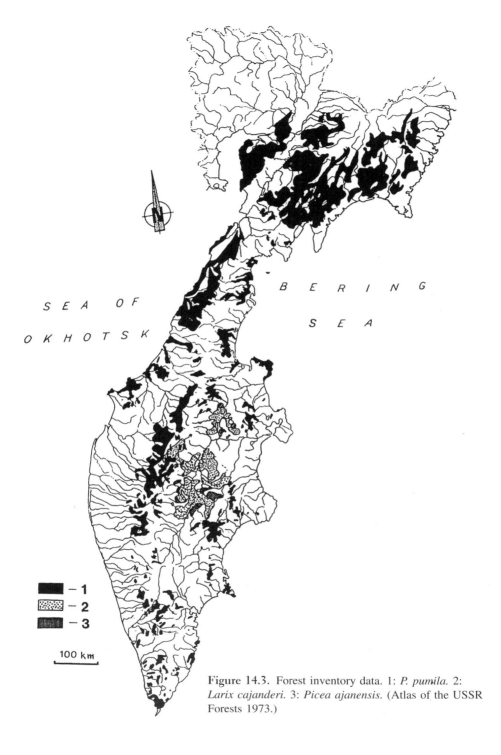

S E A   O F

O K H O T S K

B E R I N G

S E A

- 1
- 2
- 3

100 km

Figure 14.3. Forest inventory data. 1: *P. pumila*. 2: *Larix cajanderi*. 3: *Picea ajanensis*. (Atlas of the USSR Forests 1973.)

Pinus pumila is primarily a mountain species of moderately elevated sites, with the ability to adapt elsewhere. In western and central parts of its range, with a generally continental climate, it is usually pushed by interspecies competition pressure upward along the vegetation profile to the treeline. It can bear other colonizing neighbors if there is adequate sunlight and top-soil aeration, as when it grows under a sparse canopy of *Larix* spp. (larch) and *B. ermanii,* typical for the boreal zone in northeast Asia.

In Kamchatka and on the continental part of the Okhotsk seacoast, in Sakhalin, the Kuril Islands, and northern Hokkaido—with a wet and relatively cold climate—*P. pumila* grows along the whole altitudinal profile. It ranges from sea level, where it is a pioneer tree species on coastal dunes, to lowland tundra with little or no plant cover, fluvioglacial deposits, and river valleys (Semjagin 1911, cited in Tikhomirov 1949) to the upper limit of woody vegetation. The species grows in the lower canopy of subalpine *Larix* and *B. ermanii* forests and forms an independent vegetation belt around and above alpine meadows, as a pioneer species in the forest-mountain tundra ecotone and in newly formed volcanogenic ecotopes (Lipshits and Liverovsky 1937, cited in Tikhomirov 1949). Only one other species occupies the same altitudinal and habitat spectrum in northeast Asia—*Alnus kamtschatica* (dwarf alder), which often accompanies *P. pumila.*

The existence of many unique adaptive features of *P. pumila* is the result of the long-term impact of severe environmental conditions since the late Miocene and early Pliocene, the time of its establishment as a species (Khomentovsky and Egorova 1990). The strong negative impact of the Arctic Ocean and north Pacific Ocean climates on biotic systems causes the species distribution to resemble that of more northerly regions.

In Kamchatka, *P. pumila* finds specific environmental conditions. The peninsula, surrounded by cold ocean, is only linked to the continent by a narrow and boggy isthmus so that, biologically, the peninsula is an island. The peninsula is about 1,000 km north–south and at most 450 km east–west and is divided by mountain ranges nearly aligned with the meridian into an inner part with a continental climate and coastal parts with a maritime climate. It exhibits two types of vegetation zones, altitudinal and coastal (Fig. 14.4) as a combined reflection of relief structure and oceanic air masses (see also Fig. 14.5). The coastal type parallels altitudinal zonation, in the sense that moving inland from the cold and wet seashore is equal to moving downslope from the cold and wet mountain top.

In general, a dominating environmental factor in Kamchatka is the lack of warmth. Komarov (1927) accurately termed as "subalpine" the vegetation between 300 and 1,000 meters above sea level (msl), which covers the whole vegetation profile, if the coastal effect mentioned above is considered.

The macrostructure of vegetation cover is rather simple, as a result of highly intensive geodynamic processes: glaciations and transgressions; orogenic, seismic, and volcanic activity. There are three main geobotanical provinces on the Kamchatka Peninsula, each containing three or four districts (Khomentovsky et al. 1989) (Fig. 14.6). "Tundra-forest" components include *P. pumila* formations, which occur almost everywhere on the peninsula. Zones of monotypic and mixed forest are complicated by the mosaic structure of microecotopes and plant communities and are pressed by climate into narrow conjunctive belts.

The upper limit for upright tree *(Larix)* distribution in Kamchatka lies at 950–1,000 msl and for sparse subalpine *Larix* forests, at about 850 m. The upper limit for

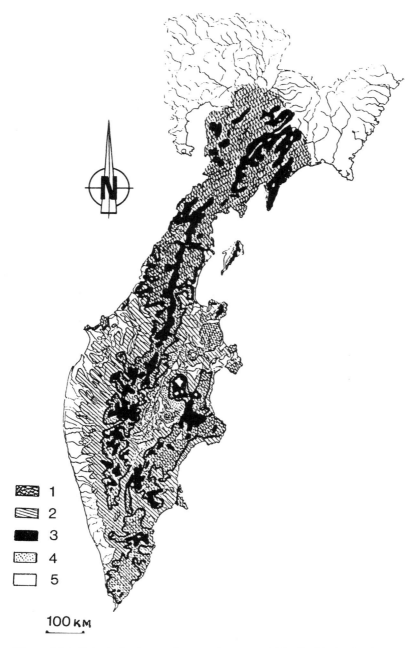

Figure 14.4. Principal scheme of vegetation zones (altitudinal belts) in Kamchatka. 1: Subalpine. 2: Boreal middle-elevation forest. 3: Alpine. 4: Inner lowland. 5: Coastal lowland. (From Lavrenko and Sochava 1954.)

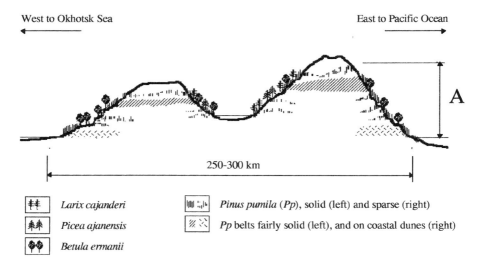

West to Okhotsk Sea   East to Pacific Ocean

250-300 km

‡‡	*Larix cajanderi*						:,	‖	*Pinus pumila* (*Pp*), solid (left) and sparse (right)
🌲🌲	*Picea ajanensis*	⁄⁄ ⁚	*Pp* belts fairly solid (left), and on coastal dunes (right)
🌲🌲	*Betula ermanii*		

Figure 14.5. Cross-section of Kamchatka Peninsula roughly indicating altitudinal vegetation belts. A = altitudinal limit of woody vegetation (~1000 m for *Larix*, ~1400 m for *Pp*). (Graphic by J. Donnette.)

creeping conifer trees *(P. pumila)* is 1,400–1,500 m, whereas in the Baikal and Honshu regions it extends higher, to 2,000 m and 3,200 m, respectively (Dmitrieva 1978, Hayashi 1960). The upper limit for dwarf deciduous trees *(A. fruticosa)* is about 100 m lower than that of *P. pumila.*

Pinus pumila, which very often lives at the limit of tolerance for woody vegetation, shows a wide variety of adaptation to conditions such as poor and frozen soils and frequent volcanic ashfalls. Given the strong interspecies competition for sunlight, soil nutrients, and oxygen, the first species to occupy bare surfaces formed by fire or volcanic activity are the winners. They must have excellent ecological flexibility with the potential to adapt to changing demands of the environment. The intensive geodynamics of the region are advantageous to a tree species such as *P. pumila,* which has all the features of a permanent pioneer. Here we describe its unique and complicated mechanism of reproduction and reforestation. The search for good explanations for its peculiarities provides new knowledge of tree survival in extreme situations and may also point the way to the solution of the northern territories' afforestation problem.

Description of Research Area

The key plots of the research area are situated near the geographic center of Kamchatka (56°N, 158°E), in the Sredinnij ("middle") mountain range between 460 and 1,200 msl. The research area covers the upper part of the conifer forest, subalpine, alpine, and above-alpine belts (Figs. 14.3 and 14.4).

The climate of the area is subcontinental, with 300–400 mm annual precipitation, moderately cold and snowy winters, and cool summers. The growing season at 400–

Figure 14.6. Geobotanical regions of Kamchatka Peninsula (from Khomentovsky et al. 1989). Provinces (in italics) and districts are as follows. *Western mountain–lowland province of stone-birch, tundra–forests.* 1: Okhotsky district; tundra, tundra–forest. 2: Western foothill–lowland district; dwarf woody vegetation, stone-birch forest. 3: Mountainous–alpine district; tundra, forest–tundra. *Central valley–submountain province; coniferous, stone-birch forests.* 4: Valley–foothill district; deciduous forest. 5: Northern district; spruce, larch forests. 6: Central district; stone-birch, larch forests. 7: Southern district; larch, spruce forests. *Eastern mountain–coastal province; stone-birch, tundra–forest.* 8: Mountainous–volcanic district; tundra, subalpine forest–tundra. 9: Eastern mountainous–coastal district; dwarf woody vegetation, stone-birch forests. 10: Pacific district; coastal and forest–tundra. Arrows show key plots (Bolgit).

500 msl is only about 100 days, from the beginning of June until the end of August. A permanent snow cover at 700 m and higher (the location of the main *P. pumila* highland vegetation belt) is established in the first half of October and usually disappears in May.

The area is mountainous with a climatic compression of its altitudinal vegetation belts. The average mountain range is around 1,500 m high. Volcanogenic basalts and andesites and Quaternary glacial, fluvioglacial, alluvial, and proluvial deposits define the territory. Glacial relief is evident, especially in upper river and creek basins.

Soils are typical for Kamchatka mountain regions: turf illuvio-humus, turf illuvio-volcanic ash, tundra illuvio-humus, tundra illuvio-humus-volcanic ash, turf primitive (Sokolov 1973 classification, with additions by N.V. Kazakov, personal communication). Soil features are defined by extended freeze periods, lower horizon anaerobiosis and repeated weak to moderate volcanic ashfall impacts (mainly from the eastern peninsula).

The vegetation of the area, reflecting its geographic position, mountain relief, and severe climate, is a spatial mosaic and is not very diverse. River valleys and lower parts of their tributaries are occupied by *Populus suaveolens* (fragrant poplar), *Chosenia arbutifolia* (Korean willow), *Larix cajanderi* (larch), *Betula kamtschatica* (white birch), and *B. ermanii* (stone birch). Two dwarf tree species, *P. pumila* and *A. kamtschatica,* are widespread along the whole vegetation profile. They occur in groups or strips along moraine tongues, river terraces, fluvioglacial deposits, watershed ridges, and slopes, and form the upper limit of woody vegetation.

Plant formations in the middle and high altitudes are almost the same (excluding *Populus* and *Chosenia*), but, as one moves upward along the narrowing valleys and watersheds, there is a change in spatial disposition from macromosaic to micromosaic. Trees more strictly follow soil temperature and drainage gradients and occupy each shelter from wind and snow damage.

Material and Methods

Data Collection

Data for *P. pumila* seed production assessment in key plots were gathered mostly from September–October 1990 at the Bolgit field base near Esso village in the Bistrinsky district of Kamchatka. The site is the Tupikin Clyuch creek basin, approximately 10 km^2. The material presented is from one slope of east-northeast exposure between 650–1,030 msl and from the nearest parts of the watershed (up to 50 m from the slope edge). Permanent plots on an existing transect and some additional points were chosen for cone crop measurement and sample collection.

The following information was registered at each site (Table 14.1): position in the landscape and ground cover of *P. pumila* clumps, type of plant community, general condition of water supply, average seasonal height of skeleton branches, and shading rate from neighboring upright trees or clumps (Box 14.1).

In each point for data and sample collection, we used 35 small randomly chosen plots of different sizes, from 1 to 15 m^3. We counted the number of mature (yr-2) cones and immature female buds (yr-1 cones) on each germinating shoot and skeleton

Table 14.1 Description of Ecotopes and *Pinus pumila* Clumps[a]

Altitude (msl)	Position in Relief	*P. pumila* Community Type[c]	Main Water Source[b]	Stand Composition[c]	*P. pumila* H avg, cm	SC(%)[d]	Shading[e] Utr	Ppc
1030	Upper part of flat watershed, shaded from S and E by ridge	Pumilae pinetum carioso-hypnoso-ericosum; with P. p. cladinosum fragments	A + S	Pp	100 (15–20)	80 unev	0	1
950	Plateau, full sun; watershed with slight slope to NNE	P. p. purum, + P. p. carioso-cladinosum fragments	A	Pp	40 (10–15)	40 unev	0	0
900	Flat watershed above creek source, shaded by ridge from S	P. p. hypnoso-carioso-ericosum	A + S	Pp + Lc (upper limit of Lc)	300	40 unev	1	1
810	Complex watershed ridge, slightly sloped to NE	P. p. hypnoso-carioso-ericosum	A	Pp + Lc	150 (35–40)	60 unev	1	2
800	Middle part of E-exposed slope of wide creek valley	P. p. hypnoso-ericosum	S + A	Pp + Lc	200	80 unev	0	1
680	Eastern border of watershed ridge with flattened top	P. p. hypnoso-carioso-ericosum	A + S	Pp + Lc	300	60 unev	2	1
650	Lower part of E-exposed slope in narrow creek valley	P. p. ericoso-sphagnosum	S	Pp + Ak	300 (40–45)	100	0	3

[a]The age of all analyzed clumps is between 150 and 260 years.
[b]Main water source. A: atmosphere. S: slope.
[c]Woodstand composition. Pp: *P. pumila*. Lc: *Larix cajanderi*. Ak: *Alnus kamtschatica*.
[d]H avg: average height (cm). In parentheses: rough estimation of annual shoot elongation (mm). SC(%): surface cover by clumps. "unev" = uneven.
[e]Shading. Utr: from upright trees of different species. Ppc: from neighboring *P. pumila* clumps. 0: no shading. 1: slight. 2: moderate. 3: heavy shading.

BOX 14.1 *Pinus pumila* Growth Form.

Pinus pumila plants cannot be treated as typical trees. They neither stand separately nor have a long single trunk; in most cases they overlap both above and below ground. We also cannot call them "clones," because we do not know the existence and level of genetic relationships. The best term to describe its collective growth form is "clump" (a term suggested by Dr. Diana F. Tomback, personal communication).

A single trunk (as traditionally understood) of *P. pumila,* as described in the text, is very short. Branches (or "trunk-branches") of the first and second order are equal physiognomically, bear a similar quantity of shoots, and form crowns that look as if they have many trunks. During the typically syngenetic way of community establishment and continuous dispersal, *P. pumila* plants form such dense cover that it is impossible to determine the real quantity of individuals without destroying the stand.

This forced us to use the only acceptable measure of cone crop or similar estimation—we measured each parameter not per single tree or clump, but per square unit, hectare, or square meter. Certainly, in a thorough evaluation, stand structure and ground cover features would also have to be taken into account.

Table 14.2 *Pinus pumila* Cone Crop in Central Kamchatka Mountains

Number of Cones[a] per Shoot per Hectare		Altitude msl						
yr-1	yr-2	650	680	800	810	900	950	1030
		Germinating shoots bearing yr-2 cones only (%)						
–	1	81	100	93	86	63	8	32
–	2	7	0	2	8	4	0	6
	Subtotals	88	100	95	94	67	8	38
		Germinating shoots bearing yr-1 cones only (%)						
1	–	4	0	0	3	25	15	30
2	–	4	0	5	0	2	11	7
3	–	0	0	0	0	0	11	0
	Subtotals	8	0	5	3	27	37	37
		Germinating shoots bearing yr-1 and yr-2 cones (%)						
1	1	4	0	0	3	3	23	10
1	2	0	0	0	0	3	4	2
2	1	0	0	0	0	0	15	6
2	2	0	0	0	0	0	5	5
3	1	0	0	0	0	0	8	1
4	1	0	0	0	0	0	0	1
	Subtotals	4	0	0	3	6	55	25

[a]yr-1: immature first-year buds. yr-2: mature second-year cones.

(trunk) branch (Table 14.2). We also recorded the number of cones damaged by insects (*Cecidomyia pumila* damage yr-1 cones, *Eupithecia abietaria* damage yr-2 cones) and by birds (*Nucifraga caryocatactes kamtschatkensis,* nutcracker).

In each site 20–100 mature cones were collected and measured before being dried. Cones and seeds were measured for 16 parameters (unpublished data), five of which are reported in this chapter: mass of cones, seeds, and nuclei; number of seed scales per cone, and total seed quantity per cone.

Samples for geographic and macroecological comparison were collected during 1990–93 at other locations on the peninsula (Fig. 14.7, Table 14.3).

Figure 14.7. Points of data collection for assessment of geographical and ecological variability of *P. pumila* seed production. 1: (Thick lines) Main mountain ranges (schematic). 2: (Stars) Major volcanoes 3: (Numbered arrows) Points of data collection (see also Table 14.3 and Figure 14.6).

Table 14.3 Description of Cone Collection Points, Shown on Fig. 14.7

Block (Altitude msl)	Collection Points	Site Description
A (650–1200)	Eastern macroslope of Sredinnij mountain range.	Inner part of peninsula; most continental in climate.
	1: 650 m.	1–5: Bolgit field base.
	2: 800 m.	1, 2, 4: *P. pumila.*
	3: 800 m.	3: among stone birch on opposite W exposure slope.
	4: 1030 m.	
	5: 1200 m.	5: on opposite W exposure slope.
	6: Korale mountain in Bystraya River valley (400 m).	6: slope of partly destroyed volcano in narrow valley, among larch and white birch.
	7: upper part of Jurtinnaya River valley (1200 m).	7: inner part of peninsula, strong continental climate; almost upper limit of dwarf pine vegetation belt in highlands.
	8: upper part of Luntos River valley (700 m).	8: as 7, but at lower border of dwarf pine highland vegetation belt.
B (400–1000)	Northeastern mountains	
	9: Povorotniaya River valley (700 m).	9: highlands dwarf pine vegetation belt; subcontinental cold climate.
	10: Ozernaya River valley (450 m).	10: relatively closed river valley, typical for mountain ranges of E Kamchatka mountain forest–tundra ecotone.
	11: Holocene lava flows on western slopes of Tolbachik Volcano (800 m).	11: lava flows covered with thick layer of volcanic ash; protected by the slope from direct Pacific wet winds impact.
	12: tundra of western slope of Ploskaya Volcano (1000 m).	12: upper limit of dwarf pine highland belt, patches among tundra.
	13: Kronotskoye Lake (400 m).	13: shore of the lake on E macroslope of Vostochnij Mountain range, open to Pacific climate impact.
	14: Uzon Volcano (500 m) caldera.	14: upper mountainous vegetation in large intrazonal locality of thermal springs.
C (150–1000)	15: saddle between Avacha and Korjaka Volcanos (1000 m).	15: upper limit of dwarf pine highland belt on S exposure slope, open to Pacific (25 km from the coast); sheltered from the north by volcanoes.
	16: Sinichkino Lake shore (150 m).	16: dwarf pine clumps on hill tops among the secondary, mature stone-birch forest, 20 km from coast.
	17: western foothill of Korjaka Volcano (400 m).	17: closed cup-like valley, protected from the north by volcano and by moraine hills from the ocean.
D (25–50)	18–20: Ossora Bay, Pacific coast.	18–20: NE coast dunes, hills, and river valley, exposed to Pacific cold, wet winds.

(continued)

Table 14.3 (*continued*

Block (Altitude msl)	Collection Points	Site Description
E (200–500)	21–26: lower part of Kichiga River basin, 20 km from Pacific.	21–26: on hills and northern slopes of wide river valley; in small creek valleys. Pacific impact less than Block D.
F (<10)	27–38: Kichiga River delta, Pacific coast.	27–38: dunes and river banks close to the water; similar to D.
G (5–20)	39: Morjovaya Bay on Pacific coast. 40: seashore dunes near Nalacheva River delta.	39: coastal dunes and cliffs open to full ocean impact, in moderate latitude. 40: dune belt of SE coast.

Results and Comments

Crown Germination Structure and
Cone Crop Estimates (Table 14.2)

On the whole, the number of skeleton branches and germinating shoots per hectare did not vary much between 650–800 m. It varied more at higher elevations, which may be explained by a higher level of acceptable insolation. This was also the most likely explanation for the steep increase in the number of germinating shoots and, correspondingly, of yr-1 and yr-2 cone crops (Khomentovsky 1994).

Even slight shading was accompanied by a change in the number of shoots (900 m and 1030 m sites). The crop of yr-1 and yr-2 cones per germinating shoot had specific dispersal features. Very large numbers of cones were produced annually at each site, especially in the lower, shaded part of the elevation profile. Intensive seed production centers "migrate" from year to year, supporting conclusions of Khomentovsky and Efremova (1991) on the existence of local seedage mosaics, which provide a continuous seed supply. This is important not only for the guaranteed reproduction and microevolutionary diversity of *P. pumila* itself, but also for all the consumers (insects, mammals, nutcrackers, and other birds) that interact with *P. pumila*.

Cone and Seed Mass and Size Variation

Mass   The micromosaic variation in mass of yr-2 cones is most evident at higher elevations. In sufficiently insolated and wind-protected sites, cone mass increases; in shaded sites or open plateaus, it decreases (Fig. 14.8). At the uppermost elevation of *P. pumila* seed-production capability (1,200 m), there is an abrupt decrease of all seed-production parameters. It is remarkable that, in general, cone and seed masses are not directly correlated with altitude, even in the case of an obviously unfavorable environment (such as the 950 m site, open plateau).

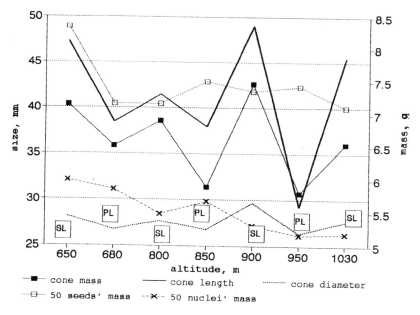

Figure 14.8. Variability of seed and cone size and mass within plot series in Bolgit field base. SL: slope. PL: plateau. Other characteristics of sites are shown in Table 14.1.

Self-regulation of seed production is also evident: The number of seed scales in the cone (with and without seeds) as well as the total number of seeds in the cone slightly increase along the elevation gradient. At the same time, seed and nuclei masses do not change, or slightly decrease. This means that seeds become smaller, but grow in greater quantity, keeping the same reproductive potential and increasing protection from environmental impacts.

Size   Structure of the landscape (slope gradient, width of valley or watershed, openness of terrain, etc.) strongly influences the microclimate mosaic. Cone length and mass, reflecting microclimatic variation, are more closely correlated with landscape structure than with elevation. Community structure is not too important here: All ecotopes, except the 950 m site, are occupied by variations of the same group of forest types, Pumilae pinetum cariosohypnoso-ericosum. Generally, cone size and mass at all elevations vary within limits known for the species (see also Data Analysis, Discussion, and Conclusion, below).

Damage Caused by Insects and Birds

Insect damage (Khomentovsky and Efremova 1991) is unlikely to cause considerable change in average-population cone size: There is almost no effect on cone diameter, and cone length decreases only up to 3–9%. However, in some cases damage by *Cecidomyia* results in nondevelopment of up to 20% of seeds in the cone. This usually does not stop development of the rest of the cone's seeds and is compensated at the population level by an abundance of seeds produced.

The level of damage caused by *C. pumila,* clearly visible as cone curvature, is rather high throughout most sites and varies depending on environmental and weather conditions to which the insects are sensitive. Damage was moderate (25–50%) in the 650 m site (slope in the narrow valley), 800 m site (slope in the wide valley), and 900 m site (wind-protected watershed at the upper limit of larch); maximal (100%) in the 810 m site (watershed ridge, partly protected from the wind by the upper slope); and minimal (8%) in the 1030 m site (weakly wind-protected and too cold for insects).

Geographical Variation of Seed Production

Material collected from many parts of the peninsula (Figs. 14.7 and 14.9, Table 14.3) reveals that seed production does not depend on site elevation. A slight increase in the number of scales (more than the number of seeds) per cone shows a larger percentage of empty seeds in highlands. Cone size and mass of seeds and nuclei slightly decrease with elevation, but here, as with all other parameters, we see only tendencies, not considerable changes. Also, no noticeable differences were related to the

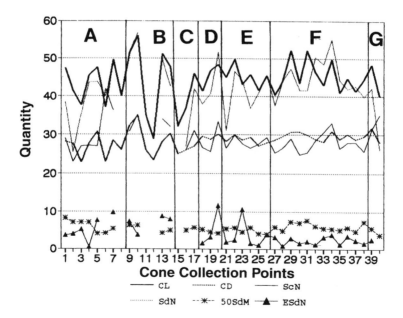

Figure 14.9. Varying parameters of *P. pumila* seed production along the Kamchatka Peninsula. x-axis: cone collection points (see also Figure 14.7 and Table 14.3). CL: cone length (mm). CD: cone diameter (mm). ScN: number of scales in the cone. SdN: number of seeds in the cone. 50SdM: mass of 50 seeds (g). ESdN: number of empty seeds in the cone. A–G: blocks from Table 14.3. All estimates are averages from 30–100 measurements, $P < 0.05$.

distance of a site from the sea, or climatic continentality of the environment. All sites showed corresponding rates of variability for most parameters. Proper detailed assessment can be done only by multifactor environmental analysis at the landscape (or microsite) level, on the basis of key-plot data. For example, site 5 shows poor microclimate conditions on the open slope at the dwarf tree's limit. However, site 4 shows a more favorable environment in a wind-protected watershed only 500 m from, and 170 m lower than, site 5, but on the opposite part of the creek valley.

Another example is found in Block B (Fig. 14.7, Table 14.3). Here we can see a mixture of impacts: more or less comfortable conditions for *P. pumila* in inner mountains facing the Pacific (sites 9 and 10), a strongly contrasting environment near volcanoes (sites 11 and 12), and warming effects of the lake or hot-spring fields (sites 13 and 14). It is evident that site quality is much more important for *P. pumila* than latitude, longitude, altitude, or large-scale distance from the coast.

Date Analysis, Discussion, and Conclusion

The main conclusion is that *P. pumila,* distributed over vast territory with numerous types of habitats, worked out such a high potential of seed production in time and space that we can speak about the environmental (abiotic and natural biotic) impact on it in terms of modification only, but not in terms of regulation. Environmental stresses—such as a sudden summer-night frost killing female buds, insect population outbreaks, etc.— sometimes can considerably lower the crop size, but do not cause a crisis in crop size: the plants have some compensating mechanisms both at the organism and population level. One mechanism is the formation of annual seed production microhabitat mosaics; another is the increase in seed number under stressful conditions.

Pinus pumila seed production in Kamchatka is not significantly dependent on altitude and general site condition; of these, landscape conditions below facies level are of greatest importance. However, this is true only when plants are not too shaded by neighboring upright trees, such as when *P. pumila* forms an independent vegetation belt or is in relatively solid fragments, as in lowland or mountain tundra, in upper forest-tundra ecotones, on seashore dunes, and in wide river valleys. When *P. pumila* grows as the lower canopy of relatively dense larch forests (more than 30% *Larix* ground cover), which have the same age and plant community origin (very often, in pyrogenic ecotopes), the *Larix* shades the dwarf pine due to its faster growth. Under these conditions *P. pumila* usually does not produce seeds, but develops vegetatively, with the help of adventitious roots, during its whole life.

The utmost upper limits of distribution (1,300–1,400 msl) are unfavorable for *P. pumila* seed production, mainly due to low temperature. Even here some cones can be found, and sometimes the crop can be abundant in wind-sheltered habitats. Phenological delay often prevents nutcrackers from finding seeds here, resulting in high-altitude seed dispersal by birds from lower sites. Trees at lower sites have more stable genotypes; this provides a better chance for successful reforestation in new areas. At the same time, *P. pumila* regeneration at high altitudes is basically provided by the remaining abundant seeds, which are spread mainly by small mammals.

Unfavorable conditions for seed production are also found along the Pacific Kamchatka seacoast on the dunes closest to the water. However, *P. pumila* seed production on the second to third dune of the coastal belt is equal in cone size and mass parameters to trees at the 900–1,000 m altitude, and in seed and nuclei parameters, to those at 650–750 m altitude in the continental-climate part of the peninsula.

Moderately favorable conditions for seed production are found in plains and foothills, on gravel-sandy alluvial, fluvioglacial, proluvial deposits (so-called "dry rivers"), burned areas, and other open ground surfaces with good drainage and full sun.

Most of the seed-productive zones, where cone crops are more or less stable from year to year, lie within 600–900 msl in inner regions of the peninsula, where the climate is either subcontinental or continental. The seed-productive zone widens and lowers toward sea level near the coast, providing proof of coastal vegetation zonation.

Considering all known features of *P. pumila* seed production, we can conclude that this species prefers open spaces, can tolerate rather severe conditions, and in many cases historically had to accommodate itself to life with upright trees. Upright trees, as species and colonizers, came geochronologically later, when the climate moderated enough for them.

Successful seed production of *P. pumila* requires the following conditions for normal development of the species: in winter, sufficient snow cover to protect buds of low-lying trees from freezing; in summer, high insolation and, in cold highland sites, good wind protection, full sun for crowns, and good drainage and aeration for root systems.

Aside from seed production, the important adaptive features of the plant in summer (still not investigated completely) are, presumably, the presence of mycorrhiza and the ability to produce adventitious roots.

The main adaptive feature of *P. pumila* in late autumn and winter is its ability to bow close to the ground and lie prostrate at low temperatures before and under snow cover. This feature provides for good chances of survival in the most extreme conditions. Strict dependence of *P. pumila* on snow-cover depth is shown by coincidence of its area with a snow-depth isocline of 40 cm and greater (Lukicheva 1964, Tstcherbakova 1964).

Productivity site mapping is needed for the practical, silvicultural estimation of current or future cone crops of *P. pumila*. An example is the scope of crop diversity. At the middle elevation site (700–800 m), with 5 g seed mass per cone, 80% germinating shoots, and 80% *P. pumila* ground cover, we can get about 110 kg seeds per hectare. At the upper limit of seed production, with 2 g seed mass per cone, 30% germinating shoots, and 10% cover, we can get about 1 kg/ha.

High variability of cone crops can also be shown by the data in Table 14.2. If we take into account only area purely covered by *P. pumila* (without considering its real distribution on the surface), we see that crop size in the 650 m site is about 6.5 kg/ha, while in the 950 m site, it is about 212 kg/ha.

Starting from 900–1,000 m elevation, which corresponds to the upper limit of upright tree distribution in Kamchatka, *P. pumila* produces an increasing amount of shoots in what may be seen as compensation for the increasing severity of abiotic environmental conditions. Photosynthetic ability increases, seed-producing potential remains at the same high level, and seed protection (thickness of seed walls) slightly strengthens. All of this occurs mostly in the subalpine belt and indicates that *P. pumila* is primarily a subalpine plant.

Coincidence of these processes with the upper limit of upright tree distribution indicates the existence of some kind of temperature-threshold trigger. It may be the sum of effective temperatures, above which the development of upright tree forms is possible, and below which the development of prostrate forms only is possible.

We believe that *P. pumila* and other stone pines have high evolutionary potential. This is revealed in the prominent and widespread polymorphism of reproductive organs and their functional flexibility (Pravdin and Iroshnikov 1982). Ecological flexibility results from keeping all important generative parts safe for effective reproduction (Table 14.4). It is noticeable not only from within-species similarity of parameters, measured in different geographical areas and site conditions (Efremova and Ivliev 1972, Krylov et al. 1983, Rylkov and Skvortsov 1984), but also from the fact that one of the main parameters, mass of nuclei and seeds, is very stable not within species only (as we showed above), but also in other stone pines (Blada and Popescu 1992).

It appears that *P. pumila*'s seed-production potential (which is one of the species' most important characteristics) exceeds the possibilities of its realization in this region. Habitat diversity in Kamchatka is evidently only a part of the whole scope acceptable for the dwarf pine. With this in mind, we should look for possible centers of its origin in regions with a broader spectrum of sites than Kamchatka.

We think that *P. pumila* is a middle-altitude species, connected in its origin and contemporary existence with the influence of wet Pacific air masses, but growing better in relatively humid mountains than at the seashore itself. The latter can be illustrated by the fact that it reaches its highest level of polyembryony at the Pacific coast (Iroshnikov 1982). We also noticed that the strong maritime climate influenced dwarf pine ecotypical distribution and biomass production (unpublished data) at the eastern shore of Lake Baikal, the farthest west, continental region of *P. pumila* growth.

Combining our observations and conclusions with other hypotheses of *P. pumila* evolution (Tikhomirov 1949, Sochava and Lukicheva 1953, and others), we think that this dwarf tree most likely appeared as a species in Tertiary times in climatically subcontinental mountain regions of Angarida (northeast Asia), not too far from the Pacific. At Kamchatka it is dated by pollen records to 1–1.5 million years ago (Malaeva 1967, Chelebaeva et al. 1974). During Pleistocene interglacial periods and in the early and middle Holocene, *P. pumila* occupied periglacial zones and other places unacceptable for upright tree species. Later, *P. pumila* shared sites with strong com-

Table 14.4 Some Parameters of *P. pumila* Seed Production on the Kamchatka Peninsula[a]

	Cone Length (mm)	Cone Diam. (mm)	Total Scales in Cone	No. of Scales Without Seeds	No. of Seeds in Cone	Mass of 50 Seeds (g)	Mass of 50 Nuclei (g)
Average	45.0	29.0	31.2	6.4	42.4	5.2	2.7
Stand deviation	5.7	2.2	7.5	4.0	10.3	1.2	1.0

[a]Number of samples: more than 1000. $P < 0.05$.

petitors, but during the last 2–2.5 thousand years, *P. pumila* again began to spread widely, apparently due to global climatic changes (Egorova and Khomentovsky 1988) favorable for boreal Pacific plant species preferring a cool, humid environment.

With its wide, strong reproductive potential, Siberian dwarf pine serves as a "permanent pioneer" in each acceptable, plant-free place—fresh volcanic ash fields, burned tundra, proluvial "dry river" deposits, coastal dunes, etc. If not disturbed by human activity, this unique creeping tree-cover will continue to provide ecosystem stability for vast regions of severe northeastern Eurasia.

Acknowledgments   I am very grateful to A.N. Smetanin, O.A. Chernyagina, A.S. Valentsev, V.A. Churikova, and other colleagues who helped collect samples, and to I.G. Khomentovskaya and J.V. Savenkova for taking numerous measurements.

References

Atlas of the USSR forests. 1973. Moscow.

Blada, I., and N. Popescu. 1992. Variation in size and weight of cones and seeds in four natural populations of Carpathians *Pinus cembra*. Materials for Internat. Workshop "Subalpine Stone Pines and Their Environment," St. Moritz. Manuscript.

Chelebaeva, A.I., A.E. Shantser, I.A. Egorova, and E.G. Lupikina. 1974. Cenozoic deposits of the Kurilo-Kamchatka area. Pages 31–57 *in* I. V. Luchitsky, ed. Kamchatka, Kuril Islands and Commander Islands (Series on History of Relief Development.) Nauka, Moscow.

Dmitrieva, E.V. 1978. *Pinus pumila* at the southern limit of its range: mensurational and synecological characteristics. Botanicheskii Zhurnal. **63(9)**:1352–1358. (In Russian.)

Efremova, L.S., and L.A. Ivliev. 1972. On dwarf stone pine seed production at Kamchatka. Part 2. Pages 158–159 *in* Utilization and reproducing of forest resources in the Far East. Abstracts of the all-USSR conference. Khabarovsk. (In Russian.)

Egorova, I.A., and P.A. Khomentovsky. 1988. Siberian dwarf pine as an indicator of volcanic activity. Volcanology and Seismology **6**:82–88. (In Russian.)

Grossett, H.E. 1959. *Pinus pumila* (Pall.) Rgl.: Materials on its biological study and economical utilization. Proceedings on the study of the fauna and flora of the USSR, edited by Moscow Soc. of Natural History. New series. Sect. of Botany. Moscow. No. **12 (XX)**. (In Russian.)

Hayashi, Y. 1960. Taxonomical and phytogeographical study of Japanese conifers. Norin-Shuppan, Tokyo.

Iroshnikov, A.I. 1982. Seed production and seed quality of conifers in northern and mountainous regions of Siberia. Pages 98–117 *in* Seed production of Siberian conifers (collection of papers). Novosibirsk, Russia. (In Russian.)

Khomentovsky, P.A. 1994. A pattern of *Pinus pumila* (Pall.) Regel seed production ecology in the mountain of central Kamchatka. Pages 67–77 *in* Schmidt, W.C. and F.K. Holtmeier (compilers). Proceedings of International Workshop "Subalpine Stone Pines and their Environment." USDA For. Serv. Tech. Rep. INT-GTR-309. Ogden, Utah.

Khomentovsky, P.A., and I.A. Egorova. 1990. An essay of *Pinus pumila* (Pall.) Rgl. formation history at Kamchatka in late Cenozoic. Pages 60–68 *in* Proceeding of international symposium "Boreal forests: state, dynamics, anthropogenic impact," held in Arkhangelsk, Russia. Part 3. Moscow. (In Russian.)

Khomentovsky, P.A., and L.S. Efremova. 1991. Seed producing and cone-feeding insects of *Pinus pumila* (Pall.) Regel at Kamchatka peninsula: aspects of coexistence. Pages 316–320 *in* Y.N. Baranchikov, W.J. Mattson, F.P. Hain, and T. L. Payne, eds. Forest Insects Guilds: Patterns of Interaction with Host Trees. U.S. Dept. Agric. For. Serv., Gen. Tech. Rep. NE-**153**: 316–320.

Khomentovsky, P.A., and I.G. Khomentovskaya. 1990. Geographical variability of *Pinus pumila* seed producing at Kamchatka. Pages 47–55 *in* V.N. Vinogradov, ed. Questions of Kamchatka Geography. Petropavlovsk-Kamchatsky, Asia-Russia. No. 10. (In Russian.)

Khomentovsky, P.A., N.V. Kazakov, and O.A. Chernyagina. 1989. Tundra-forest zone at Kamchatka: problems of conservation and management. Pages 30–46 *in* J.P. Mikhailov and V.M. Parfenov, eds. Problems of Nature Management in Taiga Zone. Institute of Geography. Irkutsk, Russia. (In Russian.)

Komarov, V.L. 1927. Flora of Kamchatka peninsula. USSR Acad. of Sci. Publ., Leningrad. Vol. 1. (In Russian.)

Krylov, G.V., N.K. Talantsev, and N.F. Kozakova. 1983. Siberian pine (*Pinus sibirica* Du Tour). Forest Industry Publ. Moscow. (In Russian.)

Lavrenko, E.M., and V.B. Sochava, eds. 1954. Geobotanical map of the USSR. Scale 1:4,000,000. Botanical Institute of the USSR Acad. of Sciences. (In Russian.)

Lukicheva, A.N. 1964. Distribution areas of Asian plants (a map). Page 112 *in* I. P. Gerasimov, ed. Geographical atlas of the world. Moscow. (In Russian.)

Malaeva, E.M. 1967. Kamchatka vegetation development in Pliocene-Pleistocene. Pages 78–170 *in* Development of Siberian and far-eastern vegetation in Quaternary. Moscow. (In Russian.)

Pravdin, L.F., and A.I. Iroshnikov. 1982. Genetics of *Pinus sibirica* Du Tour, *P. koraiensis* Sieb. et Zucc. and *P. pumila* Regel. Annales Forestales. Vol. **9(3)**:79–123.

Rylkov, V.F., and N.I. Skvortsov. 1984. Dwarf stone pine cone and seed parameters at low cone crops. Inform. Bull. of Chitinsky Information Center. Chita, Russia. Pages 72–84. (In Russian.)

Sochava, V.B., and A.N. Lukicheva. 1953. On geography of dwarf stone pine. Reports of the USSR Academy of Sciences **90(60)**: 1163–1166. (In Russian.)

Sokolov, I.A. 1973. Volcanism and soil formation (with Kamchatka as an example). Nauka, Moscow. (In Russian.)

Tikhomirov, B.A. 1949. Siberian dwarf pine, its biology and utilization. Materials on the study of the fauna and flora of the USSR, issued by Moscow Soc. of Natural History. New series. Sect. of Botany. Moscow. No. 6 (XIX). (In Russian.)

Tstcherbakova, E.J. 1964. An average of maximal decade snow depth (a map). Page 220 *in* I.P. Gerasimov, ed. Geographical Atlas of the World. Moscow. (In Russian.)

15

Robert Ornduff

The *Sequoia sempervirens* (Coast Redwood) Forest of the Pacific Coast, USA

Since its discovery in the late 18th century, scientists and lay persons alike have been impressed by the magnificence of the stately *Sequoia sempervirens* (coast redwood) and the dense forests it forms. Yet this conifer is one of the world's most restricted tree species, occurring only along a very narrow portion of the Pacific coast of California and Oregon. Within this narrow range, however, *S. sempervirens* exerts a powerful ecological influence in its ecosystem. This chapter describes the morphological features of *S. sempervirens,* its geographical distribution and the factors that limit this distribution, other organisms that occur with it, ecological dynamics of its forests, its fossil record, its economic values, and its conservation status. It also poses some questions concerning *S. sempervirens* ecology that remain to be answered.

Sequoia sempervirens (D. Don) Endlicher, a member of the bald cypress family (Taxodiaceae), is naturally restricted to the coastal fog belt of the coast ranges of California from southern Monterey County northward to the Oregon border; from there it extends slightly northward into Curry County, in the extreme southwestern portion of Oregon (Fig. 15.1). This very long-lived tree grows rapidly, achieving heights of over 100 m, and may form pure stands where no other tree species can compete successfully. The crown is typically pyramidal or narrow, and the slender branches spread horizontally or droop (Fig. 15.2). The tapered trunk may reach over 6 m in diameter at the base and has a characteristic stringy, spongy, reddish, ridged bark to 3 dm (nearly 1 foot) thick. The trunk may be markedly flanged or buttressed at its base. The species lacks a tap root, but produces an extensive "fan" of horizontal roots just below the soil surface. The evergreen leaves and the branchlets that bear them persist for up to four years, then are dropped from the tree to form a thick litter beneath. Leaves of the lower branches are sharp-pointed, linear, bright green, to 2.5 cm long, have short petioles, are two-ranked, and form flat sprays (Fig. 15.3a). Leaves on the upper branches often are short, awl-shaped, to 1 cm long, and are densely and spirally arranged on the branches (Fig. 15.3c). These upper leaves strik-

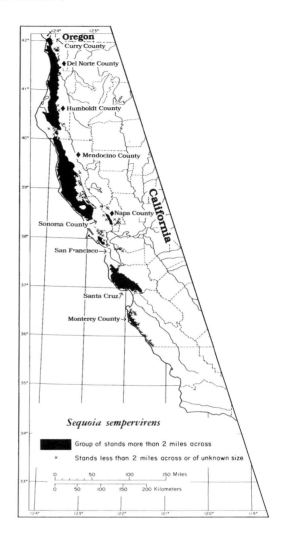

Figure 15.1. Distribution of *S. sempervirens* (coast redwood) in California and southwestern Oregon. (From Griffin and Critchfield 1972.)

ingly resemble those of *Sequoiadendron giganteum,* giant sequoia, of the western slopes of the Sierra Nevada in California.

Sequoia sempervirens is monoecious and wind-pollinated, with small, stalked, green, nearly globose or ovoid male cones 2–5 mm long. These are borne singly in axillary or terminal positions on the branchlets. Pollen is shed between late November and early March. The mature female cones are more or less spherical, reddish brown, 15–35 mm long and to 1.3 cm wide, with 15–25 shield-shaped, spine-tipped woody scales (Fig. 15.3a, c). Seeds mature during the first autumn and are shed beginning in September. They are elliptic in outline, light brown, with a narrow spongy wing, and are approximately 5–6 mm long (Fig. 15.3b). Seeds lack dormancy and germination takes from 3 to 6 weeks. Seed viability is variable and may be as low as 1% (Olson et al. 1990). Ger-

Figure 15.2. Habit of *S. sempervirens* in an area that has been logged; the standing trees are referred to as "culls." (From Jepson 1923.)

mination percentages thus may be poor. W.J. Libby (personal communication) reports seed viabilities of up to 60% in some natural populations and suggests that viability levels may be controlled in part by environmental conditions.

Distribution

Sequoia sempervirens is restricted to the coastal fog belt of the Pacific coast of central and northern California and extreme southwestern Oregon (Griffin and Critchfield 1972; Fig. 15.1). This fog results from a combination of factors that develop during the summer. A coastward shift in the Pacific High during the spring months is associated with northwesterly winds that churn upwellings of cold water from the lower depths of the Pacific Ocean along the California–Oregon coast. As a consequence, offshore summer ocean temperatures may be lower than those in the winter. The westerly winds, in their long course over the ocean, pick up considerable quantities of moisture. When these moisture-laden winds reach the coastal zone of cold water, the moisture condenses and forms the heavy persistent fogs that hang over much of the California coast during late spring and summer.

   Sequoia sempervirens ranges from just north of Salmon Creek Canyon in the Santa Lucia Mountains of extreme southern Monterey County of California (lat. 35° 41′N) in an interrupted pattern northward in the Coast Ranges to the Chetco River in Curry County, southwestern Oregon, about 20 km north of the California border, where three small groves occur (lat. 42° 09′N). Two of the Oregon groves are near the Chetco River, and the third is along the Winchuck River (Sudworth 1967). The total north–south range covers roughly 725 km (Fig. 15.1). Significant gaps in the range occur in the vicinity of Monterey Bay, the southern San Francisco Bay region, north-

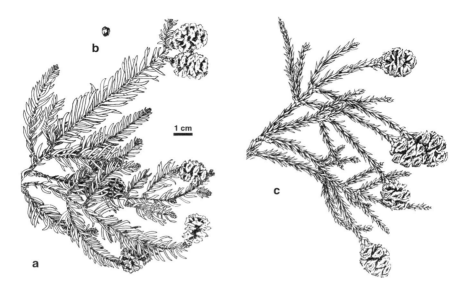

Figure 15.3 a. Upper branch of *S. sempervirens* showing needle-like leaves and mature female cones. b. Seed of *S. sempervirens*. c. Upper branch of *S. sempervirens* showing scale-like leaves and mature female cones. (From Jepson 1923.)

ern Marin and southern Sonoma counties, and southern Humboldt County, all in California (Sudworth 1967, Griffin and Critchfield 1972). The majority of trees occur within 50 km of the coastline; the most interior populations occur on Howell Mountain in Napa County north of San Francisco Bay, about 72 km from the coast, but in an area under coastal influence. *Sequoia sempervirens* forests reach their best development in Humboldt and Del Norte counties of northern California. It has been estimated that at the time of European settlement of what is now California and Oregon, *S. sempervirens* occupied approximately 800,000 hectares (ha) (Zinke 1977). Although many old-growth stands of *S. sempervirens* have been cut for lumber, the overall range of the species has been reduced little as a result of logging. Many areas once occupied by old-growth *S. sempervirens* forests now support a mixture of old-growth, second-growth, and third-growth forests (Rydelius and Libby 1993).

Sequoia sempervirens grows on canyon slopes and river flats on the seaward side of the Coast Ranges in California (Jepson 1910); in extreme northern California and in Oregon, the species is said to be more typical of slopes than of river bottoms (Franklin and Dyrness 1973). At the southern margin of its range, the species is said usually to occur in canyons, because the summer fogs are channeled up these canyons (Henson and Usner 1993). The species grows best in wet, well-drained, highly fertile areas along the coast. The biomass of the *S. sempervirens* forest type is extraordinary, having been estimated at a maximum of 2,300 metric tons per ha (Franklin and Dyrness 1973). The elevational range of the tree is from sea level to about 1,000 m. On river flats the forest is commonly composed almost exclusively of *S. sempervir-*

ens, although other species of trees grow in the well-insolated gallery forests lining the water courses. These include *Alnus* spp. (alders), *Umbellularia californica* (California bay, sometimes called California laurel or pepperwood), and *Acer macrophyllum* (big-leaf maple). On slopes above these flats, as conditions become hotter and drier, *S. sempervirens* may grow mixed with *U. californica, Lithocarpus densiflorus* (tan oak, sometimes called tanbark oak), *Pseudotsuga menziesii* (Douglas fir), and *Arbutus menziesii* (madrone).

Sequoia sempervirens rarely grows along the immediate coast, since the trees are not tolerant of strong, salt-laden maritime winds. Soils of this coastal strip may also be unfavorable for this species' growth (Olson et al. 1990). From Marin County to Sonoma County, California, the coastward side of the *S. sempervirens* forest may be occupied by *Pinus muricata* (bishop pine), *P. menziesii,* or other conifers. From Sonoma County, California, northward, the vegetation on the coastward side of the redwood belt may be composed of these conifers plus *Picea sitchensis* (Sitka spruce), *Abies grandis* (grand fir), and *Tsuga heterophylla* (western hemlock). In the southern portion of its range, the coastward side of the *S. sempervirens* forest may be occupied by a mixture of various hardwood species, *P. menziesii,* and even grasslands or scrublands. At the interior margin of its range, *S. sempervirens* may grow mixed with various hardwood species more typical of drier, hotter inland sites; as conditions become hotter and drier to the interior, *S. sempervirens* eventually gives way to hardwood forests, *P. menziesii* forest, various valley and foothill woodland types, chaparral, or valley grassland. Zinke (1977) presented an informative set of diagrams showing the occurrences of major tree species associated with *S. sempervirens* along a series of west to east transects from the southern to the northern portion of its range (Fig. 15.4).

Ecological Factors Influencing Distribution

Factors limiting the distribution of *S. sempervirens* appear to differ in various portions of its range. The coastward distribution of this species appears to be limited chiefly by a lack of deep, rich, moist soils as well as by the presence of persistent, strong, onshore, salt-laden winds. The northern distribution in Oregon may be limited in part by lack of suitable topography for some distance north of the Chetco River, by too-low winter temperatures (Sternes 1960), and by competition from other coastal conifers such as *P. sitchensis, Thuja plicata* (western red cedar), and *T. heterophylla.* Rainfall totals outside extreme southwestern Oregon may also be too low for *S. sempervirens,* since the mean annual precipitation of southwestern Oregon is high, but drops sharply inland and to the north (Sternes 1960). The eastern edge of the *S. sempervirens* range likely is determined by too-dry and too-hot summers and a lack of the moderating influence of summer fogs. The southern range of the species may be limited by unsuitable substrates and steep coastal topography, too-high summer temperatures, and too-low annual precipitation.

The distribution of *S. sempervirens* appears to be static: There is no evidence that the range of the species is (naturally) expanding or contracting at present. Rydelius and Libby (1993) have commented on the "paradox that redwoods, once established, are among the least likely of tree species to be killed by insects and disease, but

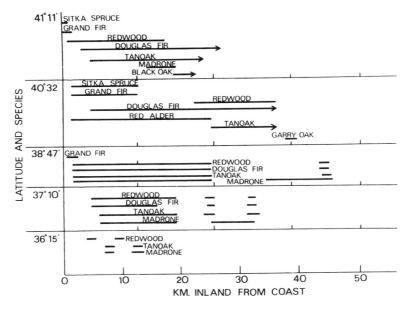

Figure 15.4. The occurrence of major tree species of north coastal forests on transects inland from the ocean at various latitudes, arranged from north to south. Arrows indicate that the species extend further inland. (From Zinke 1977.)

redwood germinants and first-year seedlings are unusually susceptible to soil pathogens." It is possible that the disease-prone nature of seedlings may be important in limiting the natural range of this species.

So-called "glades" or "balds" are characteristic of the *S. sempervirens* belt; these are treeless areas dominated by grasses and forbs, often very extensive in area, and usually on slopes. These balds are believed to result from soils with a high clay content and a relatively high pH (Zinke 1977), conditions that favor herbaceous plants over woody ones. Balds may also result from unstable and/or seasonally waterlogged soils; neither favor the establishment of trees. The ecotones between these balds and the adjacent *S. sempervirens* forest often resemble truncated versions of the vegetation types found to the east or west of the redwood belt.

Precipitation in the *S. sempervirens* belt ranges from 64 to 310 cm annually, and most of it falls during the winter months. Snow is uncommon and of short duration. However, the prolonged summer fogs that may persist unbroken for days serve to moderate air temperatures and increase atmospheric humidity. The fog-drip resulting from the condensation of fog on the foliage of *S. sempervirens* may add 25 cm or more effective summer precipitation to the soil under the trees (Johnson 1994). Soils commonly are derived from marine sandstone, but other parent materials are locally significant, as are alluvial deposits (Zinke 1977, Olson et al. 1990). High-site soils for the species are Xerochrepts, Haploxerults, and Haplohumults of various series

(Olson et al. 1990). *Sequoia sempervirens* avoids serpentine, of which there are extensive outcrops in its range. Mycorrhizal associates of *S. sempervirens* are vesicular-arbuscular fungi of the Glomales; there are probably several taxa involved, but these have not been identified taxonomically (Molina 1994, T. Bruns, personal communication). Howell (1970) and Shuford and Timossi (1989) described substrate requirements and floristic associates of *S. sempervirens* in Marin County. Zinke (1977) and Johnson (1994) listed additional plant species associated with *S. sempervirens* elsewhere in its range.

Sequoia sempervirens is a long-lived, massive tree. The growth rate is exceptionally rapid: up to 42 $m^3/ha^{-1}/yr^{-1}$ in natural populations (Zinke 1977) or more than 50 $m^3/ha^{-1}/yr^{-1}$ in a French plantation (W.J. Libby, personal communication). Specimens frequently exceed 1,000 years in age, with the oldest known individual 2,200 years old. The record height for the species is reported as 111 m (Johnson 1994) and specimens over 60 m tall are common, particularly on deep, moist soils in the northern portion of the range. Specimens in the southern portion of the range reach only about 60 m in height (Henson and Usner 1993).

Dominance

Sequoia sempervirens provides an excellent illustration of the ecological concept of dominance: a dominant species is one whose presence restricts the distribution and abundance of other plant species through its effect on the local environment. *Sequoia sempervirens* exerts dominance in several ways. The trees cast considerable shade, with the result that only the most shade-tolerant herbs and shrubs can grow and reproduce under them. In addition, the litter beneath the trees is often many centimeters thick, composed mostly of branchlets that have fallen from the trees throughout the year. Seeds that fall on this litter sift down through it until they reach a point several inches below the litter surface; if they germinate, the seedlings may die before their foliage reaches the poorly lit litter surface. While one might expect that under cool, moist conditions the litter would decompose rapidly, apparently decomposition is a fairly slow process, perhaps due to a lack of compaction, the resins and tannins present in the plant tissues, and high soil acidity. Dominance is also exerted via root competition: the shallow root system of *S. sempervirens* is pervasive under the trees.

Sequoia sempervirens is tolerant of fire, in part due to the fire resistance of the very thick bark on the lower trunks and in part because of its ability to produce vegetative sprouts from the trunk following a fire. The species is tolerant of prolonged flooding and the resultant siltation, provided the deposited sediments are free-draining (Marden 1993). Individuals can produce new roots from the trunk immediately below the raised soil surface after several feet of silting. Potential tree competitors cannot tolerate such siltation; thus, *S. sempervirens* stands occurring on repeatedly flooded alluvial flats may harbor no other tree species. Once past the seedling stage, *S. sempervirens* is highly disease-resistant. This combination of ecological traits makes *S. sempervirens* a very successful competitor under the distinctive, restricted climatic regime to which it is so well suited.

Community Status

In a pioneer classification of the vegetation and plant communities of California, Munz and Keck (1949a, b) recognized Redwood Forest as a distinct plant community belonging to the Coniferous Forest vegetation type. They indicated a growing season of 6–12 months, with 200–350 frost-free days, and little diurnal or seasonal temperature change. Mean summer maxima were given as 68–84°F (20–29°C) and mean winter minima as 33–40°F (1–4.5°C). As typical plant associates of *S. sempervirens,* they listed *P. menziesii, Myrica californica* (wax-myrtle), *L. densiflorus, Gaultheria shallon* (salal), *Rhododendron macrophyllum* (California rose-bay), *Oxalis oregana* (redwood sorrel), *Vancouveria planipetala* (as *V. parviflora,* inside-out flower), *Polystichum munitum* (western sword fern), and *Whipplea modesta* (modesty). In Oregon, most of these plant associates are present in the *S. sempervirens* forest (Franklin and Dyrness 1973). While the ranges of these associated species include the range of *S. sempervirens,* they are all more extensively distributed than is *S. sempervirens.*

Mason (1947) pointed out that if the geographic ranges of plant species associated with *S. sempervirens* are plotted,

> not a single associated species approximates a complete coincidence of geographic area with that occupied by the redwood. . . . Here is a forest whose unity of concept is based upon the total range of a single species, *Sequoia sempervirens,* together with those associated species whose geographic ranges chance to overlap into its area. . . . It has reality only by virtue of its definition in terms of the occurrence of *Sequoia sempervirens.*

In addition, Mason emphasized changes in the species composition of the flora associated with *S. sempervirens* through time; he noted, for example, that *Vaccinium ovatum* (California huckleberry) is probably a post-Tertiary entrant into *S. sempervirens* forests.

Some of the understory associates of *S. sempervirens* appear to thrive in the conditions imposed by this tree: *O. oregana,* for example, flowers under poor light intensities, produces large colonies via clonal growth, and appears to be unable to grow in areas subjected to full sun during much of the day. In contrast, adults of *P. munitum* tolerate full sun, yet this fern also thrives in the shade of *S. sempervirens.* Commonly, some of the shrubby and herbaceous species growing under *S. sempervirens* do not flower or fruit; such plants probably became established when light conditions under the canopy were more favorable, and persist as nonreproductive individuals because of the subsequent development of suboptimal light intensities. Light gaps resulting from fallen trees are locally important for establishment of seedlings of *S. sempervirens* and associated understory species, and the trunks of the fallen trees serve as "nurseries" for these seedlings. Some observers (Olson et al. 1990) have indicated that the age structure of *S. sempervirens* stands suggests fairly frequent seedling establishment.

Classification

Ornduff (1974) adopted a modified version of the Munz-Keck system and included their Redwood Forest in North Coastal Forest, referring to the region of North

Coastal Forest occupied by *S. sempervirens* as a "phase" of North Coastal Forest. He characterized *S. sempervirens* forests as "esthetically very pleasing but rather monotonous botanically."

In the southern portion of its range, three "phases" of *S. sempervirens* forest have been recognized (Henson and Usner 1993): Pure Redwood Forest, Redwood–Riparian Forest, and Redwood–Mixed Hardwood Forest. The first phase occurs on moist, north-facing slopes or adjacent to streams. The understory is poor, with few or sometimes no vascular plants present. Redwood–Riparian Forest occurs along canyon bottoms where running water is present year-round; because of the increased light conditions caused by the corridor of water, as well as enhanced soil moisture levels, here *S. sempervirens* occurs with hardwood species such as *L. densiflorus, Platanus racemosa* (western sycamore), *Alnus rhombifolia* (white alder), *A. macrophyllum,* and *U. californica.* The herbaceous and shrubby flora of this phase is relatively species-rich. Redwood–Mixed Hardwood Forest occurs on slopes and represents an ecological transition between Redwood Forest and an adjacent mixed hardwood forest more tolerant of warm, dry conditions. Here, *L. densiflorus, A. menziesii, U. californica,* and *Quercus agrifolia* (coast live oak) grow mixed with *S. sempervirens.* These species associations also occur elsewhere in the range of *S. sempervirens,* being duplicated at least as far north as Marin County, California. Other plant associates of *S. sempervirens* were listed by Olson et al. (1990) and Waring and Major (1964).

Another description of California plant communities, sponsored by the California Department of Fish and Game, was a provisional one authored by Holland (1986). It recognized Alluvial Redwood Forest and Upland Redwood Forest. Alluvial Redwood Forest occurs away from the immediate coast on alluvial flats with deep, well-drained soils, primarily in Del Norte and Humboldt counties, California. This forest is often shrouded in fog, is periodically flooded, but is also occasionally burned. It was described as "moderately dense" and similar to the Western Hemlock Forest better developed on the coast to the northward; however, it is less dense and less diverse floristically than the latter.

Upland Redwood Forest contains *S. sempervirens* that are shorter than those in the alluvial community, and which grow with several tree species and understory shrubs that are rare or absent from alluvial sites. The effects of summer drought on plant growth are greater here than on alluvial flats, and the effects of fire may be more evident. Soils are drier, shallower, and more eroded than under alluvial conditions, and on exposed slopes the trees may be severely wind-trimmed. Tree species associated with *S. sempervirens* in this community are *A. macrophyllum, Quercus chrysolepis* (canyon live oak), *L. densiflorus, P. menziesii, U. californica,* and *Torreya californica* (California nutmeg). Shrubs include *G. shallon, R. macrophyllum, Rubus* spp. (blackberries) and *V. ovatum.* This community occurs from Curry County in southwestern Oregon to southern Monterey County in California. It intergrades with several other forest communities or with chaparral in various parts of its range, and may occur adjacent to and intermixed with Alluvial Redwood Forest.

Holland and Keil (1989) recognized Coast Redwood Forest as a single community, but acknowledged the ecological and floristic differences that occur throughout the range of this species. They emphasized the narrow ecological tolerance of this tree, noting that it cannot survive prolonged freezing temperatures or prolonged summer drought. They also suggested that the importance of summer fog-drip is overestimated by some ecologists. These authors noted that in areas where fires have been suppressed, a dense understory of *L. densiflorus, U. californica, A. macrophyllum,* and *A. menziesii* may develop, and that this understory enhances the flammability of such areas and increases the intensity of fires. An excellent concise account of the floristics and general ecology of *S. sempervirens* forests was given by Johnson (1994).

Successional Status

The successional status of *S. sempervirens* is unclear and may vary from site to site (Franklin and Dyrness 1973). Because the trees are very long-lived and can propagate vegetatively, successional status is not easy to observe. Often, seedling establishment is not frequent; it and juvenile growth are best in moderately well-lit areas with exposed mineral soil. Such areas develop as a result of landslides, flooding, fire, or fallen trees (as well as from human disturbances such as logging, land clearance, road building, etc.). Olson et al. (1990) suggested that the presence of *S. sempervirens* of several ages in natural stands indicates episodic seedling establishment. This issue merits further study.

Some plant ecologists believe that *S. sempervirens* is successional and dependent on fires and floods for perpetuation (Stone and Vasey 1968). Flooding, however, would not constitute a significant environmental factor away from riparian situations. In some areas, *T. heterophylla* and *L. densiflorus* are considered to be climax species in the *S. sempervirens* belt (Franklin and Dyrness 1973). Waring (according to Franklin and Dyrness 1973) considered *S. sempervirens* to be a climax species although he "acknowledges" the significance of fire and flooding in the reproduction of this species. Likewise, Olson et al. (1990) stated that "the redwood forest is a climax type."

Future Research

Zinke (1977) discussed successional events in *S. sempervirens* and associated forest types following natural or human-induced disturbances and outlined several areas of future research needed in order to understand the ecological relationships of this species. He stated that "there needs to be more study of the autecological requirements of the various species that comprise" the forest vegetation of the northern California coast, noting that most of the coniferous tree species in this region reach their southern range limit in California. He attributed this to the presumption that the southern limits of these conifers are determined by the "greater aridity and warmer temperatures" that occur in this region, but wondered whether these limits are due to occasional extremes of drought, or to gradients in available moisture. He also asked at what stage in the species' life history are limiting factors operative: at the seedling level, or during later growth phases? In regard to *S. sempervirens,* Zinke noted that various environmental factors work together to limit the natural range of this species, but the nature of their interaction is not understood. He wondered whether the range of the species is extending, or still retreating. The past role of fires in determining the distribution of *S. sempervirens* and associated vegetation types needs to be clarified, and the nature of present human impacts on these coastal forests remains to be understood fully. Zinke suggested that effective management plans for stands of *S. sempervirens* currently under protection may require managed-fire cycles in plans spanning 500–1,000 years.

The low seed viability common in *S. sempervirens* has long been known, yet apparently the basis for this condition is not understood. This may be related in part to meiotic disturbances or genetic anomalies associated with polyploidy in this conifer. It is unknown whether *Sequoia* has been hexaploid since its first appearance in the fossil record. Some insights into the time of origin of polyploidy in the genus might be gained by measurements of epidermal cell size in fossils, because cell size often increases with an increase in ploidy level. Measurements of cell sizes in fossil remains of *Sequoia* extending over the full historical and geographical occurrence of the genus would be informative.

Vegetative Reproduction

Sequoia sempervirens is unusual for a conifer in its ability to produce vegetative shoots from the trunk following injury (Sudworth 1967, Rydelius and Libby 1993). After a tree has been mechanically injured, cut by logging, or severely burned by fire, vigorous and numerous shoots develop from the base of the tree, or if the crown has been burned badly, they develop from the crown as well. Some trees produce large thickets of shoots around their bases even in the absence of injury; the cause of this is unknown. In areas that have been logged, trees of considerable size may be present in circles around the trunk diameter of the cut mother tree. Since these "offspring," which may number a dozen or more, occupy roughly the same area as their parent, it is unclear that this behavior has a positive adaptive value. Sudworth (1967), however, remarked that trees originating by such vegetative means "are long-lived and produce large trees of good form."

Other Organisms

Sequoia sempervirens forests harbor many terrestrial fungi, but relatively few of these are large, fleshy species. Of the latter, *Caulorhiza umbonata* is "perhaps the most distinctive redwood-lover" (Arora 1986); Arora lists several species and genera of fleshy fungi typical of *S. sempervirens* forests. The ranges of many mammal species include the *S. sempervirens* forests, but no mammal species appear to be restricted to these forests. In California, *Sorex bendiri* (marsh shrew), *S. pacificus* (Pacific shrew), *Scapanus orarius* (coast mole), *Arborimus albipes* (white-footed vole), *A. longicaudus* (red tree vole), and *Zapus trinotatus* (Pacific jumping mouse) all have ranges that more or less coincide with that of *S. sempervirens*. Some of these mammals, however, may be more typical of forest margins or gaps than of the forests themselves (Jameson and Peeters 1988). The sole remnant population of *Cervus elaphus roosevelti* (Roosevelt elk) in California is in the *S. sempervirens* forests of Humboldt County.

The insect fauna of the Californian coastal coniferous forests is rich, but since these forests represent southern extensions of forests that are much better developed to the north, it is not surprising that few species of insects are known to be endemic to these forests in California (Powell and Hogue 1979). Among amphibians, *Plethodon dunni* (Dunn's salamander) and a color form of *Aneides flavipunctatus* (black salamander) appear to be restricted to *S. sempervirens* forests (Stebbins 1972). The avifauna of *S. sempervirens* forests is not a notably rich one. In California, these forests are important nesting sites for the otherwise maritime *Brachyramphus marmoratus* (marbled murrelet), whose nesting habits were unknown until 1974 (Cogswell 1977). The uncommon *Strix occidentalis caurina* (northern spotted owl) occurs in coastal coniferous forests of California and the Pacific Northwest. Beginning in the late 1970s, this owl became the center of still-continuing heated debates centering on the degree to which logging practices in its habitat jeopardize its populations, and whether conservation agencies' plans for ensuring long-term viability of the bird are sufficient (Thomas et al. 1990).

Geological History

Sequoia sempervirens has an extensive fossil history extending back roughly 100 million years to the late Cretaceous and continuing up to the Pleistocene (Axelrod 1976). The genus is known only in fossil remains from the northern hemisphere; these have been found in southern Alaska, western and central North America, Greenland, Spitsbergen, western Europe, central and southern Asia, and Japan. In texts published before the late 1940s, the geological history of *Sequoia* is inaccurately represented due to an understandable and interesting confusion between redwoods now assigned to two different genera, *Sequoia* and *Metasequoia* (Sand 1992).

Sequoia sempervirens forest as we know it today must be viewed as a floristic association of relatively recent origin, whose restricted range has been recently occupied, whose climatic regime is of recent origin, and which has become increasingly depauperate floristically over the past tens of millions of years. Many of these changes are associated with a dramatic deterioration in the climatic regime over much of the northern hemisphere during the Tertiary. In western North America there has been a gradual increase in mean summer temperatures, decrease in annual precipitation, and over much of this region, restriction of rains to the winter months; doubtless, these climatic changes were responsible for a radical diminution in the range of *S. sempervirens* and a loss of many associated species from its forests (Raven and Axelrod 1978).

Taxonomic Relationships

The family Taxodiaceae in the restricted sense contains ten genera, all but three consisting of a single species each (Mabberley 1993). These genera occur in eastern Asia, North America, and, oddly, Tasmania, with no genus shared among any of these regions. Recent studies employing different lines of evidence do not agree completely on the closest relatives of *Sequoia,* but it is probable that its closest relative is giant sequoia, *Sequoiadendron giganteum,* and that these two genera in turn have as their closest relative the (now) Asian *Metasequoia* (Brunsfeld et al. 1994).

Genetics

Sequoia sempervirens is a hexaploid with $n = 33$ (Stebbins 1948). The origin of this unusual (for conifers) high chromosome number is unknown. Although it has been suggested that the species may represent an allopolyploid (i.e., a polyploid containing the chromosome sets of two or more species) derived from hybridization between *Metasequoia* and another unspecified genus, this is unlikely (Sclarbaum and Tsuchiya 1984). The levels of genetic variation throughout the limited range of *S. sempervirens* are relatively high (W.J. Libby, personal communication). Inbreeding effects in *S. sempervirens* seem to be rather slight, though under conditions of stress they may become more evident (Libby et al. 1981). Superior genotypes can be propagated by means of tissue culture and are of economic significance in re- or afforestation (Rydelius and Libby 1993). Artificial hybrids of *S. sempervirens* with *Sequoiadendron, Taxodium,* and *Cryptomeria* have been reported (Olson et al. 1990), but these reports need confirmation.

Economic and Conservation Considerations

Sequoia sempervirens has been harvested for lumber on a commercial scale at least since the 1820s, when lumbering commenced in the hills behind Oakland, California, on San Francisco Bay. The California gold rush that began in 1848 spurred considerable demand for lumber, and by 1860 there were at least 300 lumber mills in California; most of these were in Mendocino and Humboldt counties in the northwestern portion of the state and were devoted largely to milling the lumber of *S. sempervirens*. Logging practices during the 19th century have been termed "criminally wasteful" (Barbour et al. 1993); cut stumps 3 or more meters high were left after trees had been cut, and various postharvesting practices led to substantial land erosion.

If old-growth stands are defined as those that have never been subjected to any logging practices, there are now in California approximately 24,300 ha (60,000 acres) of old-growth *S. sempervirens* stands protected in state parks; 8,100 ha (20,000 acres) protected in Redwood National Park (founded in 1968); and about 600 ha (1,500 acres) protected in county parks. Approximately 77,000 additional ha (190,000 acres) in these parks are occupied by *S. sempervirens* that are not old growth. Although old-growth stands that are being harvested will soon be gone, managed second-growth stands are periodically harvested and will continue to produce marketable timber indefinitely. There are about 2,800 ha (7,000 acres) of old-growth *S. sempervirens* in California owned by a private lumber company, and 405 ha (1,000 acres) of old-growth forest owned by various individuals. It is estimated that about 364,000 ha (900,000 acres) of *S. sempervirens* stands are under some form of active management for lumber harvesting, and that an additional 324,000 ha (800,000 acres) have been converted to other uses such as agriculture or urbanization.

In Oregon, around 525 ha (3,300 acres) of old-growth *S. sempervirens* forest are within the boundaries of Siskiyou National Forest. In that state, an additional 5,670 ha (14,000 acres) of land occupied by *S. sempervirens* are in public or private hands and are subject to logging. By the time the San Francisco-based Save-the-Redwoods

League was formed in 1918, about two-thirds of the merchantable *S. sempervirens* in California had been logged. This organization has been the prime mover in the preservation of natural areas with *S. sempervirens,* and during its 76 years of existence has contributed over $76,000,000 to state and federal agencies for the preservation of stands of this tree in public parks (figures supplied by John Dewitt, Executive Director of the Save-the-Redwoods League; see also Dewitt 1993).

The timber yield of *S. sempervirens* varies according to the ages (hence sizes) of the trees in the stands. Yields range from between 10,500 and 14,000 m³/ha on alluvial flats to 742 m³/ha on "low" sites with smaller trees (Olson et al. 1990). Young, managed stands can be highly productive, but conversion of old-growth stands to young managed stands is difficult because of negative net growth for a decade following logging (Olson et al. 1990). Clearcutting is suggested as a method for making the conversion economical (Olson et al. 1990). Managed forests may be treated aerially with herbicides to eliminate tree competitors of *S. sempervirens;* unfortunately, this practice also eradicates a number of the herbs and shrubs normally associated with the tree.

The *S. sempervirens* region of California is an important tourist attraction, especially during the summer, and is a significant economic asset for the region. That the public today can visit and enjoy the magnificent redwood forests reflects the foresight of those in governmental and private agencies who worked for so many decades to set aside these ancient forests as a legacy for future generations.

Acknowledgments I thank Bill Libby for his careful and constructive reading of an early draft of this chapter, Tom Bruns for information on mycorrhizal associates, and John Dewitt for providing detailed information on the conservation status of *S. sempervirens.*

References

Arora, D. 1986. Mushrooms demystified, 2nd edition. Ten Speed Press, Berkeley, CA.

Axelrod, D.I. 1976. History of the coniferous forests, California and Nevada. Univ. Cal. Publ. Bot. **70**:1–62.

Barbour, M., B. Pavlik, F. Drysdale, and S. Lindstrom. 1993. California's changing landscape. California Native Plant Society, Sacramento, CA.

Brunsfeld, S.J., P.S. Soltis, D.E. Soltis, P.A. Gadek, C.J. Quinn, D.D. Strenge, and T.A. Ranker. 1994. Phylogenetic relationships among the genera of Taxodiaceae and Cupressaceae: evidence from *rbc*L sequences. Syst. Bot. **19**:253–262.

Cogswell, H.L. 1977. Water birds of California. Univ. of California Press, Berkeley.

Dewitt, J.B. 1993. California redwood parks and preserves, 3rd edition. Save-the-Redwoods League, San Francisco.

Franklin, J.F., and C.T. Dyrness 1973. Natural vegetation of Oregon and Washington. U.S. Dept. Agric. For. Serv. Gen. Tech. Rep. PNW-8. U.S. Gov. Printing Office, Washington, DC.

Griffin, J.R., and W.B. Critchfield. 1972. The distribution of forest trees in California. U.S. Dept. Agric. For. Serv. Res. Paper PSW-82/1972.

Henson, P., and D.J. Usner. 1993. The natural history of Big Sur. Univ. of California Press, Berkeley.

Holland, R.F. 1986. Preliminary descriptions of the terrestrial natural communities of California. Dept. Fish and Game, Sacramento.

Holland, V.L., and D.J. Keil. 1989. California vegetation. El Corral Publications, California Polytechnic State Univ., San Luis Obispo.

Howell, J.T. 1970. Marin flora, second edition with supplement. Univ. of California Press, Berkeley.

Jameson, E.W., Jr., and H.J. Peeters. 1988. California mammals. Univ. of California Press, Berkeley.

Jepson, W.L. 1910. The silva of California. The Univ. Press, Berkeley.

———. 1923. The trees of California. Associated Students Store, Univ. of California, Berkeley.

Johnson, V.R. 1994. California forests and woodlands. Univ. of California Press, Berkeley.

Libby, W.J., B.G. McCutchan, and C.I. Millar. 1981. Inbreeding depression of selfs of redwood. Silvae Genetica **30**:15–25.

Mabberley, D.J. 1993. The plant-book. Cambridge Univ. Press, Cambridge, UK.

Marden, M. 1993. The tolerance of *Sequoia sempervirens* to sedimentation, East Coast region, New Zealand. N. Z. Forestry, November:22–24.

Mason, H.L. 1947. Evolution of certain floristic associations in western North America. Ecol. Monogr. **17**:201–210.

Molina, R. 1994. The role of mycorrhizal symbioses in the health of giant redwoods and other forest ecosystems. Pages 78–81 *in* P.S. Aune, coordinator. Proceedings of the symposium on giant sequoias: their place in the ecosystem and society. U.S. Dept. Agric. For. Serv. Gen. Tech. Rep. PSW-GTR-151. Pac. Southwest Res. Sta., Albany, CA.

Munz, P.A., and D.D. Keck. 1949a. California plant communities. El Aliso **2**:87–105.

——— . 1949b. California plant communities–supplement. El Aliso 2:199–202.

Olson, D.F., Jr., D.F. Roy, and G.A. Walters. 1990. *Sequoia sempervirens* (D. Don) Endlicher. Pages 541–551 *in* R.M. Burns and B.H. Honkala, coordinators. Silvics of North America, Volume 1, conifers. U.S. Dept. Agric. Serv. Agric. Handbook 654.

Orenduff, R. 1974. Introduction to California plant life. Univ. of California Press, Berkeley.

Powell, J.A., and C.L. Hogue. 1979. California insects. Univ. of California Press, Berkeley.

Raven, P.H., and D.I. Axelrod. 1978. Origin and relationships of the California flora. Univ. Cal. Publ. Bot. **72**:1–134.

Rydelius, J.A., and W.J. Libby. 1993. Arguments for redwood clonal forestry. Pages 159–168 *in* M.R. Ahuja and W.J. Libby, eds. Clonal forestry II, conservation and application. Springer-Verlag, Heidelberg.

Sand, S. 1992. The dawn redwood. Am. Horticulturist, October issue:40–44.

Sclarbaum, S.E., and T. Tsuchiya. 1984. A chromosome study of coast redwood, *Sequoia sempervirens* (D. Don) Endl. Silvae Genetica **32/33**:56–62.

Shuford, W.D., and I.C. Timossi. 1989. Plant communities of Marin County. Special Publication Number 10 of the California Native Plant Society, Sacramento.

Stebbins, G.L. 1948. The chromosomes and relationships of *Metasequoia and Sequoia*. Science **108**:95–98.

Stebbins, R.C. 1972. Amphibians and reptiles of California. Univ. of California Press, Berkeley.

Sternes, G.L. 1960. Climates of the states: Oregon. U.S. Department of Commerce Weather Bureau climatography of the United States No. 60–35. U.S. Govt. Printing Office, Washington, DC.

Stone, E.C., and R.B. Vasey. 1968. Preservation of coast redwood on alluvial flats. Science **159**:157–161.

Sudworth, G.B. 1967. Forest trees of the Pacific slope. Dover Publications, New York.

Thomas, J.W., E.E. Forsman, J.B. Lint, E.C. Meslow, B.R. Noon, and J. Verner. 1990. A conservation strategy for the northern spotted owl. U.S. Gov. Printing Office, Washington, DC.

Waring, R.H., and J. Major. 1964. Some vegetation of the California coastal redwood region in relation to gradients of moisture, nutrients, light, and temperature. Ecol. Monogr. **34**:167–215.

Zinke, P. 1977. The redwood forest and associated north coast forests. Pages 679–698 *in* M. Barbour and J. Major, eds. Terrestrial vegetation of California. John Wiley, New York.

16

Ingrid Olmsted & Rafael Durán García

Distribution and Ecology of Low Freshwater Coastal Forests of the Yucatán Peninsula, Mexico

Tropical freshwater coastal forests on the Yucatán Peninsula of Mexico (Fig. 16.1) are dominated by trees and shrubs capable of growing in areas that may be inundated almost all year with water depths up to 1 meter. The same species have to be able to withstand severe droughts for several months. In general, the forests are of low stature and are dominated by microphyllous, evergreen species, with twisted trunks, stratified crowns, and several dominant species with spines (Olmsted and Durán 1986).

We have encountered at least six communities of this forest type on the Yucatán Peninsula, each dominated by a different tree species. The dominance by one tree species instead of several, as is common in other tropical, dry forest types on the Yucatán Peninsula (Olmsted and Durán 1990), puts greater emphasis on the environmental factors that may be responsible for this situation.

Limestone karst underlies these forests, which grow on a variable cover of gley soils and marl. The geologic explanation for the occurrence of these forests is not known; however, they only occur in depressions, which are often elongate and parallel the coast.

These low, inundated forests—the literal translation of the Spanish *selva baja inundable* (Lot-Helgueras 1983, Olmsted et al. 1983, Olmsted and Durán 1986, Lot-Helgueras and Novelo 1990, and Olmsted 1993)—are principally distributed along the coast in the Mexican states of Quintana Roo, Yucatán, Campeche, and Tabasco (Lundell 1934, Standley and Record 1936, Miranda 1958, Rico Gray 1982, and Flores 1984). Similar forests also occur in the Peten region of northern Guatemala (Gonzalez Fuster, personal communication) and Belize (Dobson 1973). Lot-Helgueras and Novelo (1990) concentrate their descriptions and analyses on the states of Veracruz and Tabasco in southeastern Mexico, while our observations are concentrated on the northern Yucatán Peninsula.

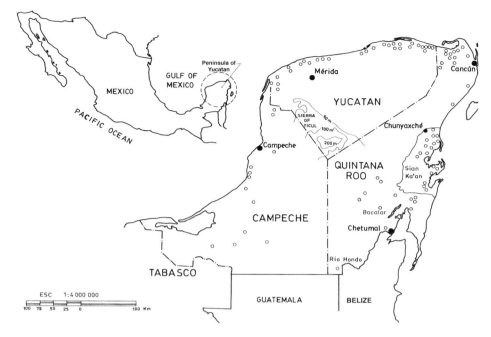

Figure 16.1. Map of Yucatán Peninsula. Open dots: approximate sites of low freshwater coastal forests (forests vary in size and shape). Black dots: towns and villages.

The distribution of these forests has been reduced during the last 100 years, especially due to habitat alteration for cattle grazing. Historically, the community dominated by logwood, *Haematoxylon campechianum* L., was cut heavily between 1600 and 1900 by the English, Spanish, and Germans and shipped to Europe for use as a natural black dye (Dobson 1973, Dachary and Arnaiz 1983). According to Dobson, the importance of *H. campechianum* to England was cause for the establishment of the colony of British Honduras (Belize).

This chapter describes the distribution and ecology of the low-stature freshwater coastal forests of the Mexican portion of the Yucatán Peninsula. We make observations for the whole peninsula and report the analysis for an area in the Sian Ka'an Biosphere Reserve.

General Description of the Yucatán Peninsula

Most of the Yucatán Peninsula has a subhumid, hot climate (Aw) according to W. Koeppen and modified by Garcia (1964). Annual precipitation usually averages between 1,000 and 1,500 mm, falling mostly in the summer. More humid climates exist in Tabasco, and the very northwestern tip of the peninsula has a tropical, dry climate with about 600 mm of annual precipitation. During the summer and fall, hurricanes may arrive from the Caribbean and during the winter, northerly storms may reach the peninsula from the Gulf.

The Yucatán Peninsula is a great limestone platform of about 175,000 square km, covered by carbonate rocks and sediments of Tertiary to Holocene age (Weidie 1985). Physiographically, the Yucatán Peninsula is part of the Gulf Coastal Plain Province of Mexico. The surficial carbonates display a variety of karst features. One of the four physiographical regions of the peninsula is the eastern block fault district (Isphording 1975). This area extends parallel to the Caribbean coast and is characterized by north-northeast trending ridges and depressions that reflect the occurrence of horst and graben blocks of the Rio Hondo Fault Zone. Isphording (1975) noted the alignment of streams, lakes, and swales in this region, such as Rio Hondo, Lake Bacalar, and Lake Chunyaxche (Fig. 16.1).

Recent applications of remote-sensing data in northeastern Quintana Roo (Southworth 1985) suggest an extension of the Holbox fracture system to about 100 km south of the north coast. This fracture system is observed as water-filled swales or linear depressions. It is in this fault system as well as the Rio Hondo fault system, the eastern fault block district, and the Reforma-Campeche fault system that most of the coastal freshwater forests are found.

The Yucatán Peninsula has a flat topography, with no surficial streams. Rainwater infiltrates rapidly through the porous limestone and flows subterraneously toward the coast. Elevations along the coast are usually not higher than 5–15 m mean sea level (msl); the highest point on the peninsula measures approximately 300 m in the Sierra de Ticul.

Distribution of the Low Freshwater Coastal Forest

Figure 16.1 shows the distribution of the low freshwater coastal communities. Because this type of forest occurs in small extensions in a lot of different places and because of the scale of this map, it is very difficult to show exact distributions. The points on the map, therefore, show the general locations of forests but not their extent.

The seven forest communities we will describe are located along a 10- to 40-km strip inland from the coast of the peninsula; however, they also occur further inland, especially in the Peten region of Guatemala.

The size of the forest stands ranges from 1 hectare (ha) or less to several hundred ha. The communities may be circular, in concentric rings of forest and marsh, or in elongate formations as mentioned earlier.

Ecological Aspects

We have observed, on various trips and during studies over the last six years, six communities of the low freshwater coastal forest and one palm swamp. The six forested communities belong to the same vegetation type, the low inundated forest. Each community is dominated by a different species (Table 16.1). The "mucal" community is dominated by the shrub *Dalbergia glabra;* the "bucidal" forest is dominated by the tree *Bucida spinosa* (Fig. 16.2). The "pucteal" forest is dominated by *B. buceras* (common Mayan name *pucté*), and *H. campechianum* (common Spanish name *tinte,* Fig. 16.3) is the dominant tree of the "tintal." Two other communities, which have

Table 16.1 Summary of the Seven Communities of the Low Freshwater Forest on the Yucatán Peninsula

Community Type	Dominant Species	Dominant Life Form	Common Name	Figure
Low inundated forest				
Mucal	*Dalbergia glabra* (Miller) Standley	shrub	*muk*	n.a.
Bucidal	*Bucida spinosa* (Northrop) Jennings	tree	(none)	Fig. 16.2
Puctea	*B. buceras* (L.)	tree	*pucté*/black olive	n.a.
Tintal	*Haematoxylon campechianum* L.	tree	*palo de tinte*/logwood	Fig. 16.3
Unnamed (observed only)	*Erythroxylum confusum* Britton	tree	*kancab'ché*	Fig. 16.4
Unnamed (observed only)	*Conocarpus erecta* L.	tree	*botoncillo*	n.a.
Palm swamp				
Tasistal	*Acoelorrhaphe wrightii*[a]	palm	*tasiste*/paurotis palm	n.a.

n.a.: not available
[a](Griseb. & H. Wendl. ex Griseb.) H. Wendl. ex Beccari

Figure 16.2. Drawing of *Bucida spinosa,* of the bucidal forest, with branch showing leaves and spines (drawings by Mauro Gomez J.).

Figure 16.3. Drawing of *Haematoxylon campechianum* of the tintal forest.

not been studied but only observed, are each dominated by *Erythroxylum confusum* (*kancab'ché* in Maya; Fig. 16.4) and *Conocarpus erecta* (*botoncillo* in Spanish). The palm swamp "tasistal" is dominated by the palm *Acoelorrhaphe wrightii* (*tasiste* in Maya, paurotis palm in English).

We carried out an ecological study of the coastal forest in the Sian Ka'an Biosphere Reserve in Quintana Roo, Mexico (Olmsted and Duran 1986). We describe five communities: four inundated forests and a palm swamp, which is a closely associated vegetation type.

Environmental Factors

The forests primarily occur on marl and gley soils. Marl is found in areas with deeper water and longer periods of inundation. Organic gleysols are found in areas that have shorter hydroperiods. Marl contains a high degree of calcium carbonate, is low in nutrients, and has poor drainage. Marls are often covered by a layer of periphyton, an association of blue-green algae and diatoms (Gleason and Spackman 1974). The organic gley-

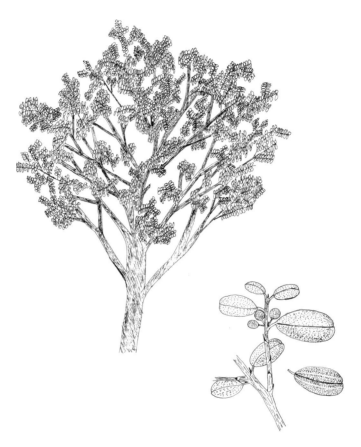

Figure 16.4. Drawing of *Erythroxylum confusum* with branch bearing leaves and fruit. Detached leaf shows abaxial side.

Table 16.2 Soil Depths and Periphyton Cover in the Five Communities Studied

Community	Soil Type	Depth (cm) x ± 1sd[a]	Periphyton (%)
Tasistal	marl	17.5 ± 3.78[b,c]	97
Mucal	marl and gleysol	9.0 ± 3.43	94
Bucidal	marl and gleysol	9.8 ± 2.59	24
Tintal	gleysol	8.9 ± 0.77	none
Pucteal	gleysol	11.8 ± 1.64	none

[a]sd, standard deviation.
[b]Significantly different from mucal and bucidal.
[c]n = 12/community; N = 60.

sols are higher in nutrients and better drained than the marls. Table 16.2 presents the type of soil and depth as well as periphyton cover associated with each community.

The low freshwater coastal forest occurs in depressions that may become inundated for various periods of the year. The hydroperiods and water depths (Fig. 16.5A,B,C) vary from one community to another depending on elevation and soil. Because the rains are unpredictable in time and space, hydroperiods may vary significantly from year to year, as shown in Table 16.3 for three of the five communities.

Figure 16.6 is a profile of vegetation, elevation, and soil depth of a tintal. The latter demonstrates the small elevation change of 30–40 cm necessary to make the difference between essentially a marsh and an inundated forest.

Floristic Composition and Structure

In each of the four forest communities in the Sian Ka'an Biosphere Reserve, we sampled four plots each of 200 m². We found 110 species, of which 65 were trees or shrubs, 25 were epiphytes, and 20 were herbs. The distribution of species by life form in each habitat is shown in Table 16.4. The number of tree species from tasistal to pucteal occurs along a gradient of decreasing hydroperiod, from the tasistal and mucal communities (longest inundation) to the bucidal (intermediate) to the tintal and pucteal (shortest). In comparison with other tropical forests, the number of species represented in these forests is low. Meave del Castillo (1983) mentions 82 tree species in a 2,500-m² area of tall evergreen rainforest in Bonampak, Chiapas, Mexico. Durán García (1986) found 80 species of trees and shrubs in a low subdeciduous forest in Quintana Roo, Mexico. The distribution of the orchids and bromeliads follows the same gradient as the trees. A list of the species found in each forest type is given in the Appendix.

The tasistal is a combination of marsh and palm swamp, with a greater cover of graminoids than trees. The palms measure up to 4 m. The mucal is a dense community of trees, shrubs, and graminoids where the tree/shrub component covers more ground and measures up to 5 m. The low stature of trees and abundance of shrubs gives this community a shrubby aspect. Bucidal, tintal, and pucteal forests are dominated by trees that measure up to 9 and 12 m. The densest tree community is found in the tintal forest with 13,733 individuals/ha, followed by the pucteal (10,000 individuals/ha) and bucidal forests (5,000 individuals/ha; Fig. 16.7).

Table 16.3 Mean Number of Days per Year with Surface Flooding, for Three Communities Measured Biweekly for Two Years, 1983–1985

	Tasistal 5 stations ± 1sd[a]	Mucal 6 stations ± 1sd	Bucidal 3 stations ± 1sd
1983–1984	$x = 315.2 \pm 22$	$x = 317 \pm 19$	$x = 122 \pm 26$[b]
1984–1985	$x = 260 \pm 46$	$x = 252 \pm 11$	$x = 185 \pm 14$

[a]Standard deviation (sd) refers to stations.
[b]Data only from November 1983 to June 1984.

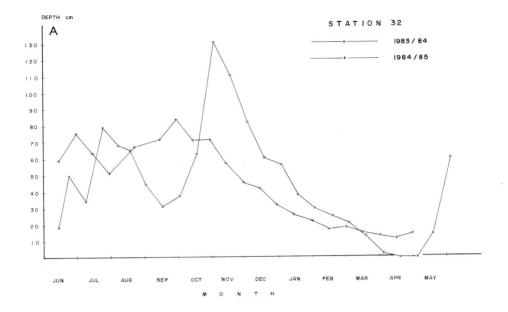

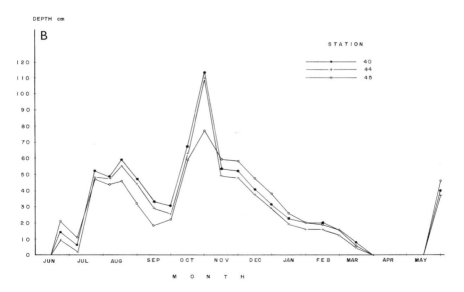

Figure 16.5. Depths of water above soil surface. A. Tasistal 1983–1985. B. Mucal 1983–1984.

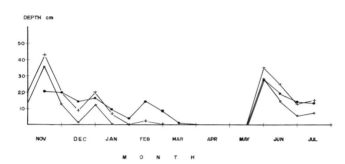

Figure 16.5. (*continued*) C. Bucidal 1983–1984.

Table 16.5 presents the 15 most important tree species in each community, giving the importance value based on relative cover, density, and frequency. Some of the trees presented in this table are more typical of noninundated forests. They are *Manilkara zapota, Metopium brownei,* and *Coccoloba floribunda* and occur in the least inundated forests.

Of all the species mentioned in Table 16.5, seven have spines of various kinds (*H. campechianum, B. spinosa, Dalbergia glabra, B. buceras, Randia aculeata, A. wrightii,* and *Jacquinia aurantiaca*). Eleven of the 15 most important trees are microphyllous. *Manilkara zapota, Me. brownei, Co. floribunda,* and *Croton niveus* have larger leaves. This microphylly may be an adaptation to the severe changes from inundation to drought. Twelve species are evergreen.

Table 16.6 presents the values of Sorensen's similarity index (Sorensen 1948) for the four forest communities based on tree species alone. The most similar communities are pucteal and bucidal (72%) and pucteal and tintal (65%). The results indicate that the vegetation type (low inundated forest) is the same for the bucidal, tintal, and pucteal

Table 16.4 Number of Species by Life Form in the Five Communities

Life Form	Total	Tasistal	Mucal	Bucidal	Tintal	Pucteal
Tree/shrub	65	2	11	32	35	49
Epiphytes	25	1	7	15	17	19
Herbs	20	9	14	6	1	1
Total	110	12	32	53	53	69

Table 16.5 Importance Values[a] of the 15 Most Important Tree Species (> 2.0 m)

Species	Tasistal	Mucal	Bucidal	Tintal	Pucteal
Acoelorrhaphe wrightii	175(1)[b]	58(2)	—	—	—
Dalbergia glabra	—	152(1)	11(9)	—	3(25)
Bucida spinosa	—	—	107(1)	—	—
Haematoxylon campechianum	—	—	9(12)	45(1)	—
Byrsonima bucidaefolia	—	—	14(6)	45(1)	17(4)
Bucida buceras	—	—	2(23)	—	55(1)
Erythroxylum confusum	—	19(4)	26(2)	20(3)	22(3)
Randia aculeata	—	—	18(3)	11(6)	14(5)
Cameraria latifolia	—	11(7)	12(7)	11(5)	8(12)
Jacquinia aurantiaca	—	12(6)	2(22)	(37)	(33)
Malpighia lundellii	—	15(5)	5(13)	10(9)	5(21)
Gymnopodium floribundum	—	—	4(17)	19(3)	7(15)
Croton niveus	—	—	—	13(4)	24(2)
Coccoloba floribunda	—	—	15(4)	11(7)	5(18)
Eugenia buxifolia[c]	—	—	(25)	5(16)	14(6)

[a]Importance value = relative cover + relative density + relative frequency (Curtis 1959).

[b]Numbers in parentheses are the ranks in each community.

[c]*Manilkara zapota, Ateleia gummifera,* and *Metopium brownei* follow in importance.

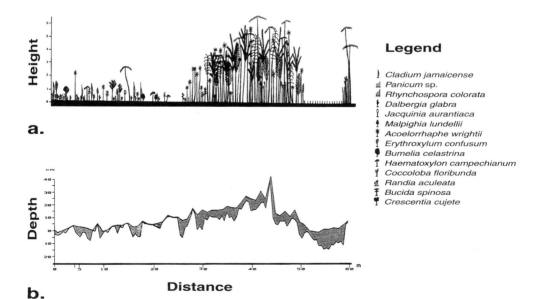

Legend

- Cladium jamaicense
- Panicum sp.
- Rhynchospora colorata
- Dalbergia glabra
- Jacquinia aurantiaca
- Malpighia lundellii
- Acoelorrhaphe wrightii
- Erythroxylum confusum
- Bumelia celastrina
- Haematoxylon campechianum
- Coccoloba floribunda
- Randia aculeata
- Bucida spinosa
- Crescentia cujete

a.

b.

Distance

Figure 16.6. A. Vegetation profile for a disturbed mucal/palm swamp community and a tintal (the taller vegetation) during dry season. B. Elevation and soil profile (shaded) for disturbed mucal/palm swamp during dry season.

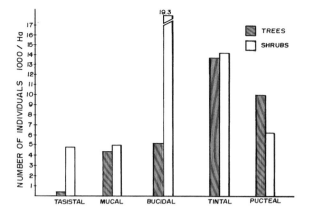

Figure 16.7. Density of trees and shrubs in the five communities.

communities (with 54–72% similarity). The similarity of the shrubby mucal community to the forested communities diminishes according to the hydroperiod gradient.

We found ten epiphytic species of bromeliads, one cactus, and 14 species of orchids in the five communities (Olmsted and Durán 1986). With regard to the distribution of epiphytes, the bucidal community is host to most individuals observed, especially of the species of *Tillandsia* in the Bromeliaceae, followed by the tintal forest, in which we found more orchid species and individuals.

Ecological Characteristics of Dominant Tree Species

The species described here were selected on the basis of their uniqueness in this vegetation type, not necessarily because of their significance as indicated by the importance values.

Bucida spinosa

Bucida spinosa (Fig. 16.2), which dominates in the bucidal, has a restricted distribution in Quintana Roo, Cuba, and the Bahamas. Within Quintana Roo, the only Mexi-

Table 16.6 Sorensen's Similarity Index for Four Communities, Based on Number of Tree Species

	Mucal	Bucidal	Tintal	Pucteal
Mucal				
Bucidal	53			
Tintal	31	54		
Pucteal	29	72	65	

can state in which it grows, *B. spinosa* was until a few months ago only known from the Sian Ka'an Biosphere Reserve. However, the authors found discontinuous populations in the remote wetlands of northeastern Quintana Roo. It is found inland up to 50 km from the coast. *Bucida spinosa* was reported for the first time in 1981 on the Yucatán Peninsula (Cabrera et al. 1981).

Bucida spinosa has very small linear leaves in fascicles and many spines. Its architecture is typical of the Combretaceae, in horizontal strata. It is evergreen, monoecious, and flowers from June to September. The seeds float. It is host to more epiphytes than any of the other tree species in this forest (Dejean et al. 1995).

Haematoxylon campechianum

Logwood or *tinte,* as *H. campechianum* (Fig. 16.3) is commonly called in English and Spanish, respectively, was of great economic importance during Spanish and English dominance of the international seas. It was appreciated for its dye and was highly sought after in European markets before the arrival of synthetic colors (Dobson 1973, Contreras Sánchez 1990).

This leguminous, monoecious tree has an original distribution on the Yucatán Peninsula, including Belize and Guatemala. Because it was so important during the 18th and 19th centuries, it was introduced to many of the West Indian islands (Standley and Steyermark 1946). Heavy exploitation of *H. campechianum* on the Yucatán Peninsula, particularly in Campeche, Tabasco, and the north- and southeast of Quintana Roo, has made this species less common in some places than it used to be. Some of its habitat has been reduced to make room for cattle grazing in Tabasco and Campeche. In Sian Ka'an, our main study area, *B. spinosa* is more abundant than *H. campechianum.*

The tree is evergreen, with small prickly spines, and it flowers for a short period during the dry season, synchronously and massively (Laura Perez del Valle, personal communication). It is pollinated by *Trigona* spp., *Apis mellifera,* and *Scoptotrigona pectoralis,* which is the most abundant of the pollinators. It is self-compatible, and the seeds are anemophilous (wind-dispersed). According to Perez del Valle, germination is rapid, and the seeds have no latent period. However, the authors have never found many young plants in the field.

Bucida buceras

Bucida buceras, a widely distributed tree of the Combretaceae, is very similar in aspect to *B. spinosa;* however, it has a wider representation in the semi-evergreen tropical forest than in the low freshwater forest. It has larger, clustered leaves, fewer spines, and attains heights of up to 12 m in the low forest, but may reach 40 m in the tall semi-evergreen forest. It is monoecious. We have found the two *Bucida* species growing in proximity in the bucidal, but never in the pucteal. *Bucida buceras* is economically important as a timber tree in other parts of Mexico and as a cultivated street tree in Florida in the USA. In its tall form, it is distributed in Mexico, Central

America, and the West Indies, but in its shorter pucteal form, it is mostly found in Quintana Roo.

Erythroxylum confusum

One of three species of this genus in Quintana Roo, *E. confusum* (Fig. 16.4) of the Erythroxylaceae occurs in all communities, but is abundant in the bucidal and tintal. We have found it dominant in low inundated forests, but did not study it because of the remoteness of the sites. It has fissured bark as does *B. spinosa* and is an important phorophyte for epiphytes. The leaves are somewhat areolate, with faint lines delimiting areas on the abaxial side. It is monoecious, deciduous, and is distributed on the Yucatán Peninsula and the West Indies.

Malpighia lundellii

In Quintana Roo we have only found *M. lundelli* in the low freshwater forest. Its distribution is restricted to the Yucatán Peninsula, including Belize and Guatemala, and has also been reported in Palo Verde, Costa Rica (Janzen 1982). It is probably not a well-known species. *Malpighia lundellii* is not abundant, but it is constant in all communities of the low freshwater forest on the Yucatán Peninsula. It is host to a leafless miniature orchid *(Harrisella prorrecta)*. The species is monoecious and deciduous.

Cameraria latifolia

Cameraria latifolia (Fig. 16.8) belongs to the Apocynaceae and has small dark-green leaves and white latex of great toxicity. It is monoecious, deciduous, and flowers between May and August. The stout branches bear a multitude of tiny twigs. Its architecture of a short, stout trunk with stout branching provides support for tank bromeliads.

   Cameraria latifolia has a mostly Caribbean distribution, including the Yucatán Peninsula, Cuba, Jamaica, and Puerto Rico (Woodson, 1940). The authors have never found it anywhere other than the low freshwater forest.

Conclusion

The low freshwater coastal forest of the Yucatán Peninsula seems to be located mostly in the swales and linear depressions of various fault systems (Isphording 1975, Southworth 1985).

   The communities of the low freshwater coastal forest on the Yucatán Peninsula are distributed according to elevation gradient and hydroperiod, from the longest hydroperiod to the shortest, in the following order: tasistal–mucal–bucidal–tintal–pucteal.

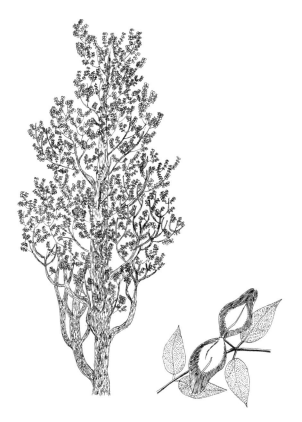

Figure 16.8. Drawing of *Cameraria latifolia*.

The tasistal, or palm swamp, has an important marsh component, whereas the mucal is rather shrubby in aspect, with the tallest individuals up to 5 m. The tasistal is widely distributed along the coast of the Yucatán Peninsula, while the mucal is not as common.

Tintal and bucidal are the most common of the low freshwater forests. The tintal has a wide distribution on the whole peninsula, while the bucidal only occurs in Quintana Roo. The pucteal, with the shortest hydroperiod, is mostly found in Quintana Roo.

Vertical structure of the low freshwater forest is rather low, with the tallest trees not surpassing 12 m. The tree density of the communities follows the hydroperiod gradient; the tasistal has the fewest trees and lowest number of species and the tintal and pucteal have the largest number of individuals and species.

The dominant tree species of the different communities of this forest have a pronounced Caribbean distribution. Sorensen's similarity index shows that the forested bucidal, tintal, and pucteal communities belong to the same vegetation type; however, as the importance values for their dominant species indicate, they are different communities. The mucal community varies quite a bit from the tintal and pucteal communities due to its shrub component and low species richness in comparison to the other communities. Gramineae and Cyperaceae are most represented in the tasistal and mucal, with a few grasses also occurring in the bucidal.

The abundance of epiphytes has been noted more for the number of individuals than the number of species. Bromeliads are more abundant and noticeable than orchids in the mucal and bucidal. The extreme exposure to light of *Tillandsia* in this situation seems to be a similar adaptation of species of the same genus in a *Taxodium* (cypress) forest in Florida (Benzing and Renfrow 1971). *Tillandsia bulbosa,* which is abundant on *B. spinosa,* seems to qualify as an "atmospheric epiphyte" according to Benzing (1980 and 1990).

The tremendous changes from inundation for long periods of time to extreme drought require the species living in this forest to cope with stress situations. The occurrence of some of the dominant trees *(B. spinosa, C. latifolia, M. lundellii)* in the low freshwater coastal forest suggests a special adaptation to this habitat.

Because this forest is often inundated and located in remote areas, it has been spared from the development that has proceeded at a great rate in other forest types, in Quintana Roo in particular.

Acknowledgment   We thank Mauro Gomez J. for the drawings of the plant species.

Appendix: List of Plant Species in Each Community

Species	Tasistal	Mucal	Bucidal	Tintal	Pucteal
Acanthaceas					
Bravaisia tubiflora			X		X
Alismataceae					
Sagittaria lancifolia	X				
Anacardiaceae					
Metopium brownei (chechen negro, poisonwood)			X	X	X
Apocynaceae					
Cameraria latifolia (chechen blanco)		*X*	*X*	*X*	*X*
Plumeria obtusa					X
Bignoniaceae					
Crescentia cujete (calabash tree, *jíca*)		X	X		
Bromeliaceae					
Aechmea bracteata			X	X	X
Catopsis berteroniana		X			
Tillandsia baileyi			X		X
T. balbisiana		X	X	X	X
T. brachycaulos			X	X	X
T. bulbosa		X	X	X	X
T. dasyliriifolia		X	X	X	X
T. fasciculata				X	
T. flexuosa		X	X	X	
T. streptophylla		X	X		X

(continued)

Appendix (continued)

Species	Tasistal	Mucal	Bucidal	Tintal	Pucteal
Cactaceae					
Selenicereus donkelaarii*				X	
Celastraceae					
Rhacoma gaumeri*				X	X
Chrysobalanaceae					
Chrysobalanus icaco	X	X	X	X	
Combretaceae					
Bucida buceras (black olive, pucté)			X		X
B. spinosa			X		X
Convolvulaceae					
Ipomoea sagittata	X	X			
Cyperaceae					
Cladium jamaicense	X	X			
Eleocharis cellulosa		X			
Fuirena sp.			X		
Rhynchospora colorata		X	X		
R. holoschoenoides	X	X	X		
R. microcarpa	X	X			
R. nervosa			X	X	
R. tracyi	X	X			
Ebenaceae					
Diospyros cuneata					X
Erythroxylum confusum (kancab'ché)	X	X	X	X	X
E. rotundifolium				X	X
Euphorbiaceae					
Croton cortesianum			X		
C. niveus				X	X
Sebastiania adenophora*					X
Flacourteaceae					
Xylosma flexuosum				X	X
Gentianaceae					
Nymphoides indica	X	X			
Gramineae					
Dichanthelium dichotomum		X			
Eragrostis elliottii		X			
Panicum condensum		X			
P. tenerum	X	X			
P. virgatum	X	X			
Leguminosae					
Ateleia gummifera			X	X	X
Bauhinia divaricata					X
Chamaecrista glandulosa		X	X		
Dalbergia glabra (muk)	X	X	X	X	X
Haematoxylon campechianum (palo de tinte)			X	X	X
Leg. 3			X		
Lysiloma latisiliquum					X
Mimosa bahamensis				X	X
Pithecellobium albicans					X
P. recordi					
Stylosanthes humilis				X	

(continued)

Appendix *(continued)*

Species	Tasistal	Mucal	Bucidal	Tintal	Pucteal
Lentibulariaceae					
Utricularia sp.	X				
Liliaceae					
Smilax sp.		X		X	X
Loranthaceae					
Psittacanthus americanus			X		
Lythraceae					
Cuphea utriculosa				X	
Malpighiaceae					
Byrsonima bucidaefolia (nance ugrio, sak paj)			X	X	X
Malpighia lundellii	X	X	X	X	X
Malvaceae					
Hampea trilobata					X
Malvaviscus arboreus					X
Menispermaceae					
Hyperbaena winzerlingii				X	
Myricaceae					
Myrica cerifera (wax-myrtle)		X	X	X	
Myrsinaceae					
Ardisia escalonioides					X
Myrtaceae					
Calypthranthes pallens			X	X	
Eugenia axillaris (white stopper)			X	X	X
E. buxifolia				X	
E. winzerlingii *			X	X	X
Eug. 3					X
Myrt. 1					X
Myrt. 2			X	X	
Myrt. 3			X	X	
Myrt. 4			X	X	
Nymphaeaceae					
Nymphaea ampla	X	X			
Ochnaceae					
Ouratea nitida		X	X		
Olacaceae					
Schoepfia schreberi					X
Onagraceae					
Ludwigia octovalvis	X				
Orchidaceae					
Brassavola nodosa			X	X	X
B. cucullata					X
Encyclia boothiana			X	X	X
E. belizensis			X	X	X
E. cochleata			X		X
E. nematocaulon				X	X
Epidendrum nocturnum			X		
E. stanfordianum			X		
Harrisella prorrecta				X	
Notylia barkeri					X
Oncidium ascendens				X	X
Polystachia sp.				X	X

(continued)

Appendix *(continued)*

Species	Tasistal	Mucal	Bucidal	Tintal	Pucteal
Rhyncholaelia digbyana	X		X	X	X
Pleurothallis tikalensis					X
Schomburgkia tibicinis		X	X	X	X
Palmae					
Acoelorrhaphe wrightii (paurotis palm, *tasiste*)	X	X	X		
Passifloraceae					
Passiflora foetida	X	X			
Polygalaceae					
Polygala paniculatum		X	X		
Polygonaceae					
Coccoloba acuminata					X
C. cozumelensis (sak boob)					X
C. diversifolia (pigeon plum)			X		X
C. floribunda			X	X	X
Gymnopodium floribundum			X	X	X
Rubiaceae					
Borreria verticillata			X	X	
Guettarda elliptica			X	X	
Morinda royoc					X
Psychotria nervosa					X
Randia aculeata		X	X	X	X
Sapotaceae					
Bumelia celastrina (saffron plum)				X	X
Bumelia sp.				X	
Bumelia sp.					X
Chrysophyllum caimito				X	
Manilkara zapota			X	X	X
Simaroubaceae					
Picramnia antidesma					X
Solanaceae					
Lycium sp.			X	X	
Theophrastaceae					
Jacquinia aurantiaca	X	X	X	X	X
Typhaceae					
Typha dominguensis		X			
Umbelliferae					
Centella asiatica	X	X			
Verbenaceae					
Phyla nodiflora	X	X			

*Endemic species for the Yucatán Peninsula (Estrada-Loera 1991).

References

Benzing, D.H. 1980. The biology of bromeliads. Mad River Press, Eureka, CA.

———. 1990. Vascular epiphytes. Cambridge University Press, New York.

Benzing, D.H., and A. Renfrow. 1971. The biology of the atmospheric bromeliad *Tillandsia circinnata* Schlecht. I. The nutrient status of populations in South Florida. Am. J. Bot. **58**:867–873.

Cabrera, E., M. Sousa, and O. Tellez. 1981. Inventario de recursos vegetales de Quintana Roo. Reporte final del proyecto de investigación realizado para la Comisión del Plan Nacional Hidraulico. Centro de Invest. de Quintana Roo (CIQRO) e Instituto de Biología de la Univ. Nacional Autonoma de México (UNAM).

Contreras Sánchez, A. 1990. Historia de una Tintorea olvidada. El proceso de explotación y circulación del palo de tinte 1750–1807. Univ. Autónoma de Yucatán, Mérida, México.

Curtis, J.T. 1959. The vegetation of Wisconsin. An ordination of plant communities. Univ. of Wisconsin Press, Madison.

Dachary, A.C., and S.M. Arnaiz. 1983. Estudios socioeconomicos preliminares de Quintana Roo, Sector Agropecuario y Forestal. CIQRO, Pto. Morelos, Q. Roo.

Dejean, A., I. Olmsted and R.R. Snelling. 1995. Tree-epiphyte-ant relationships in the low inundated forest of Sian Ka'an Biosphere Reserve, Quintana Roo, Mexico. Biotropica 27:57–70.

Dobson, N. 1973. A history of Belize. Longman Caribbean Ltd. Trinidad and Jamaica.

Durán García, R. 1986. La vegetación de la Selva Baja subcaducifolia con *Pseudophoenix sargentii* en Quintana Roo. Facultad de Ciencias, UNAM, Mexico.

Estrada-Loera, E. 1991. Phytogeographic relationships of the Yucatán Peninsula. Journal of Biogeography 18:687–697.

Flores, J.S. 1984. Vegetación insular de la Península de Yucatán. Bol. Soc. Bot. Mex. 45:23–37.

García, E. 1964. Modificaciones al sistema de clasificación climática de Koeppen. Univ. Nacional Autónomade México, Instituto de Geografía. México.

Gleason, P.J., and W. Spackman, Jr. 1974. Calcareous periphyton and water chemistry in Everglades. Pages 148–181 *in* P.J. Gleason, ed. Environments of South Florida: present and past. Memoir 2. Miami Geol. Soc., Coral Gables, FL.

Isphording, W.D. 1975. The physical geology of Yucatan. Gulf Coast Assoc. Geol. Soc. Trans. Vol. 25:231–262.

Janzen, D.H. 1982. Costa Rican natural history. The Univ. of Chicago Press. Chicago and London.

Lot-Helgueras, A. 1983. La vegetación acuática del Sureste de México, Ciencia y Desarrollo 51. July–August. CONACYT, México.

Lot-Helgueras, A., and A. Novelo. 1990. Forested wetlands in Mexico. Pages 287–298 *in* A.E. Lugo, M. Brinson, and S. Brown, eds. Forested wetlands. Elsevier, New York.

Lundell, C.L. 1934. Preliminary sketch of the phytogeography of the Yucatan Peninsula. Carnegie Inst. Wash. Publ. 436:257–321.

Meave del Castillo, J.A. 1983. Estructura y composición de la selva alta perennifolia en los alrededores de Bonampak, Chiapas. Tesis Fac. Ciencias. UNAM, México.

Miranda, F. 1958. Estudios acerca de la vegetación. Pages 215–271 *in* E. Beltran, ed. Los recursos naturales del Sureste y su aprovechamiento, Tomo II. I.M.R.N.R., México.

Olmsted, I. 1993. Wetlands of Mexico. Pages 637–677 *in* D.F. Whigham, D. Dykyjova, and S. Hejny, eds. Wetlands of the world. Vol. 1. Kluwer Academic Publishers, Dordrecht, The Netherlands.

Olmsted, I., and R. Durán García. 1986. Aspectos ecologicos de la selva baja inundable de la Reserva Sian Ka'an, Quintana Roo, Mexico. Biotica 11:151–179.

———. 1990. Vegetación de Sian Ka'an. Pages 1–12 *in* D. Navarro L. and J. G. Robinson, eds. Diversidad Biologica en la Reserva de la Biosfera de Sian Ka'an, Quintana Roo, Mexico. CIQRO, Chetumal, Q. Roo and Program of Studies in Tropical Conservation, Univ. of Florida, Gainesville.

Olmsted, I., A. Lopez-Ornat, and R. Durán García. 1983. Vegetación de Sian Ka'an. Reporte preliminar, in Sian Ka'an, CIQRO-SEDUE, Pto. Morelos, Q. Roo y Mexico.

Rico-Gray, V. 1982. Estudio de la vegetación de la zona costera inundable del NO de Campeche, Mexico: Los Petenes. Biotica **7**:171–188.

Sorensen, T. 1948. A method of establishing groups of equal amplitude in plant sociology based on similarity of species content. Det Kong Danske Videusk. Selsk-Biol. Skr. Copenhagen. **5**:1–34.

Southworth, C.S. 1985. Application of remote-sensing data, Eastern Yucatan. Pages 12–18 *in* W.C. Ward, A.E. Weidie, and W. Back, eds. Geology and hydrogeology of the Yucatan and quarternary geology of Northeastern Yucatan Peninsula. New Orleans Geological Society, Univ. New Orleans, New Orleans.

Standley, P.C., and S.J. Record. 1936. The forest and flora of British Honduras. Publication 350, Bot. Series, Vol. XII. Field Museum of Natural History. Chicago.

Standley, P. C., and J.A. Steyermark. 1946–1977. Flora of Guatemala. Fieldiana: Botany. Field Museum of Natural History. Chicago.

Weidie, A.E. 1985. Geology of Yucatan platform. Pages 1–12 *in* W.C. Ward, A.E. Weidie, and W. Back, eds. Geology and hydrogeology of the Yucatan and quarternary geology of Northeastern Yucatan Peninsula. New Orleans Geological Society, Univ. New Orleans, New Orleans.

Woodson, Jr., R.E. 1940. The apocynaceous flora of the Yucatan Peninsula. Botany of the Maya area. Misc. Papers XIV–XXI. Pub. 522. Carnegie Inst. of Washington, Washington, DC.

17

William H. McWilliams, John B. Tansey,
Thomas W. Birch, & Mark H. Hansen

Taxodium–Nyssa (Cypress–Tupelo) Forests along the Coast of the Southern United States

The *Taxodium–Nyssa* (cypress–tupelo) association inhabits freshwater wetlands across the southern United States. *Taxodium–Nyssa* forests are a prevalent subset of lowland forests on sites inundated for most of the year. The association occurs on a continuum from cypress-dominated to tupelo-dominated, but pure stands of either species are common. Pure stands of tupelo often follow clearcutting of cypress (Putnam 1951).

Taxodium distichum (L.) Rich. (baldcypress) and *Nyssa aquatica* L. (water tupelo) are the principal species of the *Taxodium–Nyssa* association (Figs. 17.1–17.4), although *T. distichum* var. *nutans* (Ait.) Sweet (pondcypress), *N. sylvatica* var. *biflora* (swamp tupelo), and *N. ogeche* (Ogeechee tupelo) are common in some areas. The occurrence of associate species varies by the degree of inundation and type of swamp (or soil; Braun 1950). The number of associates increases in fringe areas where swamp and upland forests merge (Mitsch and Ewel 1979, Larson et al. 1981).

Little (1971) and Mattoon (1915) have described the botanical and commercial ranges of cypress and the botanical range of *N. aquatica*. Ecology and silvics of *Taxodium–Nyssa* forests are discussed by Putnam (1951), Langdon (1958), Putnam et al. (1960), Wharton et al. (1982), Ewel and Odum (1984), Dunn and Sharitz (1987), and Burns and Honkala (1990). Several studies of individual community structure, productivity, and nutrient cycling have been conducted, of which Hall and Penfound (1939 and 1943), Conner and Day (1976), Schlesinger (1978), Brown (1981), and Conner et al. (1981), are good examples. However, information on the current extent and character of *Taxodium–Nyssa* forests across their range is lacking.

This chapter describes *Taxodium–Nyssa* forests with emphasis on the coastal southern United States and has three objectives:

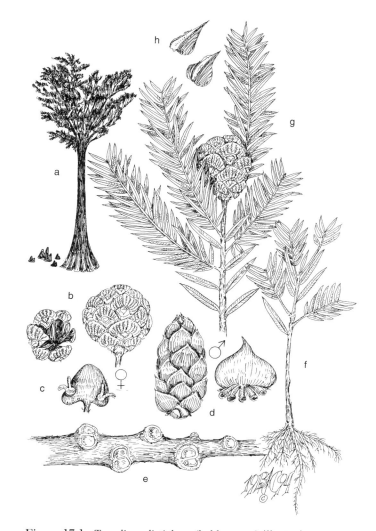

Figure 17.1. *Taxodium distichum* (baldcypress) illustration. a: Tree silhouette (30–37 m) with knees at left. b: Open and closed cone (2.0–2.5 cm). c: Female flower (2 mm). d: Cluster (2.5 mm) and individual (1.0 mm) male flowers. e: Twig (12 cm). f: Seedling (30 cm, taproot x 1/4). g: Foliage with cone (25 cm). h: Seed (12 mm). (Original illustration by Bruce L. Cunningham.)

1. To present the current distribution of *T. distichum* and *N. aquatica*;
2. To estimate and characterize the area of the *Taxodium–Nyssa* association; and
3. To examine the broad species composition of the *Taxodium–Nyssa* association.

We conclude with a discussion of trends in the area of *Taxodium–Nyssa* forests.

Figure 17.2. *Taxodium distichum* stand in Atchafalaya basin, southern Louisiana.

Materials and Methods

Study Region

The Coastal South study region stretches along the coast of the southern United States from Delaware to Galveston Bay in Texas (Fig. 17.5). In keeping with the objectives of this chapter, the inland boundary was set at approximately 250 km from the Atlantic and Gulf coastlines. The study region is located from 25°N to 40°N latitude and covers part or all of six physiographic sections of the Atlantic (or Coastal) Plain Province described by Fennemann (1938). The region, which contains all or part of 11 states, is coincident with the Coastal Plain Province along the Atlantic coast and occupies the lower portion of the Province along the Gulf coast.

Most of the Coastal South has a warm-temperate climate. However, the Floridian section and a band extending along the Gulf coastline have a subtropical climate, and the extreme southern portion of the Floridian section has a tropical climate. Annual rainfall averages from 1,000 to 1,300 mm over the region.

Source Data

Source data are from successive large-scale forest inventories conducted by the U.S. Department of Agriculture, Forest Service, Forest Inventory and Analysis (FIA) project (USDA Forest Service 1992). The FIA inventories use a two-phase sample design consisting of aerial photo samples and sample locations visited on the ground (Birdsey 1990). At each sample location, trees at least 2.5 cm in diameter at breast height (dbh) are sampled using various cluster designs and combinations of fixed- and variable-radius plots. Some species are grouped by genus (e.g., *Fraxinus* spp.) for consistency in species identification. For this study, tree-level data for all sample locations where a cypress or *N. aquatica* was tallied were pooled into a single data set. All the forest inventories had been completed since 1984.

The basal area (BA) of dominant and codominant stems within the dominant size

Figure 17.3. *Nyssa aquatica* (water tupelo) illustration. a: Tree silhouette (25–30 m). b: Cluster of male flowers (2.0 cm). c: Female flower
(1.5 cm). d: Twig (10 cm). e: Seedling (45 cm). f: Foliage (25 cm) and
fruit (2.5 cm). g: Seed (2.0–2.5 cm). (Original illustration by Bruce L.
Cunningham.)

class of stems for each sample location was used to classify the lowland forest associations. The use of stems in the dominant size class overcomes the problem of basing
the assignment on smaller stems that occur in gaps within the sampled area. Such
stems, though dominant in small areas, are not part of the main canopy. The size
classes used to partition sample trees were based on dbh: 2.5 to 10.0 cm, 10.0 to
25.0 cm, and 25.0 cm and larger. Size classes assigned to each sample location also

Figure 17.4. Late successional *N. aquatica* stand, central Mississippi.

proved useful as proxies for stage of stand development in analyzing the results. For stands classified as *Taxodium–Nyssa,* the contribution of cypress to the total BA of the two principal species was used to rank sample locations on the continuum from pure cypress to pure tupelo.

Geographic distribution and species composition were analyzed by using total-tree dry weight as the measure of importance. Dry weight, excluding foliage and roots, was estimated by using regression equations developed by Clark et al. (1985).

Results

Distribution

The geographic distributions of *Taxodium* and *N. aquatica* biomass are shown in Fig. 17.6. Each dot represents 100,000 metric tons of total tree biomass, and hence small concentrations are not included in the distributions.

Taxodium is found throughout the study region from east Texas to Delaware. The warm, humid climate of the Coastal South provides optimum conditions for the development of cypress stands (Kennedy 1972). *Taxodium* is not limited to a particular range of soil pH, but rather is found on soils ranging from acidic to alkaline (Davis

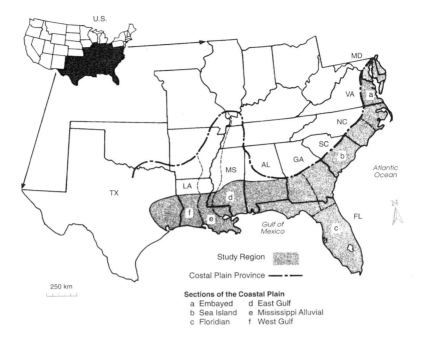

Figure 17.5. Location of the study region within the Coastal Plain Province (adapted from Fennemann 1938). TX: Texas. LA: Louisiana. MS: Mississippi. AL: Alabama. GA: Georgia. SC: South Carolina. NC: North Carolina. VA: Virginia. MD: Maryland. FL: Florida. Delaware is shown to the east of Maryland.

1943) and is common in the permanent swamps of the region. Concentrations occur in the Mississippi Alluvial, Floridian, and Sea Island sections. *Taxodium* is abundant over extensive areas of the alluvial swamps of the lower Mississippi River, as well as along the Pee Dee, Santee, Savannah, Apalachicola, and Mobile Rivers. It is found in nonalluvial swamps such as the Dismal and Okefenokee, and in the many river swamps of Florida (Mattoon 1915). *Taxodium distichum* occurs in the floodplains of large rivers, while *T. distichum* var. *nutans* is common in upland swamps (Penfound 1952). The natural range of *Taxodium* extends farther southwest in Texas than the study region boundary, but this portion of the range was excluded because no FIA data have been collected there.

Nyssa aquatica typically thrives in alluvial soils common in low, wet flats or sloughs, and in deep swamps (Burns and Honkala 1990). It is abundant in the Mississippi Alluvial, East Gulf, and Sea Island sections, and is noticeably absent from the Floridian section. The FIA data show *N. aquatica* occurring as far north as the Eastern Shore of Maryland, farther than Little's (1971) natural range.

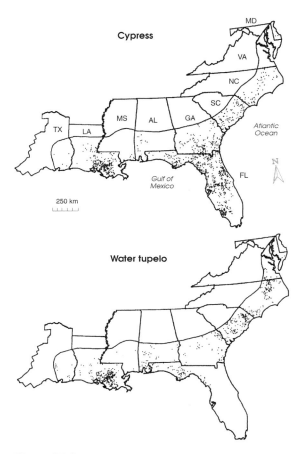

Figure 17.6. Distribution of *Taxodium* spp. and *N. aquatica* in the Coastal South. Each dot represents 100,000 metric tons of total dry weight for trees 2.5 cm and larger at breast height. Source data are from forest inventories (1984–1990) conducted by the U.S. Department of Agriculture, Forest Service, Forest Inventory and Analysis project (USDA Forest Service 1992).

Forest Area

Forests cover 81.3 million ha in the southern United States (Table 17.1). Upland forests, which predominate across most of the South and account for 84% of the total forest area, include *Pinus* (pine), *Quercus–Pinus* (oak–pine), and *Quercus–Carya* (oak–hickory) associations. These three associations account for 37%, 17%, and 46% of the upland forest area, respectively. Lowland forests constitute 16% of the total forest area. By FIA definition, lowland forests are a subset of the wetlands ecosystem that includes areas that are at least 10% stocked with tree species. Lowland forests

Table 17.1 Area of Forest Land by Association and Region for the Southern United States[a]

	Area (10^3 ha)		
Association	Total	Coastal South	Inland South
Upland			
Pinus (pine)	25,402.1[b]	14,082.1	11,320.0
Quercus–Pinus (oak–pine)	11,607.0	4,691.3	6,915.7
Quercus–Carya (oak–hickory)	31,485.8	6,307.9	25,177.9
Total	68,494.9	25,081.3	43,413.6
Lowland			
Taxodium–Nyssa (cypress–tupelo)	1,653.0	1,491.4	161.6
Other lowland	11,177.8	6,512.7	4,665.1
Total	12,830.8	8,004.1	4,826.7
All types	81,325.7	33,085.4	48,240.3

[a]The southern United States includes Alabama, Arkansas, Delaware, Florida, Georgia, Kentucky, Louisiana, Maryland, Mississippi, North Carolina, Oklahoma, South Carolina, Tennessee, East Texas, and Virginia, as well as counties in southeast Missouri and southern Illinois in which *Taxodium–Nyssa* forests occur.
[b]Source data are from the most recent forest inventories conducted by the U.S. Department of Agriculture, Forest Service, Forest Inventory and Analysis project (USDA Forest Service 1992).

range from moist- to wet-site associations, with the former being the most abundant (McWilliams and Rosson 1990). *Taxodium–Nyssa* is the most common wet-site association, and *Taxodium–Nyssa* forests total 1.6 million ha or 13% of the lowland forest (2% of the South's forest).

The Coastal South study region contains 33.1 million ha of forest, or 41% of the South's total. The Inland South is composed of areas in the southern United States outside the Coastal South study region. Compared to the Inland South, the Coastal South has more pine and lowland forests. Sixty-two percent of the South's lowland forest is located within the Coastal South study region. Nearly all of the South's *Taxodium–Nyssa* forests (90%) are along the coast. The *Taxodium–Nyssa* stands of the Coastal South total 1.5 million ha, and most of the *Taxodium–Nyssa* stands outside the Coastal South are located within the Mississippi Alluvial Plain to the north of the study region.

Classification of stands according to the percent of cypress highlights the monotypic dominance that is common within the *Taxodium–Nyssa* association. If we consider stands to be monotypic, in which either of the principal species accounts for 80% or more of the BA of both species combined, then 59% of the stands are cypress-dominated, 20% are tupelo-dominated, and 21% have mixed dominance (Fig. 17.7).

Figure 17.8 shows the distribution of *Taxodium–Nyssa* area by size class. The proportion of *Taxodium–Nyssa* in the 2.5 to 10.0 cm size class is an indicator of the association's ability to replace stands in larger size classes and provide habitat for early seral-stage wildlife species. The proportion of stands in this early stage of stand

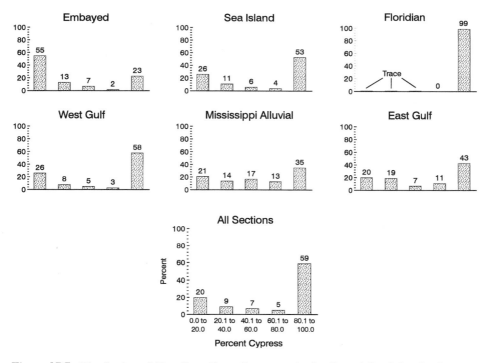

Figure 17.7. Distribution of *Taxodium–Nyssa* forest area in the Coastal South by physiographic section and percent cypress. Source data are from citation in Figure 17.6.

development is very low (less than 10%) in all physiographic sections to the south and west of the Sea Island section.

Composition of Canopy

Broad composition traits are illustrated by ranking the ten most common canopy trees of the *Taxodium–Nyssa* association for the Coastal South and making comparisons among physiographic sections (Table 17.2). As pointed out by Putnam et al. (1960) and McKnight et al. (1981), wet-site forests tend to support fewer tree species than moist-site forests. *Taxodium–Nyssa* illustrates this tendency well, because the principal trees contribute over 60% of the association's total biomass. The ten most abundant trees contribute roughly 90%. Just over half of the biomass of the principal trees is cypress, which is divided equally between *T. distichum* and *T. distichum* var. *nutans*. The most common associates include *Fraxinus* spp. (ash), *N. sylvatica* var. *biflora* (swamp tupelo), *Acer rubrum* and *A. rubrum* var. *drummondii* (red maple), and *Quercus laurifolia* (laurel oak).

The importance of the principal trees of the *Taxodium–Nyssa* association is similar in four of the six physiographic sections (about 60%), while the Floridian and Mississippi Alluvial Plain sections have slightly higher concentrations (69 and 75%, respectively). The relative importance of *Taxodium* and *N. aquatica* follows geographic patterns. *Tax-*

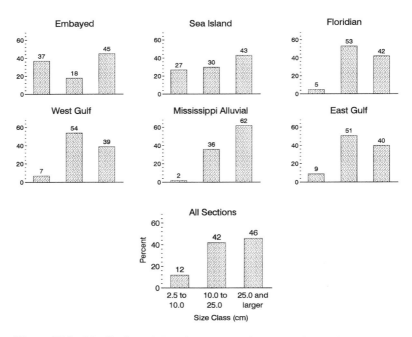

Figure 17.8. Distribution of *Taxodium—Nyssa* forest area in the Coastal South by physiographic section and size class. Source data are from citation in Figure 17.6.

odium clearly dominates in the Floridian section and shares dominance in the Sea Island, East Gulf, and Mississippi Alluvial Plain sections. Of the two cypress varieties, *T. distichum* is most important in the Embayed, Mississippi Alluvial Plain, and West Gulf sections; it shares dominance with *T. distichum* var. *nutans* in the Sea Island and East Gulf sections. *Taxodium distichum* var. *nutans* is most important in the Floridian section. *Nyssa aquatica* dominates in the Embayed section, is very important in the West Gulf section, and is least important in the Floridian section.

Discussion

Sources of Variability

In light of this study's baseline estimate of 1.6 million ha for the *Taxodium–Nyssa* association, some discussion of the estimate's variability and factors that will affect the future status of the association is relevant. The statistical error associated with the estimate of *Taxodium–Nyssa* forest area is 0.6%. More important sources of variability arise from computational assumptions made in developing the estimate. Perhaps the most significant source, in terms of variability, was the use of size classes to classify the lowland forest associations. Sensitivity analysis showed that varying these classes resulted in differences of up to 100,000 ha, or about 6.0%.

Table 17.2 Percent of Total-Tree Dry Weight by Physiographic Section for the Ten Most Important Tree Species in *Taxodium–Nyssa* Forests of the Coastal South[a]

Scientific and Common Name	Physiographic Section						
	Coastal South	Embayed	Sea Island	Floridian	East Gulf	Miss. Alluvial	Gulf
	Percent dry weight						
Nyssa aquatica (water tupelo)	27	49	30	2	29	37	37
Taxodium distichum (baldcypress)	18	10	12	12	13	38	24
Taxodium distichum var. *nutans* (pondcypress)	17	1	9	55	19	0	0
Fraxinus spp. (ash)	7	8	7	8	7	7	2
Nyssa sylvatica var. *biflora* (swamp tupelo)	6	9	10	6	8	*[b]	2
Acer rubrum and *A. rubrum* var. *drummondii* (red maple)	6	8	6	5	2	8	2
Quercus laurifolia (laurel oak)	4	1	9	4	4	*	2
Liquidambar styraciflua (sweetgum)	2	3	5	*	2	*	4
Quercus lyrata (overcup oak)	1	1	2	*	1	*	7
Salix spp. (willow)	1	1	*	*	1	4	1
Total-tree dry weight (10^6 t)	246.5	23.9	56.0	55.1	52.4	51.3	7.3
Number of other tree species sampled	68	25	38	24	46	19	23

[a]Source data are from the most recent forest inventories conducted by the U.S. Department of Agriculture, Forest Service, Forest Inventory and Analysis project (USDA Forest Service 1992).
[b]*: Less than 1 percent.

Trends

The historical decline of lowland hardwood forests has been well documented (Sternitzke 1976, MacDonald et al. 1979, Turner and Craig 1980, Abernethy and Turner 1987). The periods of most rapid loss occurred in the 1940s and from the 1960s to early 1970s. Recent FIA results suggest that declines in the area of lowland forests have slowed, at least for the present (McWilliams and Rosson 1990). Although there are no estimates specifically for *Taxodium–Nyssa,* it has likely followed these same general trends of lowland forests.

The most influential factors affecting the lowland forest landscape have been clearing and draining for agricultural use. Urban expansion and oil exploration (mainly in the lower Mississippi Alluvial Plain) have also depleted swamp areas. The impact of these factors has been less in recent years because of poor markets for agricultural and petroleum products, legislation that protects wetland habitat (Environmental Protection Agency 1980), and federal guidelines for defining wetlands (Federal Manual 1989). Heightened public awareness of wetland loss has resulted in the increased purchase of wetlands by public agencies and privately funded organizations such as the Nature Conservancy. Also, recent federal legislation has had a strong negative influence on conversion of wetlands to agricultural use (Heimlich et al. 1989) and has offered incentives for the reforestation of cropped wetlands (Sims 1989).

The effects of rising sea level and problems with regeneration are factors of great concern. Permanent flooding that results from rising water levels is not an immediate threat to *Taxodium–Nyssa* forests, but the potential for increased salinity could have a negative long-term effect, particularly in the lower Mississippi Alluvial Plain (Conner and Day 1989). A lack of natural regeneration in permanently flooded swamps will continue to forestall the replacement of existing stands following harvest or natural disturbance (Conner and Toliver 1990). The problem is compounded in the lower Mississippi Alluvial Plain because of the damage inflicted on planted cypress seedlings by *Myocastor coypus* (nutria; Blair and Langlinais 1960, Conner and Day 1989).

Conclusion

Although trends imply that decreases in the area of *Taxodium–Nyssa* forests have slowed, the volatility of agricultural and oil markets, the continued threat of rising sea levels and urban expansion, and difficulties with regeneration underscore the need for continued long-term monitoring. Large-scale forest inventories can be used to assess overall changes in the extent and character of the *Taxodium–Nyssa* association, thus providing support to more intensive research initiatives.

Acknowledgments   We gratefully acknowledge James F. Rosson, Jr. for his suggestions for estimating the area of the *Taxodium–Nyssa* association.

In Memory   The authors extend heartfelt appreciation to the late John B. Tansey, whose leadership in using large-scale inventory data to address wetland forest characterization and many other issues has provided us with inspiration and guidance in conducting this and other studies. His energy and expertise are sorely missed. His creativity will continue to inspire us.

References

Abernethy, Y., and R.F. Turner. 1987. U.S. forested wetlands: 1940–1980. Bioscience **37(10)**:721–727.

Birdsey, R.A. 1990. Forest change—an assessment of nationwide forest inventory methods. Pages 119–125 *in* V.J. LaBau and T. Cunia, eds. State of the art methodology of forest inventory: a symposium proceedings, State Univ. New York, Syracuse. U.S. For. Serv., Gen. Tech. Rep. PNW-GTR-263.

Blair, R.M., and M.G. Langlinais. 1960. Nutria and swamp rabbits damage *T. distichum* plantings. J. Forestry **58**:388–389.

Braun, L.E. 1950. Deciduous forests of eastern North America. Hafner, New York.

Brown, S. 1981. A comparison of the structure, primary production, and transpiration of cypress ecosystems in Florida. Ecol. Monogr. **51**:403–427.

Burns, R.M., and B.H. Honkala. 1990. Silvics of North America, Volume 1, Conifers. Volume 2, Hardwoods. U.S. For. Serv., Agric. Handb. No. 654.

Clark, A. III, D.R. Phillips, and D.J. Frederick. 1985. Weight, volume, and physical properties of major hardwood species in the Gulf and Atlantic coastal plains. U.S. For. Serv., Res. Paper SE-250.

Conner, W.H., and J.W. Day, Jr. 1976. Productivity and composition of a baldcypress-water tupelo site and a bottomland hardwood site in a Louisiana swamp. Am. J. Bot. **63(10)**:1354–1364.

———. 1989. Response of coastal wetland forests to human and natural changes in the environment with emphasis on hydrology. Pages 34–43 *in* D.D. Hook and R. Lea, eds. The forested wetlands of the Southern United States. Orlando, FL. U.S. For. Serv., Gen. Tech. Rep. SE-50.

Conner, W.H., and J.R. Toliver. 1990. Long-term trends in bald-cypress (*Taxodium distichum*) resource in Louisiana (U.S.A.). For. Ecol. and Man. **33/34**:543–557.

Conner, W.H., J.G. Gosselink, and R.T. Parrondo. 1981. Comparison of the vegetation of three Louisiana swamp sites with different flooding regimes. Am. J. Bot. **68**:320–331.

Davis, J.H., Jr. 1943. The natural features of southern Florida, especially the vegetation and the Everglades. Fla. Geol. Surv., Geol. Bull. 25.

Dunn, C.P., and R.R. Sharitz. 1987. Revegetation of a *Taxodium-Nyssa* forested wetland following complete vegetation destruction. Vegetatio **72**:151–157.

Environmental Protection Agency. 1980. Section 404(b)(1) guidelines for specification of disposal sites for dredged or fill material (40 CFR part 230). Federal Register **45**:85344–85357.

Ewel, K.C., and H.T. Odum, eds. 1984. Cypress swamps. Univ. of Florida Press, Gainesville.

Federal manual for identifying and delineating jurisdictional wetlands. 1989. An interagency publication. U.S. Dept. Inter. Fish Wildl. Serv., U.S. EPA, U.S. Dept. Army, and U.S. Soil Cons. Serv.

Fennemann, N.M. 1938. Physiography of eastern United States. 1st Edition. McGraw-Hill Book Company, NY and England.

Hall, T.F., and W.T. Penfound. 1939. A phytosociological analysis of a cypress–gum forest in southeastern Louisiana. Am. Midl. Nat. **21**:378–395.

———. 1943. Cypress–gum communities in Blue Girth Swamp near Selma, Alabama. Ecology **24**:208–217.

Heimlich, R.E., M.B. Carey, and R.J. Brazee. 1989. Beyond swampbuster: a permanent wetland reserve. J. Soil and Water Cons. **44(5)**:445–450.

Kennedy, H.E., Jr. 1972. Baldcypress. U.S. For. Serv., FS-218.

Langdon, O.G. 1958. Silvical characteristics of baldcypress. U.S. For. Serv., Southeastern For. Exp. Sta., Paper No. 94.

Larson, J.S., M.S. Bedinger, C.F. Bryan, S. Brown, R.T. Huffman, E.L. Miller, D.G. Rhodes, and B.A. Touchet. 1981. Transition from wetlands to uplands in southeastern bottomland hardwood forests. Pages 225–273 *in* J.R. Clark and J. Benforado, eds. Wetlands of bottomland hardwood forests. Elsevier Scientific, Amsterdam, Oxford, NY.

Little, E.L. 1971. Atlas of United States trees, Vol. 1. Conifers and important hardwoods. U.S. For. Serv., Misc. Pub. No. 1146.

MacDonald, P.O., W.E. Frayer, and J.K. Clauser. 1979. Documentation, chronology, and future projections of bottomland hardwood habitat losses in the lower Mississippi Alluvial Plain. Vol. 1. U.S. Dept. Int. Fish Wildl. Serv.

Mattoon, W.R. 1915. The southern cypress. U.S. For. Serv. Bulletin No. 272.

McKnight, J.S., D.D. Hook, O.G. Langdon, and R.L. Johnson. 1981. Flood tolerance and related characteristics of trees of bottomland forests of the Southern United States. Pages 29–69 *in* J.R. Clark and J. Benforado, eds. Wetlands of bottomland hardwood forests. Elsevier Scientific, Amsterdam, Oxford, NY.

McWilliams, W.H. and J.F. Rosson, Jr. 1990. Composition and vulnerability of bottomland hardwood forests of the Coastal Plain Province in the south central United States. For. Ecol. Man. **33/34**:495–501.

Mitsch, W.J., and K.C. Ewel. 1979. Comparative biomass and growth of cypress in Florida wetlands. Am. Midl. Nat. **101**:417–426.

Penfound, W.T. 1952. Southern swamps and marshes. Bot. Rev. **18**:413–446.

Putnam, J.A. 1951. Management of bottomland hardwoods. U.S. For. Serv., Southern For. Exp. Station, Occasional Paper 116.

Putnam, J.A., G.M. Furnival, and J.S. McKnight. 1960. Management and inventory of southern hardwoods. U.S. For. Serv., Agric. Handb. No. 181.

Schlesinger, W.H. 1978. Community structure, dynamics, and nutrient cycling in the Okefenokee cypress swamp-forest. Ecol. Monogr. **48**:43–65.

Sims, D.H. 1989. CRP extended to hardwoods for some bottomland applications. For. Farmer **49(1)**:13.

Sternitzke, H.S. 1976. Impact of changing land use on the Delta Hardwood forests. J. Forestry **74**:24–27.

Turner, R.E., and N.J. Craig. 1980. Recent areal changes in Louisiana's forested wetland riparian habitat. Louis. Acad. Sci. **43**:61–68.

USDA Forest Service. 1992. Forest Service Resource Inventories: an overview. U.S. Forest Service, Forest Inventory, Economics, and Recreation Research. Washington, DC.

Wharton, C.H., W.M. Kitchens, and T.W. Sipe. 1982. The ecology of bottomland hardwood swamps of the southeast: a community profile. U.S. Dept. Int. Fish and Wildl. Serv., FSW/OSB-81/37.

18

William H. Conner

Impact of Hurricanes on Forests of the Atlantic and Gulf Coasts, USA

Hurricanes are common, but unpredictable, occurrences on the Atlantic and Gulf coasts of the southeastern United States. There is evidence that 160,000 to 320,000 hurricanes have occurred in the area of the Florida keys during the past two million years (Ball et al. 1967). They occur about once every 20 years in south Florida (Lugo et al. 1976) and are even more common on the Gulf coast (Conner et al. 1989). South Carolina experienced 38 hurricanes between 1700 and 1983, or approximately one every seven years (Dukes 1984). Hurricane paths have been plotted for several time periods. Figure 18.1 shows hurricanes plotted for the years 1886–1963. As can be seen in the figure, the Gulf coast has been hit by more hurricanes than the Atlantic coast.

High winds are usually associated with hurricanes, but other forces include tidal storm surges and torrential rains. These latter forces are often more destructive than wind alone (Baker 1978). Hurricane winds cause defoliation, breakage, and windthrow in forests with the severity of damage related to storm intensity, forest structure, and soil conditions (Weaver 1989). The storm surge (also called storm tide, hurricane tide, or tidal wave) is a mound of water pushed ashore by the hurricane. In coastal locations, the combined flooding and pounding by waves can cause great damage (Baker 1978). Heavy rains associated with hurricanes can be both destructive and beneficial. Much damage has been caused in areas where large amounts of rain have caused flooding of tributaries and major streams. In contrast, crops in the southeastern United States have been saved from drought more than once by hurricane rains (Simpson and Riehl 1981). Tropical cyclone-related rainfall contributes about 15% to mean seasonal precipitation in the Gulf of Mexico region, and up to 30% or more in some areas (Cry 1967, Meeder 1987).

Despite the interest that hurricanes generate immediately following their destruc-

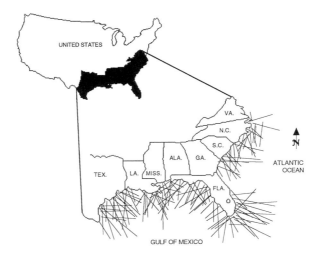

Figure 18.1. Hurricanes crossing the Atlantic and Gulf coasts, 1886–1963 (after Baker 1978).

tive landfall, very little is actually known about their long-term impacts on forested ecosystems, mainly because of the difficulty in planning hurricane-related research (Lugo et al. 1983). It is difficult to study an area in anticipation of a hurricane, since it cannot be predicted where or when a hurricane will occur. After a hurricane occurs, it takes time to design a study and get money allocated for that study. It is often easier to study something else and hope that others will follow up on the questions that have been raised. In this chapter, I attempt to synthesize what we know about the impacts of hurricanes on forests of the southeastern United States. Special emphasis will be placed on Hurricane Hugo (1989), as it may provide the best opportunity for long-term ecological research in the coming years.

Impact to Forests

The ecological impacts of hurricanes on freshwater forested wetlands and upland coastal species are not well documented. The most often-reported statistics include area affected, severity and type of damage, number of trees damaged, growing stock volume lost, and economic value of the timber lost. It has been shown repeatedly that coastal economies recover quickly following a major hurricane (Janiskee 1990), but little information exists on the recovery of forest ecosystems (Conner et al. 1989).

Damage to forest types is generally greatest to pine and least to swamp, with hardwood forests intermediate (Touliatos and Roth 1971). However, there are exceptions, as in Hurricane Camille (1969) in Mississippi, where Hedlund (1969) reported that both pine and hardwood were equally vulnerable. During Hurricane Donna (1960), *Pinus elliottii* (slash pine) was one of the most resistant species in Florida (Craighead and Gilbert 1962). A positive correlation between annual growth of *P.*

elliottii on barrier islands along the Mississippi coast and hurricanes has even been reported (Stoneburner 1978). In addition, Stoneburner found that hurricane-induced washover deposits stimulated germination of pine seedlings by reducing the competitive shrub understory and exposing the mineral soil surface. Vogel (1980) hypothesized that hurricanes have replaced fire as a dominant perturbation that creates proper conditions for *P. elliottii* establishment.

In swamp forests dominated by *Taxodium distichum* (baldcypress), *Nyssa aquatica* (water tupelo), and *Nyssa sylvatica* var. *biflora* (swamp blackgum), hurricanes are capable of defoliating, topping, and overturning trees, but it is the defective and hollow trees that usually break, and windthrow of these wetland species is generally rare (Craighead and Gilbert 1962, Duever et al. 1984, Hook et al. 1991). (For more on *Taxodium-Nyssa* forests, see McWilliams et al., this volume.) Windthrow of bottomland species is more common and may be related to shallow rooting in moist, soft soil (Hedlund 1969, Gunter and Eleuterius 1973). In south Florida, new leaf growth was unusually rapid for several species of trees and many species flowered a second time immediately following Hurricane Donna in 1960 (Vogel 1980).

One aspect of hurricane-induced rains that has generally been overlooked is their role in inducing high export of organic matter. Day et al. (1977) reported that Hurricane Carmen (1974) caused litterfall to occur two months early in swamp forests of the Barataria Basin, Louisiana, and that a large pulse of carbon, nitrogen, and phosphorus was flushed from the area to the Barataria estuary following the storm. The material exported after the storm represented 20–30% of the total export for the year and the authors suggested that this input was important in stimulating the productivity of the Barataria estuary. (See also Conner and Day, this volume.)

Even though mangrove forests represent a small portion of the total forested area of the southeastern United States, there is probably more information on hurricane impacts for this forest type than for any other. Mangrove species include *Avicennia germinans* (black mangrove), *Languncularia racemosa* (white mangrove), and *Rhizophora mangle* (red mangrove). Lugo and Snedaker (1974) found differences in susceptibility between individual species of mangrove. Hurricanes may, to some extent, control species composition (Craighead and Gilbert 1962) and keep mangrove forests in a juvenile successional state (Lugo and Snedaker 1974). Net production over the long term is increased (Lugo et al. 1976), but maximum biomass appears to be limited by hurricanes.

Hurricane Hugo

Hurricane Hugo made landfall on the night of September 21–22, 1989, with the eye of the storm passing just north of Charleston, South Carolina. Estimated maximum sustained winds were 54 meters per second at Charleston and destructive winds continued to do damage 325 km inland (Janiskee 1990). Although wind damage was widespread, a storm surge averaging approximately 3 m above sea level also caused major damage. Overall, Hugo was the costliest hurricane in history with nearly $7 billion in damage on the United States mainland alone (Case and Mayfield 1990).

The center of the hurricane followed a path as shown in Figure 18.2, causing damage to 1.8 million ha of forest (Hook et al. 1991). This represents more area than

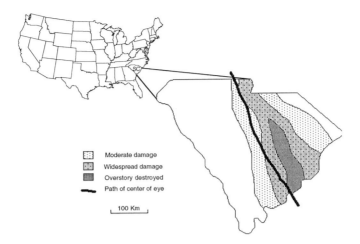

Figure 18.2. Map of South Carolina showing the path of Hurricane Hugo and the area of damage.

that affected by Hurricane Camille (1969), Mount St. Helens' eruption (1980), and the 1988 Yellowstone fires combined. A nine-county area experienced a loss of 70–90% of its older, taller timber (Janiskee 1990). The most extreme forest damage occurred within the area impacted by the eyewall, an area of very strong winds surrounding the hurricane eye. Pine and hardwood species exhibited little differences in their resistance to wind damage (Hook et al. 1991). Outside of the eyewall, there was generally less damage to the bottomland and swamp species than to pines (Sharitz and Putz 1990, Hook et al. 1991), although in some areas oaks were damaged heavily. The major short-term effect to bottomland and swamp species was the loss of foliage and small branches. Sharitz and Putz (1990) also reported that trees that had previously suffered wind damage were more susceptible to new damage.

In one way, the point of landfall and subsequent damage will be very beneficial to the scientific world. Directly in the path of the storm were the Francis Marion National Forest and the Santee Experimental Forest (Fig. 18.3), where forest management/resource studies have been underway for over 50 years. Since the storm, the United States Department of Agriculture (USDA) Forest Service has implemented a program of research to learn about the storm's effects on the forest and to develop appropriate means to restore this ecosystem. Projects include

- Recovery of wetland functions in watersheds;
- Ecophysiological and hydrologic studies of wetland forests and forest tree species impacted by tidal surge;
- Mitigation of salvage logging effects on wetland soils and wetland site productivity;
- Characterization of storm stress effects on trees and subsequent attack by insects and disease;
- Recovery of wildlife populations; and
- Wetland hardwood ecophysiology, restoration ecology, and silviculture.

Figure 18.3. Map of South Carolina showing the location of study areas in relation to the path of Hurricane Hugo.

Hobcaw Forest, just 90 km northeast of Charleston, South Carolina, is another area of active research since the 1960s. Post-Hugo research efforts by Clemson University faculty are focusing on the recovery of wetland forests after inundation by storm surge. Studies include seedling and watershed response to saltwater, natural and artificial regeneration of impacted areas, productivity and nutrient cycling studies, soil and hydrologic changes, and modeling efforts. Other efforts in the state include long-term study plots in the Congaree Swamp National Monument (Rebecca Sharitz, personal communication) and Four Hole Swamp (Norman Brunswig, personal communication).

Hurricane Andrew

Hurricane Andrew passed over south Florida on August 24, 1992, partially or completely defoliating a 50-km-wide swath across the state (Pimm et al. 1994). The hurricane then proceeded across the Gulf of Mexico and made landfall on August 26th in south central Louisiana, where it caused moderate to severe damage to over 51,000 ha (Doyle et al. 1994). Preliminary observations of damage to forests within both states found that cypress suffered only modest damage, while bottomland hardwoods, pines, and mangroves exhibited greater amounts of damage (Gorham et al.

1994, Loope et al. 1994, Smith et al. 1994). Only 1–2% of the cypress trees in Florida, and <5% in Louisiana, suffered major damage. Over 38% of the bottomland hardwood forests in Louisiana were severely damaged, while 20–30% of large trees on tropical hardwood hammocks in Florida were extensively damaged (downed or had large branches broken off). Fortunately, there exist some long-term studies in the Florida and Louisiana impacted areas that will help expand our knowledge of hurricanes and ecosystem disturbances due to natural episodic events (Pimm et al. 1994).

Conclusions

Hurricanes occur frequently and are thus a recurring aspect of coastal forest development, but their impacts have not been studied extensively. Only now are scientists beginning to recognize the importance of natural catastrophes as critical to the dynamics of ecosystems (Boucher 1990). When considered in the broader context of coastal ecosystem functioning, hurricanes could be considered a periodic disordering stress that causes alteration of the biological and physical structure, elimination of some habitats and creation of others, and high material fluxes (Conner et al. 1989). Since coastal forests have developed in areas prone to hurricanes, it is likely that these forests have developed mechanisms to reestablish themselves rapidly following disturbance, as has been suggested for rainforests (Boucher 1990). Research efforts initiated after Hurricanes Hugo and Andrew will help provide a better understanding of how coastal forests respond to and recover from large-scale disturbance.

References

Baker, S. 1978. Storms, people and property in coastal North Carolina. University of North Carolina Sea Grant Publication UNC-SG-78–15, Raleigh.

Ball, M.M., E.A. Shinn, and K.W. Stockman. 1967. The geologic effects of Hurricane Donna in south Florida. J. Geology 75:583–597.

Boucher, D.H. 1990. Growing back after hurricanes. BioScience 40:163–166.

Case, B., and M. Mayfield. 1990. Atlantic hurricane season of 1989. Monthly Weather Review 118:1165–1177.

Conner, W.H., J.W. Day, Jr., R.H. Baumann, and J.M. Randall. 1989. Influence of hurricanes on coastal ecosystems along the northern Gulf of Mexico. Wetlands Ecol. Manage. 1:45–56.

Craighead, F.C., and V.C. Gilbert. 1962. The effects of Hurricane Donna on the vegetation of southern Florida. Q. J. Fla. Acad. Sci. 25:1–28.

Cry, G.W. 1967. Effects of tropical cyclone rainfall on the distribution of precipitation over the eastern and southern United States. U.S. Dept. of Commerce, Washington, DC, ESSA Prof. Paper No. 1.

Day, J.W., Jr., T.J. Butler, and W.H. Conner. 1977. Productivity and nutrient export studies in a cypress swamp and lake system in Louisiana. Pages 255–269 in M.L. Wiley, ed. Estuarine processes, Volume. II. Academic Press, NY.

Doyle, T.W., L.E. Gorham, and B.D. Keeland. 1994. Wind damage relationships of Hurricane Andrew on forested wetlands of southern Louisiana. ASB Bull. 41:83.

Duever, M.J., J.E. Carlson, and L.A. Riopelle. 1984. Corkscrew Swamp: a virgin cypress strand. Pages 334–348 *in* K.C. Ewel and H.T. Odum, eds. Cypress swamps. Univ. of Florida Press, Gainesville.

Dukes, E.K. 1984. The Savannah River Plant environment. E.I. du Pont de Nemours & Co., DP-1642. Savannah River Laboratory, Aiken, SC.

Gorham, L.E., B.D. Keeland, and T.W. Doyle. 1994. Patterns of primary versus secondary damage caused by Hurricane Andrew in the Atchafalaya Basin of southern Louisiana. ASB Bull. **41**:84.

Gunter, G., and L.N. Eleuterius. 1973. Some effects of hurricanes on the terrestrial biota, with special reference to Camille. Gulf Res. Repts. **4(2)**:174–185.

Hedlund, A. 1969. Hurricane Camille's impact on Mississippi timber. Southern Lumberman **219 (2728)**:191–192.

Hook, D.D., M.A. Buford, and T.M. Williams. 1991. Impact of Hurricane Hugo on the South Carolina coastal plain forest. J. Coastal Res. SI-**8**:291–300.

Janiskee, R.L. 1990. "Storm of the century": Hurricane Hugo and its impact on South Carolina. Southeastern Geog. **30**:63–67.

Loope, L., M. Duever, A. Herndon, J. Snyder, and D. Jansen. 1994. Hurricane impact on uplands and freshwater swamp forest. BioScience **44**:238–246.

Lugo, A.E., and S.C. Snedaker. 1974. The ecology of mangroves. Ann. Rev. of Ecol. and Systematics **5**:39–54.

Lugo, A.E., M. Sell, and S.C. Snedaker. 1976. Mangrove ecosystem analysis. Pages 113–145 *in* B.C. Patten, ed. Systems analysis and simulation in ecology. Academic Press, NY.

Lugo, A.E., M. Applefield, D.J. Pool, and R.B. McDonald. 1983. The impact of Hurricane David on the forests of Dominica. Can. J. For. Res. **13**:201–211.

Meeder, J.F. 1987. Variable effects of hurricanes on the coast and adjacent marshes: a problem for land managers. Pages 337–374 *in* N.V. Brodtmann, ed. Fourth water quality and wetlands management conference proceedings. New Orleans.

Pimm, S.L., G.E. Davis, L. Loope, C.T. Roman, T.J. Smith III, and J. Tilmant. 1994. Hurricane Andrew. BioScience **44**:224–229.

Sharitz, R.R., and F.E. Putz. 1990. Damage from Hurricane Hugo to the Congaree Swamp National Monument. Unpublished manuscript. Aiken, SC.

Simpson, R.H., and H. Riehl. 1981. The hurricane and its impact. Louisiana State Univ. Press, Baton Rouge.

Smith, T.J., III, M.B. Robblee, H.R. Wanless, and T.W. Doyle. 1994. Mangroves, hurricanes, and lightning strikes. BioScience **44**:256–262.

Stoneburner, D.L. 1978. Evidence of hurricane influence on barrier island slash pine forests in the northern Gulf of Mexico. Am. Midl. Nat. **99**:234–237.

Touliatos, P., and E. Roth. 1971. Ten lessons from Camille. J. For. **69**:285–289.

Vogel, R.J. 1980. The ecological factors that produce perturbation-dependent ecosystems. Pages 63–94 *in* J. Cairns, Jr., ed. The recovery process in damaged ecosystems. Ann Arbor Science, Ann Arbor, MI.

Weaver, P.L. 1989. Forest changes after hurricanes in Puerto Rico's Luquillo mountains. Interciencia **14**:181–192.

19

William H. Conner & John W. Day, Jr.

The Effect of Sea Level Rise on Coastal Wetland Forests

The Mississippi Delta, USA, as a Model

Numerous reports have recently emphasized the potential impact of global warming trends on future sea level rise in coastal areas (e.g., Hoffman et al. 1983, Kerr 1991). Predictions are that many low-lying coastal areas, especially those dominated by wetlands, will be flooded as sea level rises from 30 to 50 cm over the next century. Current evidence indicates that sea level rise is leading to wetland loss in a number of coastal areas, but most research has concerned coastal marshes (Baumann et al. 1984). Little information is available on the impact of sea level rise on coastal wetland forests (Salinas et al. 1986, Conner and Day 1988b). There is a critical need to consider both the impacts of sea level rise and the possible policy responses to these impacts.

The Mississippi River Delta area of Louisiana, USA, can serve as a model of the effects of sea level rise because it has been experiencing a "relative sea level rise" of about a meter per century, caused primarily by regional subsidence. In this chapter, we discuss the effects of this rise on coastal wetland forests in the Mississippi River Delta area and use this information to consider the potential impacts of eustatic (worldwide) sea level rise on coastal wetland forests in other areas.

The Mississippi Deltaic Region

The Mississippi Delta is a large area of lakes, bays, coastal wetlands, and low-lying uplands created during the past 7,000 years by frequent channel changes of the Mississippi River (Kolb and van Lopik 1958, Morgan 1967; Fig. 19.1). As the river overflowed its banks during floods, a new layer of nutrient-rich silt was deposited. Open water areas filled in with sediment and marsh plants took root, further enhancing the rate of sediment deposition. As the land surface continued to rise with new sediment depositions, swamp

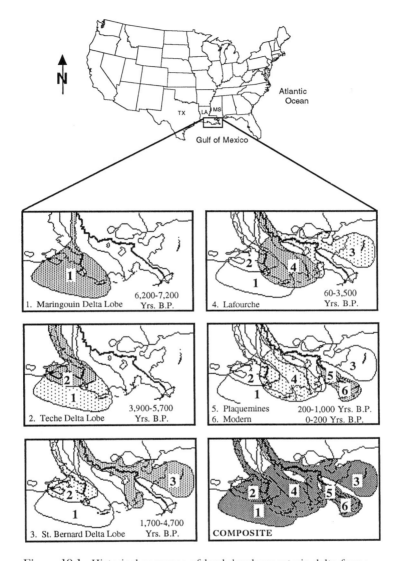

Figure 19.1. Historical sequence of land development via delta formation in coastal Louisiana. TX: Texas. LA: Louisiana. MS: Mississippi. (After Kolb and van Lopik 1958, Morgan 1967.)

and bottomland hardwood tree species often became established in freshwater areas. When the river switched courses, the land-building process continued in another part of the coastal plain while the abandoned delta eroded and subsided. Overall, channel switching created approximately 1.6 million ha of marshes and forested wetlands in Louisiana. Today, the landscape of coastal Louisiana is composed of a series of alluvial natural levee ridges along active and abandoned distributaries with vast areas (over 150,000 ha) of interdistributary forested wetlands.

Causes of Wetland Loss

During the past several decades, the long-term pattern of net land gain over the past 7,000 years has been reversed. There has been an increase in the rate of land loss in Louisiana with values as high as 100 km^2 yr^{-1} reported (Gagliano et al. 1981). This dramatic change is the result of a number of factors (including geologic, catastrophic, biologic, and human) which contribute to the inability of wetlands to maintain surface elevation in the face of rising water levels (Day and Templet 1989).

Geologic Factors

Sea level change may occur on a worldwide scale or be of more local importance, as in Louisiana, where localized downwarping occurs because of sediment and water loading of coastal sediments (Walker et al. 1987). Subsidence occurs naturally in Louisiana where great thicknesses of sediment have accumulated from Mississippi River flooding. Holocene sediment thickness may exceed 100 m in some areas of the delta region, and the rate of subsidence increases as compaction processes occur. Whereas the average eustatic rise in sea level is only 0.15 cm/yr (Gornitz et al. 1982), a number of studies have shown that the current rate of relative sea level rise (subsidence + eustatic sea level rise) in the Mississippi Deltaic Plain is between 1.0 and 1.2 cm/yr (Baumann et al. 1984, Hatton et al. 1983).

Catastrophic Factors

By far the greatest catastrophic factor to influence coastal Louisiana is the hurricane. Even though they occur at infrequent intervals, hurricanes are a normal part of the climatic regime of the Gulf coast and represent a major factor affecting coastal ecosystems (Conner et al. 1989). The degree of impact from an individual hurricane is largely controlled by size, speed, and storm path (Walker et al. 1987). Wind speed is especially important in that it determines the amount of physical damage to the forest and the height of the associated storm surge (for a more detailed description of hurricane damage to coastal forests, see Conner, this volume). While short-term impacts may be dramatic, hurricanes generally do not produce long-term detrimental impacts to natural systems and often provide net benefits along the Gulf coast (Conner et al. 1989). Hurricane impacts are often most severe and long-lasting in wetlands that have been modified by human impacts such as partial or complete impoundments where saltwater may be trapped for days after a storm surge.

Biologic Factors

Biomass production by plants (roots and aboveground) can contribute more than 50% of vertical accretion (Day and Templet 1989), thereby helping to offset subsidence. In addition, thick stands of vegetation are capable of trapping sediment as floodwaters flow over an area. In the Netherlands, Dijkema and Wolff (1980) reported that sedimentation was two to three times higher after vegetation became established. Sedi-ment input to an area builds up the elevation and, in addition, is an important source of new nutrients for vegetation (DeLaune et al. 1983). Since coastal wetland vegeta-tion grows within a narrow elevation range, a reduction in sedimentation results in lowered surface elevation and eventually death of the vegetation.

Herbivory can also cause problems in wetland areas. In Louisiana, big consumers like *Myocastor coypus* (nutria) and *Ondatra zibethicus* (muskrat) commonly consume large areas of marsh vegetation (Gosselink 1984) and cause damage to newly planted tree seedlings in cypress swamps (Conner and Toliver 1988). *Myocastor coypus* dam-age caused Louisiana Soil Conservation Service personnel to recommend that the planting of cypress be suspended until some means of *M. coypus* control were per-fected (Blair and Langlinais 1960). How much these creatures contribute to wetland loss is unknown, but there is at least one report of conversion of a southwestern Louisiana marsh area to open water following a large-scale eatout (O'Neil 1949). Fuller et al. (1984) reported that the exclusion of *M. coypus* and *O. zibethicus* from marsh areas resulted in significant increases in plant biomass and community compo-sition.

Human Factors

Human activity has dramatically changed the wetland system during the past 50 years. In the Mississippi River Delta region, an important impact has been a reduction in the amount of suspended sediment reaching the Delta (Walker et al. 1987). Flood control levees along the Mississippi and Atchafalaya rivers prevent annual spring flooding of coastal wetlands, depriving the wetlands of the normal addition of sedi-ments. In addition, dams along the upstream portions of the Mississippi and its dis-tributaries have resulted in a reduction of sediment load from 434×10^6 tons/year to 255×10^6 tons/year (Walker et al. 1987).

Canal construction for navigation, access, drainage, and oil and gas extraction also have tremendous impacts on coastal wetlands. Over 15,000 km of canals have been constructed in Louisiana's coastal zone, and these canals have changed the regional hydrology of much of the area (Day and Templet 1989). The canals with their associ-ated spoil banks restrict the flow of water into some areas and in other areas may act as channels for saltwater intrusion.

Effects of Waterlogging and Salinity on Forested
Wetland Communities

Flooding is a natural occurrence in forested wetlands. Forest productivity appears to peak at an annual flood frequency if flooding occurs during the dormant period (Gosselink et al. 1981). These seasonally flooded forests tend to be very productive (Conner and Day 1976). Studies have shown that production is closely correlated with nutrient input and flooding water volume (Brown et al. 1979, Conner and Day 1982).

The effect of flooding is dependent on duration. Short-term flooding events during the growing season seem to have very little effect on mature trees. Mitsch and Rust (1984) found very little correlation between growth of water-tolerant trees and annual flooding. However, they did find that years with a high percentage of flooding during the growing season were years of low tree growth. Johnson and Bell (1976), in a similar study, found no relationship between frequency of flooding and growth of bottomland tree species with diameters exceeding 4 cm.

Flooding that extends into the growing season or exists for extended periods can have serious effects on the survival of bottomland trees. Even though flooding may have no visible effect the first growing year, trees usually start dying the second year, and only a few species have been reported to survive three years of continuous flooding (Green 1947). Hall and Smith (1955) reported that in Tennessee none of the 39 common deciduous tree species could survive flooding if the root system was covered for more than 54% of the growing season during an eight-year period.

Flooding and submersion during the growing season can have a substantial impact on tree growth. Studies at the Southeastern Forest Experiment Station (1958) have shown that *Liriodendron tulipifera* (yellow poplar) seedlings submerged during the dormant season were relatively unaffected, while those submerged for only three days during the growing season were adversely affected. After 14 days of submersion during the growing season, only 5% of the seedlings survived. Baker (1977) studied the influence of spring submersion on the growth and survival of several bottomland tree species. He noted, as others have, that seedlings of most species lose their leaves during submersion. Intolerant species died, the growth of *Populus deltoides* (cottonwood) and *Nyssa aquatica* (water tupelo) occurred in the form of sprouts from the root collar, while *Fraxinus pennsylvanica* (green ash) was the only species studied where new growth was from the original stem.

Mature *Taxodium distichum* (baldcypress) and *N. aquatica* can survive under flooded conditions. *Taxodium distichum* is well known for its ability to grow in flooded areas (Wilhite and Toliver 1990). However, increased flooding can sometimes have serious consequences if the mean depth of flooding exceeds 60 cm (Brown and Lugo 1982). In Florida, Harms et al. (1980) found that in water from 20 to 100 cm deep, 0–16% of the *T. distichum* trees died in seven years. In water over 120 cm deep, 50% of the *T. distichum* died after four years. In Louisiana, a long-term study of *T. distichum* survival was conducted near Lake Chicot (Eggler and Moore 1961). After four years of flooding with water 60–300 cm deep, 97% of the *T. distichum* were still alive. Eighteen years after flooding, 50% of the *T. distichum* were still alive. However, most of the living trees in the deep water had dead tops (Eggler and Moore 1961). From the available data on flooding stress and *T. distichum* survival, it appears that *T. distichum* can adapt to shallow, permanent flooding (< 60 cm), and even in deep water (> 60 cm), death and decline is a gradual process (Eggler and Moore 1961, Harms et al. 1980).

Most of the available literature on salinity tolerance of coastal tree species comes from studies on the effects of salt spray on trees (Wells and Shunk 1938, Moss 1940, Little et al. 1958). With the increased interest in global warming and sea level rise in recent years, more emphasis has been placed on determining the response of coastal tree species to increased flooding and salinity. Some research has been conducted on the physiological response of wetland tree species to salinity (Pezeshki and Chambers

1986, Pezeshki 1987, Pezeshki et al. 1988). (For more on *Taxodium–Nyssa* forests, see McWilliams et al., this volume.)

History of Forested Wetland Research in Coastal Louisiana

The earliest published works on Louisiana's forested wetlands dealt with descriptions of the plant communities found in the state (e.g., Penfound and Hathaway 1938). It was not until the 1960s that studies began describing the ecology of these forests and how they respond to environmental factors. By the 1970s, the value of these systems and the role they play in the overall coastal ecosystem were just being recognized and serious ecological studies were begun (Conner and Day 1976, Conner et al. 1981). Much of the work in forested wetland ecology was done by researchers at Louisiana State University's Center for Wetland Resources (see later paragraphs describing the various types of research conducted). Some work was done through the School of Forestry, Wildlife, and Fisheries, but those studies were concerned more with silvicultural aspects of how to properly manage wetland species (Faulkner et al. 1985, Jackson and Morris 1986, Toliver et al. 1987, Dicke and Toliver 1988).

Ecology of Deltaic Forested Wetlands and the Impact of Rising Water Levels

Hydrology

Apparent water level rise in coastal Louisiana is greater than in other coastal areas of the United States. For example, Stevenson et al. (1986) reported that apparent water level rise on the Atlantic coast is 3–4 mm/yr versus 8–14 mm/yr in Louisiana (Conner and Day 1988b).

Water levels in Louisiana's coastal forests typically follow a seasonal pattern of flooding and drying with the extent of flooding depending on the elevation of the site and seasonal water budget. In the Barataria and Verret Basins (Fig. 19.2), wetland forests are very near sea level, and they are flooded almost year-round with a short dry period during late July–early August, a time when rainfall is low and evapotranspiration is high. Flooding occurs during the winter and early spring, but for most of the growing season the forest floor on the ridge areas is dry. The lower *Taxodium–Nyssa* areas are flooded for most of the year.

Analysis of the yearly average water level in both watersheds shows that there has been a significant apparent increase in water levels through time. This similarity is not surprising considering that both basins have undergone similar patterns of geologic development. Apparent water level rise was calculated to be 8.5 and 13.7 mm/yr for Barataria Basin and Verret Basin, respectively (Conner and Day 1988b).

Both basins have experienced significant increases in the total number of days flooded per year. Bottomland ridges in the Verret Basin did not experience any major flooding until the 1970s, but there has been a steady increase in the number of days flooded per year since then. Before 1970, the Verret bottomland areas were at an elevation sufficient to keep the forest floor from flooding. However, the lack of sedimentation in the area, combined with apparent water level rise, has resulted in the

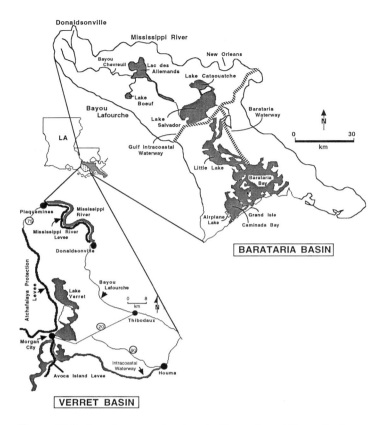

Figure 19.2. Location map showing the Barataria and Verret Basins of south Louisiana, and most commonly mentioned places or features in the text.

ridge now being at an elevation where flooding occurs frequently. The history of flooding in the swamp forests of Verret Basin is similar to the bottomland ridge described above, except that increased flooding was evident by the late 1960s.

In the Barataria Basin, the swamps have been flooded to some extent since the 1950s, but flooding has increased such that the forests are flooded almost year-round. Even during dry periods such as 1981 and 1985–86, these forests were rarely free of standing water although the total days flooded decreased during these years.

Water levels have also been measured in a forested wetland consisting of a bottomland hardwood ridge adjacent to a *Taxodium–Nyssa* forest near Thibodaux, Louisiana (Conner and Day 1989). The bottomland ridge system there is somewhat higher than the Verret ridge. There was a strong seasonality with higher water levels in the winter. While water levels in the *Taxodium–Nyssa* forest were above or at the sediment surface all year, the water level on the ridge was about one meter below the ground surface during much of the growing season.

The three forested wetland systems discussed above are located on an elevation gradient with Thibodaux higher than Verret, which is higher than Barataria. These three systems illustrate the impact of rising water levels on coastal forested wetlands. There will be progressively higher water levels and longer flooding periods. The effects of this increased flooding are discussed in the following sections.

Vertical Accretion in Wetland Forests

There is a direct relationship among flooding duration, sediment availability, and vertical accretion in Louisiana's forested wetland areas (Conner and Day 1991b). In the Lake Verret wetlands, for example, sedimentation was shown to increase with flooding frequency. Vertical accretion in drier bottomland hardwood areas was 0.03 cm/month, while in more flooded areas there was 0.06 cm/month of sediment accretion (Conner and Day 1991b). Most of the accumulated material in the driest plots appeared to be organic in nature (the result of decaying leaf litter), whereas in the more flooded plots, inorganic material seemed to be the dominant component of the sediment layer.

During the 1984–85 winter flood period, sedimentation averaged 2.7 ($\pm$ 1.2) mm on a bottomland hardwood ridge in the Verret Basin as compared to 8.8 mm and 6 mm in more flooded *Taxodium–Nyssa* forests of the Barataria and Verret Basins, respectively (Conner and Day 1988b). This difference is undoubtedly due to frequency and height of flooding experienced by each basin. There are about ten flooding events of varying duration each year in Louisiana coastal forests. Each flood event transports sediments into the forest areas. As only the higher floods cover the bottomland ridges, much less sediment is imported and deposited there.

The foregoing information indicates that vertical accretion is related to both flooding duration and frequency. High accretion rates can be sustained with rising water levels if there is regular flooding with sediment-laden water. There must be suspended sediments in the flood water, and rates of sedimentation vary from year to year depending upon the amount of flooding experienced. The mineral sediment component of vertical accretion comes from floodwaters that inundate an area during high-water periods. Extended, long-term flooding, as in impounded areas, does not result in high accretion rates. Thus, flooding alone does not ensure that vertical accretion will occur. Sediment availability and flooding have to coincide in order to help maintain surface elevation.

Although hurricane activity has been shown to be very important in the sedimentation of Louisiana's coastal marshes (Baumann et al. 1984), it does not seem to be so important in the sedimentation of forested wetlands. In swamp forests, there is little chance of large-scale resuspension of sediments because of the low percentage of open water areas compared to the large expanses of forest. Instead, most of the sediment is derived from erosion of agricultural fields surrounding wetland areas. During the late summer–early fall when hurricanes and tropical storms are most active, the fields are completely covered with vegetation (mainly with *Saccharum officinarum* [sugarcane] in south Louisiana), and erosion is undoubtedly low. During the winter and spring months, however, the vegetation-free and newly plowed fields are highly susceptible to erosion by rainfall runoff, and this is when the greatest sedimentation occurs.

Forest Composition and Structure

The forested wetland community is strongly affected by water level and drainage. *Taxodium* and *Nyssa* are characteristic of poorly drained and frequently flooded areas, while bottomland hardwoods are found on slightly higher, better-drained ridge areas. Brown (1972) stated that a 15-cm difference in wetland elevation is more significant in changing plant communities in Louisiana than 30 m in the mountains. Typically, there is an increase in the number of tree species from the most frequently flooded areas to the least flooded areas. A particular species composition may persist indefinitely as long as the flooding pattern and relative elevation remain constant from year to year.

Of wetland habitats in the state, plant diversity is greatest in the swamp forest. Over 200 species of plants have been noted in this area by the authors. *Taxodium* and *Nyssa* are the dominant trees in the swamps. *Acer rubrum* var. *drummondii* (Drummond red maple), *Fraxinus* spp. (ashes), and a number of woody shrubs such as *Itea virginica* (Virginia willow) and *Cephalanthus occidentalis* (buttonbush) are also relatively dominant. In the slightly drier areas, species like *Populus heterophylla* (cottonwood), *Salix nigra* (black willow), *Celtis laevigata* (sugarberry), *Gleditsia aquatica* (water locust), *Quercus* spp. (oaks), and *Carya* spp. (hickories) occur.

Rising water levels are altering succession in the forested wetlands of coastal Louisiana. Increased flooding results in tree death for less flood-tolerant species, leading to reverse succession toward a more aquatic assemblage. Even though *Taxodium* and *Nyssa* are capable of surviving prolonged flooding during the growing season, other tree species are not (Hook 1984). Already 25–50% of the less flood-tolerant trees have died in permanently flooded plots under study in the Verret and Barataria watersheds.

Succession modeling has been used to predict the impact of progressive water level rise on coastal wetland forests (Brody et al. 1989). The FORFLO model of bottomland hardwood succession and habitat change simulates the growth, reproduction, and tree composition in a mixed species stand. The model predicted that bottomland hardwood species will be replaced by *Taxodium* and *Nyssa* as flooding increases. The largest increase in basal area in the understory is by *Taxodium* and *Nyssa*, indicating that conditions become more favorable for their germination and survival with time. However, even *Taxodium* and *Nyssa* will decline with time, and no new trees will enter the understory.

In forests with rising water levels, tree stress is visually evident in the number of dead tops and branches. In one study, the number of visibly stressed trees increased from 8 to 53% of the total number of trees on a gradient of increasing flooding (Conner and Day 1988a). If present trends of increasing water level continue as the coastal area subsides, all but the most flood-tolerant trees *(Taxodium* and *Nyssa)* will die in the near future as a result of permanent flooding (Conner and Brody 1989).

These data indicate that sea level rise will lead to a change in forest structure. Bottomland hardwood species will die out first due to waterlogging stress. Deep swamp species will persist for longer periods of time but will eventually die and will not be replaced, as regeneration cannot occur because of permanent flooding.

Vegetation Productivity

Conner and Day (1976) were among the first to recognize that seasonally flooded systems are more productive than continuously flooded ones and, along with Mitsch and Ewel (1979), described a Shelford-type parabolic curve for productivity as a function of water regime. Hydrology appears to be the key factor influencing species productivity (Conner and Day 1976), although nutrient flux is also related to biomass production (Brown 1981). One of the difficulties in separating the factors controlling ecosystem processes in forested wetlands is that hydrology itself controls so many other factors, such as nutrient delivery and soil redox status (Brinson 1989). Fluctuating water levels are typical of forested wetlands, and periodic flooding represents an energy source bringing sediments and nutrients into the system (Conner and Day 1976). The general pattern of productivity response to flooding has been reported to be: flowing water $>$ slow-flowing water $>$ stagnant water.

Stem productivities and litterfall have been measured in three areas of the Barataria Basin of Louisiana with different flooding regimes for a number of years (Conner and Day 1992a,b). Even though the stem growth of individual *Taxodium* and *Nyssa* trees in a permanently flooded area was greater than in the other areas (presumably the result of reduced competition and greater sunlight), the lower number of trees in that area resulted in lower areal productivity. Average aboveground net productivity decreased in the order of: *Procambarus clarkii* (crayfish) pond (which is pumped down in the summer) $>$ a seasonally flooded area $>$ a permanently flooded area.

These results suggest that rising sea level will lead to a progressive decline in forest productivity. Ultimately, tree productivity will approach zero as the forest disappears. As trees die, the forests are replaced by marsh and/or open water.

Decomposition

Forested wetlands are generally characterized as having large detrital accumulations and slow decomposition rates because of anaerobic conditions (Day 1982). The rate of decomposition varies depending on the species and the conditions where the process occurs, with moisture being an important factor in determining the speed of decomposition. Decomposition of plant material is generally rapid in seasonally flooded systems (Kemp et al. 1985).

Conner and Day (1991a) found that decomposition rates were highest in areas characterized by a high rate of water turnover, high oxygen levels, a dense population of *P. clarkii* in the spring, and significant materials export. These conditions lead to rapid litter decomposition and nutrient loss. In continuously flooded and stagnant areas, there is a much slower rate of water turnover, lower macroinvertebrate populations, and lower dissolved oxygen levels, resulting in slower decomposition rates and an accumulation of organic matter. The undecomposed to partially decomposed matter is buried by newly fallen leaves or sediment. Burial is accelerated due to a high regional subsidence rate of approximately 1.0 m/century (Conner and Day 1988b). Undecomposed organic matter and associated nutrients are buried and represent a permanent loss from the system. Rising water levels will likely lead to slower decomposition, less detrital export, and higher rates of burial in coastal forested wetlands.

Nutrient Cycling

The physical and chemical nature of waters and sediments in forested wetlands is controlled to a large degree by the flooding characteristics of a site and in turn has a strong impact on the ecology of forested wetlands. The degree of waterlogging strongly affects physiochemical properties and nutrient cycling in wetland soils. There are strong seasonal and spatial changes in dissolved oxygen (DO) levels in overlying waters and in soil redox potentials. Soils of higher ridge areas are more oxidized than lower, permanently flooded areas. In permanently waterlogged sites the soil redox potential is generally reduced all year. Overlying waters in swamps are characterized by very low oxygen levels for much of the year. Only during colder months do DO levels rise appreciably. Phosphate concentrations in the sediments are inversely related to DO levels. This occurs because phosphate is more soluble under reducing conditions. These results suggest that as water levels rise and soils become waterlogged, soils will be more reduced and low dissolved oxygen will be more persistent.

Forested wetlands can act as buffers in terms of time, composition, and concentration of material entering them. In a study of a Louisiana swamp receiving agricultural runoff, it was found that the swamp retained 26% of the nitrogen (N) and 41% of the phosphorus (P) introduced annually (Day and Kemp 1985). These losses are likely due to a combination of burial, denitrification, and uptake into woody tissues. Concentrations of dissolved inorganic N, particulate N, and particulate P were lower in the swamp water than in input water, while levels of dissolved organic N and dissolved P were somewhat higher in the swamp. The chemical composition of water overlying the swamp is different from that of the input water. The ratio of particulates to dissolved fractions was much lower for swamp water than for input water. The swamp tends to maintain a relatively constant N:P ratio and thus buffers inputs. The characteristically low and relatively constant N:P ratio of water overlying the swamp contrasts sharply with water in a channel receiving agricultural runoff. Nitrogen: phosphorus ratios in channels were strongly affected by rainfall, which washed material from agricultural fields. These data suggest that as swamps die out due to rising water levels, the buffer effect will be lost and water quality in adjacent areas will decline.

Management implications

The preceding discussion indicates that rising sea level will likely lead to the progressive loss of coastal forested wetlands. Proper management, however, can sustain the life of these forests for long periods even in the face of rapid sea level rise. The key to achieving this goal is to maintain vertical accretion at a rate that is approximately equal to the rate of water level rise. This can be accomplished by enhancing suspended sediment input into forested wetlands. The sediments will directly add to accretion and also stimulate forest productivity and thus organic matter soil formation. Freshwater inputs will also help lessen the effects of saltwater intrusion. In the Mississippi Delta, wetland forests within a few kilometers of the Mississippi River are dying out. Water diversions from the river could reverse the decline and lead to the regeneration of these forests.

Conclusions

Because of a high rate of apparent sea level rise, coastal wetland forests in the Mississippi Delta can serve as a model of the effects of eustatic sea level rise in other areas. Rising water levels will lead to changes in hydrology. There will be progressively higher water levels and longer flooding periods in these forested wetlands. As flooding frequency and duration increase, accretion rates that historically built these wetlands can be sustained only if there is regular flooding with sediment-laden water. Increased flooding will lead to a change in forest structure. Bottomland hardwood species will die out rather quickly due to waterlogging stress. Deep swamp species will persist for longer periods of time but will eventually die and will not be replaced, as regeneration cannot occur when there is permanent flooding. Rising water levels will lead to a progressive decline in forest productivity until, ultimately, the forest dies out. Rising water levels will likely result in slower decomposition rates, less detrital export, and higher rates of burial in coastal forested wetlands. Higher water levels will cause wetland soils to become more waterlogged. Soils will be more reduced, and low dissolved oxygen will be more persistent. As these forests die out, their ability to buffer water chemistry will be lost and water quality in adjacent areas will likely decline. Proper sediment and water management can prolong the life of coastal forested wetlands in the face of rising sea level.

Acknowledgments A variety of funding sources have provided support for our studies in forested wetlands including Louisiana Sea Grant College Program (part of the National Sea Grant Program maintained by the National Oceanic and Atmospheric Administration, U.S. Department of Commerce), Louisiana Board of Regents, Louisiana Water Resources, Louisiana Department of Environmental Quality, Savannah River Ecology Laboratory's NERP Program, U.S. Environmental Protection Agency, and the city of Thibodaux, Louisiana. Manuscript preparation was supported by the Department of Forest Resources, Clemson University, and the Department of Oceanography and Coastal Sciences and the Coastal Ecology Institute, Louisiana State University.

References

Baker, J.B. 1977. Tolerance of planted hardwoods to spring flooding. South. J. Appl. For. **1**:23–25.

Baumann, R.H., J.W. Day Jr., and C.A. Miller. 1984. Mississippi deltaic wetland survival: sedimentation vs. coastal submergence. Science **224**:1093–1095.

Blair, R.M., and M.J. Langlinais. 1960. Nutria and swamp rabbits damage baldcypress seedlings. J. For. **58**:388–389.

Brinson, M.M. 1989. Riverine forests. Pages 87–141 *in* A.E. Lugo, M.M. Brinson, and S. Brown, eds. Ecosystems of the world, Vol. 15: forested wetlands. Elsevier Scientific, Amsterdam.

Brody, M., W.H. Conner, L. Pearlstine, and W. Kitchens. 1989. Modeling bottomland forest and wildlife habitat changes in Louisiana's Atchafalaya basin. Pages 991–1004 *in* R.R. Sharitz and J.W. Gibbons, eds. Freshwater wetlands and wildlife symposium: perspectives on natural, managed and degraded ecosystems. CONF-8603101, DOE Symposium Series No. 61, Office of Scientific and Technical Information, Oak Ridge, TN.

Brown, C.A. 1972. Wildflowers of Louisiana and adjoining states. Louisiana State University Press, Baton Rouge.

Brown, S. 1981. A comparison of the structure, primary productivity, and transpiration of cypress ecosystems in Florida. Ecol. Monogr. **51**:403–427.

Brown, S., and A.E. Lugo. 1982. A comparison of structural and functional characteristics of saltwater and freshwater forested wetlands. Pages 109–130 *in* B. Gopal, R.E. Turner, R.G. Wetzel, and D.F. Whigham, eds. Wetlands: ecology and management. International Scientific, Jaipur, India.

Brown, S., M.M. Brinson, and A.E. Lugo. 1979. Structure and function of riparian wetlands. Pages 17–31 *in* R.R. Johnson and J.F. McCormick, eds. Strategies for protection and management of floodplain wetlands and other riparian ecosystems. Gen. Tech. Rep. WO-12, U.S. Dept. Agric. For. Serv., Washington, DC.

Conner, W.H., and M. Brody. 1989. Rising water levels and the future of southeastern Louisiana swamp forests. Estuaries **12**:318–323.

Conner, W.H., and J.W. Day, Jr. 1976. Productivity and composition of a baldcypress–water tupelo site and a bottomland hardwood site in a Louisiana swamp. Am. J. Bot. **63**:1354–1364.

———. 1982. The ecology of forested wetlands in the southeastern United States. Pages 69–87 *in* B. Gopal, R.E. Turner, R.G. Wetzel, and D.F. Whigham, eds. Wetlands: ecology and management. International Scientific, Jaipur, India.

———. 1988a. The impact of rising water levels on tree growth in Louisiana. Pages 219–224 *in* D.D. Hook et al., eds. The ecology and management of wetlands, Vol. 2: Management, use, and value of wetlands. Croom Helm, England.

———. 1988b. Rising water levels in coastal Louisiana: implications for two coastal forested wetland areas in Louisiana. J. Coastal Res. **4**:589–596.

———. 1989. A use attainability analysis of wetlands for receiving treated municipal and small industry wastewater: a feasibility study using baseline data from Thibodaux, Louisiana. Center for Wetland Resources, Louisiana State Univ., Baton Rouge.

———. 1991a. Leaf litter decomposition in three Louisiana freshwater forested wetland areas with different flooding regimes. Wetlands **11**:303–312.

———. 1991b. Variations in vertical accretion in a Louisiana swamp. J. Coastal Res. **7**:617–622.

———. 1992a. Diameter growth of *Taxodium distichum* (L.) Rich. and *Nyssa aquatica* (L.) from 1979–1985 in four Louisiana swamp stands. Am. Midl. Nat. **127**:290–299.

———. 1992b. Water level variability and litterfall productivity of forested freshwater wetlands in Louisiana. Am. Midl. Nat. **128**:237–245.

Conner, W.H., and J.R. Toliver. 1988. The problem of planting Louisiana swamplands when nutria *(Myocastor coypus)* are present. Pages 42–49 *in* N.R. Holler, ed. Proc. Third Eastern Wildlife Damage Control Conference. Ala. Coop. Ext. Serv., Auburn Univ., Auburn, AL.

Conner, W.H., J.G. Gosselink, and R.T. Parrondo. 1981. Comparison of the vegetation of three Louisiana swamp sites with different flooding regimes. Am. J. Bot. **63**:1354–1364.

Conner, W.H., J.W. Day, Jr., R.H. Baumann, and J.M. Randall. 1989. Influence of hurricanes on coastal ecosystems along the northern Gulf of Mexico. Wetl. Ecol. Manage. **1**:45–56.

Day, F.P., Jr. 1982. Litter decomposition rates in the seasonally flooded Great Dismal Swamp. Ecology **63**:670–678.

Day, J.W., Jr., and G.P. Kemp. 1985. Long-term impacts of agricultural runoff in a Louisiana swamp forest. Pages 317–326 *in* P.J. Godfrey, E.R. Kaynor, S. Pelczarski, and J. Benforado, eds. Ecological considerations in wetlands treatment of municipal wastewaters. Van Nostrand Reinhold, NY.

Day, J.W., Jr., and P.H. Templet. 1989. Consequences of sea level rise: implications from the Mississippi Delta. Coastal Man. **17**:241–257.

DeLaune, R.D., C.J. Smith, and W.H. Patrick, Jr. 1983. Nitrogen losses from a Louisiana Gulf Coast salt marsh. Estuar. Coastal Shelf Sci. **17**:133–141.

Dicke, S.G., and J.R. Toliver. 1988. Effects of crown thinning on baldcypress height, diameter, and volume growth. South. J. Appl. For. **12**:252–256.

Dijkema, K.S., and W.J. Wolff. 1980. Flora and vegetation of the Wadden Sea islands and coastal areas. Report 9 of the Wadden Sea Working Group. Blakema, Rotterdam.

Eggler, W.A., and W.G. Moore. 1961. The vegetation of Lake Chicot, Louisiana, after eighteen years impoundment. Southwestern Nat. **6**:175–183.

Faulkner, P.L., F. Zeringue, and J.R. Toliver. 1985. Genetic variability among open-pollinated families of baldcypress seedlings planted on two different sites. Pages 267–272 *in* Proc. 18th South. Forest Tree Improvement Conf., Long Beach, MS. Sponsored Publ. No. 40.

Fuller, D.A., C.E. Sasser, W.B. Johnson, and J.G. Gosselink. 1984. The effects of herbivory on vegetation on islands in Atchafalaya Bay, Louisiana. Wetlands **4**:105–114.

Gagliano, S.M., K.J. Meyer-Arendt, and K.M. Wicker. 1981. Land loss in the Mississippi River Deltaic Plain. Trans. Gulf Coast Assoc. Geol. Sci. **31**:295–300.

Gornitz, V., S. Lebedeff, and J. Hansen. 1982. Global sea level trend in the past century. Science **215**:1611–1614.

Gosselink, J.G. 1984. The ecology of delta marshes of coastal Louisiana: a community profile. U.S. Fish and Wildl. Serv., Office of Biological Services, Washington, DC. FWS/OBS-84/09.

Gosselink, J.G., S.E. Bayley, W.H. Conner, and R.E. Turner. 1981. Ecological factors in the determination of riparian wetland boundaries. Pages 197–219 *in* J.R. Clark and J. Benforado, eds. Wetlands of bottomland hardwood forests. Elsevier, Amsterdam.

Green, W.E. 1947. Effect of water impoundment on tree mortality and growth. J. For. **45**:118–120.

Hall, T.F., and G.E. Smith. 1955. Effects of flooding on woody plants, West Sandy dewatering project, Kentucky Reservoir. J. For. **53**:281–285.

Harms, W.R., H.T. Schreuder, D.D. Hook, and C.L. Brown. 1980. The effects of flooding on the swamp forest in Lake Ocklawah, Florida. Ecology **61**:1412–1421.

Hatton, R.S., R.D. DeLaune, and W.H. Patrick, Jr. 1983. Sedimentation, accretion and subsidence in marshes of Barataria Basin, Louisiana. Limnol. Oceanogr. **28**:494–502.

Hoffman, J.S, D. Keyes, and H.G. Titus. 1983. Projecting future sea level rise: methodology, estimates to the year 2100, and research needs. U.S. EPA, Office of Policy and Resource Management, Washington, DC.

Hook, D.D. 1984. Waterlogging tolerance of lowland tree species of the south. South J. Appl. For. **8**:136–149.

Jackson, B.D., and R.A. Morris. 1986. Helicopter logging of baldcypress in southern swamps. South. J. Appl. For. **10**:20–23.

Johnson, F.L., and D.T. Bell. 1976. Plant biomass and net primary production along a flood-frequency gradient in a streamside forest. Castanea **41**:156–165.

Kemp, G.P., W.H. Conner, and J.W. Day, Jr. 1985. Effects of flooding on decomposition and nutrient cycling in a Louisiana swamp forest. Wetlands **5**:35–51.

Kerr, R.A. 1991. Cooling the greenhouse cheaply. Science **251**:621.

Kolb, C.R., and J. van Lopik. 1958. Geology of the Mississippi Deltaic Plain, Southeastern Louisiana. Report to the U.S. Army Engineer Waterways Experiment Station, Corps of Engineers, Vicksburg, MS. Tech. Rept. #3-483, 2 vols.

Little, S., J.J. Mohr, and L.L. Spicer. 1958. Salt-water storm damage to loblolly pine forests. J. For. **56**:27–28.

Mitsch, W.J., and K.C. Ewel. 1979. Comparative biomass and growth of cypress in Florida wetlands. Am. Midl. Nat. **101**:417–426.

Mitsch, W.J., and W.G. Rust. 1984. Tree growth responses to flooding in a bottomland forest in northwestern Illinois. For. Sci. **30**:499–510.

Morgan, J.P. 1967. Ephemeral estuaries of the deltaic environment. Pages 115–120 *in* G. Lauff, ed. Estuaries. AAAS Publ. No. 83.

Moss, A.E. 1940. Effect on trees of wind-driven salt water. J. For. **38**:421–425.

O'Neil, T. 1949. The muskrat in the Louisiana coastal marshes. Louisiana Department of Wildlife and Fisheries, New Orleans.

Penfound, W.T., and E.S. Hathaway. 1938. Plant communities in the marshlands of southeastern Louisiana. Ecol. Monogr. **8**:1–56.

Pezeshki, R.S. 1987. Gas exchange response of tupelo-gum *(Nyssa aquatica)* to flooding and salinity. Photosynthetica **21**:489–493.

Pezeshki, R.S., and J.L. Chambers. 1986. Effects of soil salinity on stomatal conductance and photosynthesis of green ash *(Fraxinus pennsylvanica)*. Can. J. For. Res. **16**:569–573.

Pezeshki, S.R., R.D. DeLaune, and W.H. Patrick, Jr. 1988. Effect of salinity on leaf ionic content and photosynthesis of *Taxodium distichum* L. Am. Midl. Nat. **119**:185–192.

Salinas, L.M., R.D. DeLaune, and W.H. Patrick, Jr. 1986. Changes occurring along a rapidly submerging coastal area: Louisiana, USA. J. Coastal Res. **2**:269–284.

Southeastern Forest Experiment Station. 1958. Annual Report for 1957. Asheville, NC.

Stevenson, J.C., L.G. Ward, and M.S. Kearney. 1986. Vertical accretion in marshes with varying rates of sea level rise. Pages 241–259 *in* D.A. Wolfe, ed. Estuarine variability. Academic Press, Orlando and London.

Toliver, J.R., S.G. Dicke, and R.S. Prenger. 1987. Response of a second-growth natural stand of baldcypress to various intensities of thinning. Pages 462–465 *in* D.R. Phillips, compiler. Proc. fourth biennial silvicultural conference. U.S. Dept. Agric. For. Serv., Gen. Tech. Rep. SE-42.

Walker, H.J., J.M. Coleman, H.H. Roberts, and R.S. Tye. 1987. Wetland loss in Louisiana. Geografiska Annaler **69A**:189–200.

Wells, B.W., and I.V. Shunk. 1938. Salt spray: an important factor in coastal ecology. Bull. Torrey Botanical Club **65**:485–492.

Wilhite, L.P., and J.R. Toliver. 1990. *Taxodium distichum* (L.) Rich. Baldcypress. Pages 563–572 *in* R.M. Burns and B.H. Honkala, compilers. Silvics of North America. Vol. 1. Conifers. U.S. Dept. Agric. For. Serv. Handb. 654. Washington, DC.

20

Wolfgang Grosse, Hans B. Büchel, & Sibylle Lattermann

Root Aeration in Wetland Trees and Its Ecophysiological Significance

Oxygen deficiency is an important stress factor for roots of trees growing in frequently or permanently flooded habitats. Due to the low diffusion rate of oxygen in water and rapid oxygen consumption by active bacterial metabolism, oxygen partial pressure equals zero in saturated soil of wet sites during waterlogged periods (Armstrong 1978; Crawford 1978, 1982). The roots are unable to meet their oxygen demand from the rhizosphere under these conditions and are restricted to internal aeration from their air-exposed regions. However, oxygen diffusion over gas transport pathways exceeding 200 mm inside the plants is not adequate for respiration in the root tissue (Armstrong 1979, Drew et al. 1985).

The formation of prop roots by mangroves, knees by baldcypresses, and aerenchyma by submerged plant parts are well-known adaptations for improving the oxygen supply of tree roots growing in wetlands and areas of varying groundwater levels (Crawford 1990). Recently, pressurized gas transport has been identified in alder species as another adaptive mechanism in wetland trees for improving the root system's oxygen supply in anaerobic root environments (Grosse and Schröder 1984, 1985; Schröder 1989; Grosse et al. 1990).

Gas Diffusion to Tree Roots

Tree species vary greatly in their capacity to supply oxygen to their root systems by diffusion. The gas diffusion rates of six different deciduous tree species, which were cultivated under non-wetland conditions and measured in their dormant state during winter, varied between 0.6 to 9.0 μl air min^{-1} (Table 20.1). The highest diffusion rate was found in *Fraxinus excelsior* (European ash; see Table 20.2 for species list).

A similar situation was found for foliated trees. Diffusion rate increased with tree

Table 20.1 Gas Flux through 6- to 12-Month-Old Trees of Various Species in Relationship to Light-Induced Temperature Gradients in the Stem

Species	Age (months)	Effusion Rate[a] (μl air min^{-1}) Dark[b]	Light[c]	Light as % Dark[d]	Ref.[e]
Leafless tree seedlings (dormant)					
Acer pseudoplatanus	6	0.6	0.7	117	(a)
Aesculus hippocastanum	6	1.1	2.6	236	(a)
Alnus glutinosa	6	1.4	5.8	414	(a)
Carpinus betulus	6	3.0	3.2	107	(a)
Fagus sylvatica	6	3.3	3.3	100	(a)
Fraxinus excelsior	6	9.0	11.1	123	(a)
Foliated tree seedlings (growing)					
Populus tremula	12	0.8	1.2	150	(b)
Tilia cordata	12	1.1	1.3	119	(b)
Betula pubescens	12	1.2	1.4	117	(b)
Alnus glutinosa	6	1.8	3.2	189	(a)
	12	4.0	10.4	261	(a)
(2-month flooded)	8	15.0	22.7	151	(a)
Taxodium distichum	12	8.4	14.9	176	(b)
Fraxinus excelsior	12	13.9	15.9	114	(b)

[a]Ethane was supplied as a tracer gas in the air surrounding the stem. Gas flux through the tree and out of the roots was measured by gas chromatography.
[b]Dark: stem shaded.
[c]Light: stem under a photon flux of 230 μmol m^{-2} s^{-1}.
[d]The percent enhancement is calculated relative to the dark rate for each species.
[e]Adapted from (a) Grosse and Schröder 1985 and (b) Grosse et al. 1992.

age and was highest in *F. excelsior,* which is known to be well adapted to growth in water-saturated soil. *Taxodium distichum* (baldcypress), which thrives in American swamps, had a relatively high gas diffusion rate as well.

Diffusion rate values of all other tree species assessed by these studies were relatively low. This is true as well for *Alnus glutinosa* (common alder), which is the predominant tree species of European floodplain forests and riverine temperate forests. These regions are characterized by periods of high water table and springtime flooding, respectively. Considering the low oxygen level of soil in this species' natural habitat, the above result demonstrates a very poor internal oxygen supply to the roots and was generally unexpected.

The situation was different when young trees were adapted to wetland conditions by being cultivated in water-saturated soil for 60 days. Besides aerenchyma formation, lenticels were hypertrophied and transformed finally into intumescences in the adapted trees. The resulting diffusion rate was about eight times higher than in nonadapted trees. The adaptation of *A. glutinosa* to wetland conditions was similar to that described for *Nyssa sylvatica* var. *biflora* (swamp tupelo) by Hook et al. (1970, 1971).

Table 20.2 Soil Saturation Indicator Values of Wetland Species Referred to in Text

	Common Name		
Scientific Name	English	German	F[a]
Acer pseudoplatanus	sycamore	Echter Berg-Ahorn	6[b]
Aesculus hippocastanum	commom horse-chestnut	Balkan-Rosskastanie	n.d.
Alnus glutinosa	common alder	Schwarz-Erle	9[c]
Alnus incana	gray alder	Grau-Erle	7
Alnus viridis	green alder	Grün-Erle	6
Betula pubescens	hairy birch	Moor-Birke	x
Carpinus betulus	hornbeam	Gemeine Hainbuche	x
Fagus sylvatica	European beech	Rotbuche	5[b]
Fraxinus excelsior	European ash	Gemeine Esche	x
Glyceria maxima	great water reed grass	Großer Schwaden	10[d]
Ilex aquifolium	holly	Gemeine Stechhülse	5[b]
Iris germanica	German iris	Deutsche Schwertlilie	n.d.
Iris pseudacorus	yellow flag iris	Sumpf-Schwertlilie	10
Nyssa sylvatica var. biflora	swamp tupelo	Wald-Tupelo	n.d.
Populus tremula	aspen	Zitter-Pappel	5[b]
Taxodium distichum	baldcypress	Sumpfzypresse	n.d.
Tilia cordata	small-leaved lime	Winter-Linde	x

Source: Ellenberg 1974.
[a]Ellenberg's indicator values of soil saturation (F) on a scale of 1–12 from arid (1) through aquatic emergent (10) to total immersion year-round (12). n.d.: no data. x: varying behavior at different regions.
[b]Not clear.
[c]Frequently flooded sites.
[d]Varying water table.

During the first adaptation period of trees to water-logging and soil anoxia, the root system may profit from an oxygen supply improved by pressurized gas transfer. When solar radiant energy is absorbed by the tree bark, air is pumped to the roots and the physical effect of thermo-osmosis increases the oxygen partial pressure in the root tissues.

Gas Transport to Tree Roots

When tree stems are warmed to several degrees above ambient levels, solar and incandescent light energy are absorbed by the brown pigmented tissue underlying the stem surface (Fig. 20.1). Air is then pressurized inside the intercellular spaces forcing ambient air into the stem, down to the roots, and to the rhizosphere by bubbles formed at the root surface (Grosse and Schröder 1984, 1986). Correlation among light intensity, temperature difference, pressurization inside the intercellular spaces, and gas flow to the roots in *A. glutinosa* is shown in Figure 20.2.

This light-induced gas flow quadruples the oxygen supply to the root system of leafless *A. glutinosa* in comparison to that of trees in the dark (Table 20.1). The stimulation is lower but still detectable in foliated *A. glutinosa* as well as in *Aesculus hippocastanum* (common horse-chestnut), *T. distichum,* and *Populus tremula* (aspen).

Figure 20.1. Photograph of the lower stem of
Alnus glutinosa (common alder), with a sector
cut out, showing that pigmentation is restricted
to the surface area.

Pressurized Gas Transport

The pressurized flow of air through plants has been described by many different terms (e.g., thermo-diffusion, thermal molecular flow, and thermal transpiration as reviewed by Grosse and Bauch, 1991), but is defined now as thermo-osmosis of gases. Generally, under isobaric conditions, a flow of gas along a temperature gradient from one bulk phase to another, separated by a porous partition, is referred to as thermo-osmosis (Haase 1969). Two prerequisites are important for this gas flow. One, a temperature gradient must be established between the two bulk phases. Two, pore diameters in the partition separating the bulk phases must be similar to or smaller than the "mean free path length" of the gas molecules in the system (e.g., 70 nm at room temperature [20°C] and standard barometric pressure [100 kPa]).

Alnus glutinosa meets both these prerequisites. When trees were irradiated by incandescent light, stems were warmed relative to the surrounding air and the resulting temperature gradient was steepened with increasing light intensities (Fig. 20.2A). The porous partition that separates the warmer stem tissue from cooler ambient air is localized in the lenticels, restricted areas of loosely arranged cells at the stem surface that allow gas exchange between the atmosphere and internal tissues. Lenticels are

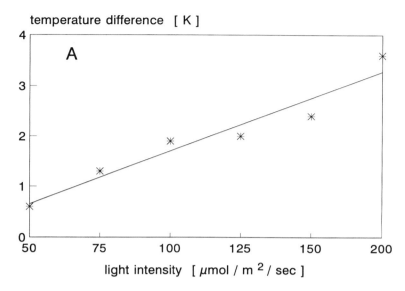

Figure 20.2. A. Temperature difference between stem interior and outside air as related to stem-irradiation intensity.

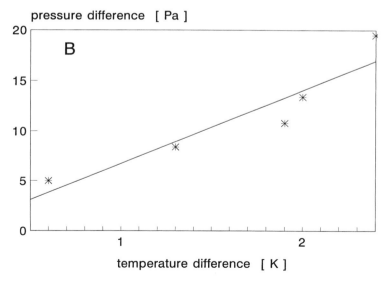

Figure 20.2. (*continued*) B. Pressure difference between stem interior and outside air as related to temperature difference.

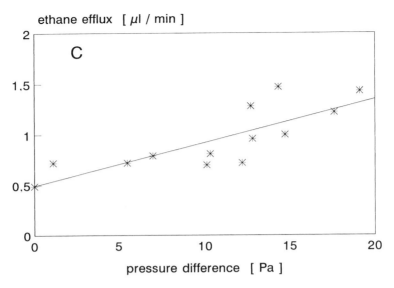

Figure 20.2. (*continued*) C. Gas flow rate (ethane is tracer gas) as related to pressure difference. Measurements were performed on one-year-old leafless *A. glutinosa* seedlings. (Adapted from Grosse and Schröder 1986.) K: Kelvin. Pa: Pascal.

formed by the phellogen, a layer of actively dividing meristematic cells. The phellogen, equipped with intercellular spaces of about 14 nm in diameter, has been identified as the thermo-osmotically active partition (Büchel and Grosse 1990). The conditions for gas diffusion in the dark and pressurized gas flow in the light are illustrated in Figure 20.3. In the dark (upper left sketch), the temperature inside the intercellular spaces of the stem equals that of the ambient air. Therefore, the mean free path length of the gas molecules is identical at both sides of the pores of the phellogen, and equal numbers of molecules move through the intercellular spaces in and out of the stem. No pressurization occurs in the intercellular system of the stem.

When a temperature gradient is established between the warmer tree stem and the ambient air by irradiation (lower left sketch), the mean free path length of the gas molecules are increased inside the tree, due to the higher temperature of the stem. This increased mean free path length prevents movement of molecules back to the outside air. By this one-way process, gas molecules accumulate in the intercellular system, causing pressurization (Fig. 20.2B), and are pumped down to the roots (Fig. 20.2C), which are more aerenchymatous (contain large intercellular air spaces) in *A. glutinosa* than in other trees (Köstler et al. 1968). Because oxygen is included in the gas flow, the root tissues should profit from this improved aeration.

Physiological Significance of Pressurized Gas Transport

Tree roots exploit the rhizosphere to meet their oxygen demand. When internal aeration is increased by pressurized gas transport, oxygen uptake rates are reduced sig-

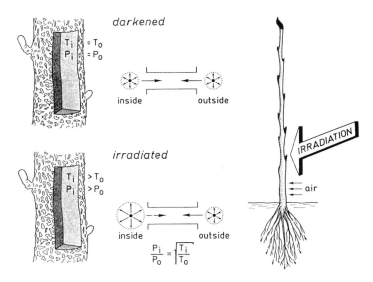

Figure 20.3. Structural properties of stems showing how exposure to light warms the tree stem interior, causing gas pressure to be higher than that of the cooler outside air as the result first of net gas influx, then of gas efflux from the root tips. Arrows point in direction of gas flow. T_i: interior temperature. T_o: outside temperature. P_i: interior pressure. P_o: outside pressure.

nificantly in tree species that are not sensitive to water-saturated soil (Table 20.3). The reduction of oxygen uptake is higher in leafless than in foliated trees, as seen in *Betula pubescens* (hairy birch) and *P. tremula,* and agrees with the lower gas transport rates detected in foliated *A. glutinosa* (Table 20.1).

Oxygen uptake measurements in two alder species (*A. glutinosa* and *A. incana,* gray alder) have shown that oxygen demand of the root system is met completely and surplus oxygen escapes into the rhizosphere when enhanced internal oxygen supply is established (Table 20.3). Because oxygen uptake in *A. viridis* (green alder) and *Ilex aquifolium* (holly) is not significantly reduced, it is clear that pressurized gas transport is an important adaptation mechanism for wetland trees. This mechanism allows survival during the initial period of flooding and soil anoxia until other adaptation mechanisms such as ethylene synthesis and aerenchyma formation may be activated.

Gas bubbles containing 1–6 vol% O_2 are observed at the roots of alder trees under pressurized gas transport conditions (Grosse et al. 1990). Therefore, we suppose that oxygen excretion is high enough to protect trees from phytotoxins that accumulate in saturated soils during periods of anoxia. Oxygen excretion may also benefit aerobic microorganisms, as shown by increased levels of nitrogen fixation by the symbiotic actinomycete *Frankia alni* associated with *A. glutinosa* (Grosse et al. 1990).

A sufficient oxygen supply is important for growth and many other physiological activities of roots. The stimulation of nutrient uptake in trees, which has been observed during periods of pressurized gas transport (Grosse and Meyer 1992), may

Table 20.3 The Effect of Pressurized Gas Transport on Oxygen Uptake by Roots of 6- to 12-Month-Old Tree Seedlings from an Aqueous Medium at 5°C Except for *Alnus glutinosa* and *A. incana* at 20°C

Species	Oxygen Uptake[a] ($\mu mol\ h^{-1}\ tree^{-1}$)		Light as % Dark[d]	Significance[e]	Ref.[f]
	Dark[b]	Light[c]			
Alnus glutinosa					
leafless	0.9	−0.3	<0	***	(a)
Alnus incana					
leafless	0.3	−0.7	<0	***	(a)
Alnus viridis					
foliated	8.0	7.6	95	n.s.	(b)
Betula pubescens					
leafless	2.2	0.7	32	**	(b)
foliated	3.3	2.5	76	**	(b)
Ilex aquifolium					
evergreen	15.9	12.9	81	n.s.	(b)
Populus tremula					
leafless	10.2	5.0	49	**	(b)
foliated	5.0	4.2	84	**	(b)
Taxodium distichum					
foliated	3.0	2.0	67	**	(b)

[a]The oxygen concentrations were measured polarographically. Gas transport conditions were created by irradiating the tree stem.
[b]Dark: stem shaded.
[c]Light: stem under a photon flux of 230 $\mu mol\ m^{-2}\ s^{-1}$.
[d]The percent reduction of the uptake rate by light is calculated relative to the dark rate for each species.
[e]The t-test values are different to a very high level with $t_p = 0.1\%$ (***), high with $t_p = 1\%$ (**), or not significant with $t_p \geq 5\%$ (n.s.).
[f]Adapted from (a) Grosse et al. 1990 and (b) Grosse et al. 1992.

result from this mechanism. Trees that were cultivated hydroponically for 72 hours had nutrient uptake rates that were significantly reduced, to about 20%, 30%, and 70% of nitrate, potassium, and phosphate, respectively, when the nutrient solution became anoxic (Fig. 20.4). This reduction was not observed when oxygen deficiency of the root tissue was avoided by enhanced internal aeration resulting from light-induced pressurized gas transport.

The improved ion uptake may have adverse effects on trees growing on ferrous-contaminated anoxic soil. The ferrous iron appears not to be detoxified by oxidation and $Fe(OH)_3$ deposition, but instead is taken up in increased amounts that can reach toxic concentrations, impairing the photosynthetic activity of alder leaves (unpublished results).

Flooding Resistance

Recent studies on *A. glutinosa* have shown that pressurized gas transport becomes ineffective with a rising water table (Grosse and Schröder 1985). This adaptation

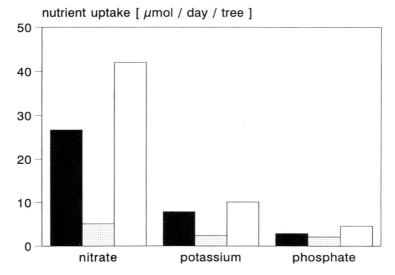

Figure 20.4. Effect of root anoxia and enhanced internal oxygen supply resulting from pressurized gas transport on nutrient uptake by one-year-old foliated *A. glutinosa.* Trees were cultivated hydroponically for 72 h at 20°C (day) and 15°C (night). Root system was ■: externally aerated; ▨: anoxic without enhanced internal oxygen supply; or □: anoxic with enhanced internal oxygen supply. (Adapted from Grosse and Meyer 1992.)

mechanism, which is highly effective for tree survival during periods of saturated soil, does not seem to be active enough to supply oxygen to the roots during flooding. The root surface of trees that were flooded up to 10 cm above ground for 20 days during summer became a deep black and did not appear healthy.

Measurements of superoxide dismutase (SOD) activity in root-tip tissue from alders cultivated in well-aerated soil, following a period of flooding, corroborate this first superficial impression of root viability. SOD is one of several enzymes that account for the anaerobic polypeptides synthesized during anoxia in tissues of plants that are successful in anaerobic environments. This enzyme protects plant tissues from self-induced post-anoxic injury resulting from highly toxic species of oxygen radicals, which are generated in larger amounts when aerobic conditions are restored (Hendry and Brocklebank 1985, Crawford et al. 1987).

When the herbaceous monocots *Iris pseudacorus* (yellow flag iris), *Iris germanica* (German iris), and the grass *Glyceria maxima* (great water reed grass) were assessed for SOD activity in their rhizomes by Monk et al. (1987), a high correlation between SOD accumulation and the degree of tolerance against longtime anoxia and post-anoxic injury was found. Large amounts of SOD accumulated during the anoxic and post-anoxic period in the most resistant *I. pseudacorus,* but a decrease of SOD activity during anoxia and the following post-anoxic period in the less tolerant grass *G. maxima* was observed. *Alnus glutinosa* is more comparable to the less resistant *I.*

Table 20.4 The Development (percent of the untreated control) of Superoxide Dismutase (SOD) Activity in *Iris pseudacorus, Iris germanica,* and *Glyceria maxima* Rhizomes as Well as *Alnus glutinosa* Roots Resulting from Anoxia, Post-Anoxic Recovery, and Light-Induced Gas Transport

	A: SOD Activity After 28 Days of Anoxia and 2 Days of Post-Anoxic Recovery			
	SOD Development (%)			
Species	Control	Anoxia (28 d)	Post-Anoxia (2 d)	Ref.[a]
Iris pseudacorus	100	1370	1420	(a)
Iris germanica	100	69	83	(a)
Glyceria maxima	100	59	29	(a)
Alnus glutinosa	100	115	80	(b)
	B: SOD Activity Affected by Light-Induced Gas Transport			
	SOD Development (%)			
Species	Control	Anoxia Light (1 d)	Aerated Light (1 d)	Ref.
Alnus glutinosa	100	118	223	(b)

[a] Adapted from (a) Monk et al. 1987 and (b) Büchel 1991.

germanica than to the resistant *I. pseudacorus* (Table 20.4, panel A), with a low accumulation of SOD during anoxia and a decrease of SOD activity soon after the restoration of oxygen. From these results we suggest that *A. glutinosa* is less adapted to soil anoxia than *I. pseudacorus,* but better adapted than *I. germanica* and *G. maxima.*

The SOD accumulation in alder roots during one day of post-anoxia shows that thermo-osmotic gas transport is able to change the low flood-resistance pattern to one of high resistance (Table 20.4, panel B). However, the high rate of *A. glutinosa* adventitious rooting should be interpreted as a further effective adaptation to frequently flooded areas.

Adventitious Rooting

Growth of adventitious roots is a common response to flooding (Gill 1975). It has been suggested that trees generate these new roots to meet their oxygen demand from the more oxygenated surface water as well as for nutrient uptake. This seems to be only partly true. The oxygen that diffuses from the nonsubmerged to the submerged part of the stem is much more important for this rhizogenesis (new root growth) than the oxygen supplied from the surface water. This phenomenon becomes evident from the assessment of rhizogenesis of three-year-old alder trees that were flooded for three weeks in August.

When the lowest 10 cm of the stems of the two *Alnus* species were flooded, many

Figure 20.5. Photograph of young tree seedlings of *A. incana* (A, B) and *A. glutinosa* (C, D) flooded for 3 weeks to approximately 10 cm above ground level. A strong reduction of adventitious rooting can be observed when the aerial parts of the tree stems are sealed by stopcock grease (B, D). Photo was taken a few minutes after removal of water.

Table 20.5 Relation of ATP Content in Root Tips and Adventitious Roots to Internal Oxygen Supply from the Aerial Shoot after 25 Days of Flooding

	ATP Content (nmol ATP g^{-1} fwt[a])	
Species	Lenticels Open[b]	Lenticels Closed[c]
Alnus glutinosa		
root tips	11 ± 8 (19)[d]	8 ± 6 (14)
adventitious roots	45[e] ± 19 (14)	16 ± 12 (2)
Alnus incana		
root tips	6 ± 4 (2)	4 ± 2 (2)
adventitious roots	35 ± 20 (2)	n.d.

Note: Values are mean ± standard deviation.

[a] fwt: fresh weight.
[b] Lenticels open = no grease blocked entry of oxygen.
[c] Lenticels closed = oxygen uptake into the stem was blocked by greasing the nonsubmerged stem surface up to 10 cm above the water level.
[d] ( ): number of tree samples.
[e] The differences to root tips of *A. glutinosa* were highly significant at P = 0.001.

adventitious roots developed (Fig. 20.5). No roots were detected on *A. incana* and very few were seen on *A. glutinosa* when oxygen diffusion from the aerial part was blocked by covering the stem surface with stopcock grease. Based on the ATP content of the root tissues (Table 20.5), we conclude that the energy supply is too low for vigorous adventitious rooting after cessation of the internal oxygen supply. A sufficiently high ATP level is guaranteed only in roots that are aerated internally through the tree. When oxygen supply to the adventitious roots is blocked, ATP content is decreased to nearly the low levels of poorly aerated submerged roots.

Evaluation of Pressurized Gas Transport

Pressurized gas transport in wetland trees, caused by thermo-osmosis of gases and as identified in different unrelated genera (Table 20.1 and Table 20.2), improves oxygen supply to the root system and consequently stimulates nutrient uptake by roots. This gas transport is not detectable in trees restricted to drier habitats and may be seen, therefore, as an adaptation mechanism of some species for growing on wet sites (Grosse and Schröder 1985, Grosse et al. 1992).

Oxygen release from roots into the rhizosphere may create oxygenated zones. Aerobic microorganisms may profit from that oxygen excretion, as shown by increased N_2 fixation activity of the symbiotic fungus *Frankia alni* (Grosse et al. 1990).

Because oxygen must travel a greater distance in older trees to get to the root tips, and gas transport activities are low in foliated trees, pressurized gas transport is unable to protect the roots of three-year-old alder trees from dangerous effects of anoxia during periods of summer flooding. Therefore, we believe that this ventilation phenomenon is only beneficial for deciduous trees in wetlands during the dormant season and for seedling establishment on water-saturated soil. Roots that rot in floods during the growing season will be replaced by adventitious roots to establish an adequate water and nutrient supply to the trees.

Acknowledgments The authors are grateful to Dr. Debashish Bhattacharya, Department of Biochemistry, Max Planck Institut, Goettingen, Germany, for reading the manuscript.

References

Armstrong, W. 1978. Root aeration in the wetland condition. Pages 269–297 *in* D.D. Hook and R.M.M. Crawford, eds. Plant life in anaerobic environments. Ann Arbor Sciences Publishers, Ann Arbor.
———. 1979. Aeration in higher plants. Pages 226–332 *in* H.W. Woolhouse, ed. Advances in botanical research. Vol. 7. Academic Press, London.
Büchel, H.B. 1991. Mechanismen der Anpassung von Pflanzen an das Wachstum im sauerstoffarmen Lebensraum. Dissertation. Univ. of Cologne, Köln.
Büchel, H.B. and W. Grosse. 1990. Localization of the porous partition responsible for pressurized gas transport in *Alnus glutinosa* (L.) Gaertn. Tree Physiol. **6**:247–256.
Crawford, R.M.M. 1978. Metabolic adaption to anoxia. Pages 119–136 *in* D.D. Hook and R.M.M. Crawford, eds. Plant life in anaerobic environments. Ann Arbor Sciences Publishers, Ann Arbor.

————. 1982. Physiological responses to flooding. Pages 453–477 *in* O.L. Lange, P.S. Nobel, C.B. Osmund, and H. Ziegler, eds. Encyclopedia of plant physiology. Vol. **12B,** Physiological plant ecology II. Springer Verlag, Berlin.

————. 1990. Studies in plant survival. Page 113 *in* D.J. Anderson, P. Greig-Smith, and F.A. Pitelka, eds. Studies in ecology. Vol. **11.** Blackwell Scientific, Oxford.

Crawford, R.M.M., L.S. Monk, and Z.M. Zochowski. 1987. Enhancement of anoxia tolerance by removal of volatile products of anaerobiosis. Pages 375–384 *in* R.M.M. Crawford, ed. Plant life in aquatic and amphibious habitats. Br. Ecol. Soc., Spec. Publ. Ser. No. **5,** Blackwell, Oxford.

Drew, M.C., P.H. Saglio, and A. Pradet. 1985. Large adenylate energy charge and ADP/ATP ratios in aerenchymatous roots of *Zea mays* in anaerobic media as a consequence of improved internal oxygen transport. Planta **165**:51–58.

Ellenberg, H. 1974. Indicator values of vascular plants in Central Europe. *In* H. Heller, ed. Scripta Geobotanica. Vol. **9,** Verlag Erich Goltze, Göttingen.

Gill, Chr. J. 1975. The ecological significance of adventitious rooting as a response to flooding in woody species, with special reference to *Alnus glutinosa* (L.) GAERTN. Flora **164**:85–97.

Grosse, W., and C. Bauch. 1991. Gas transfer in floating-leaved plants. Vegetatio **97**:185–192.

Grosse, W., and D. Meyer. 1992. The effect of pressurized gas transport on nutrient uptake during hypoxia of alder roots. Bot. Acta **105**:223–226.

Grosse, W. and P. Schröder. 1984. Oxygen supply of roots by gas transport in alder-trees. Z. Naturforsch. **39C**:1186–1188.

————. 1985. Aeration of roots and chloroplast free tissues of trees. Ber. Deutsch. Bot. Ges. **98**:311–318.

————. 1986. Plant life in anaerobic environments, the physical basis and anatomical requirements. Review. Ber. Deutsch. Bot. Ges. **99**:367–381.

Grosse, W., S. Sika, and S. Lattermann. 1990. Oxygen supply of roots by thermo-osmotic gas transport in *Alnus glutinosa* and other wetland trees. Pages 246–249 *in* D. Werner and P. Müller, eds. Fast growing trees and nitrogen fixing trees. Gustav Fischer Verlag, Stuttgart, New York.

Grosse, W., J. Frye, and S. Lattermann. 1992. Root aeration in wetland trees by pressurized gas transport. Tree Physiol. **10**:285–295.

Haase, R. 1969. Thermodynamics of irreversible processes. Pages 191–214. Addison-Wesley, Reading.

Hendry, G.A.F. and K.J. Brocklebank. 1985. Iron-induced oxygen radical metabolism in waterlogged plants. New Phytol. **101**:199–206.

Hook, D.D., C.L. Brown, and P.P. Kormanik. 1970. Lenticels and water root development of swamp tupelo under various flooding conditions. Bot. Gaz. **131**:217–224.

————. 1971. Inductive flood tolerance in swamp tupelo (*Nyssa sylvatica* var. *biflora* [Walt.] Sarg.). J. Exp. Bot. **22**:78–89.

Köstler, J.N., F. Bruckner, and H. Bibelriether. 1968. Die Wurzeln der Waldbäume. Verlag Paul Parey, Hamburg.

Monk, L.S., K.V. Fagerstedt, and R.M.M. Crawford. 1987. Superoxide dismutase as an anaerobic polypeptide. Plant Physiol. **85**:1016–1020.

Schröder, P. 1989. Characterization of a thermo-osmotic gas transport mechanism in *Alnus glutinosa* (L.) Gaertn. Trees **3**:38–44.

21

Aimlee D. Laderman & Rachel L. Donnette

Coastal Forest Management and Research

Ecosystem management and research are closely linked. Although research provides the guidelines to govern sensible management, management problems indicate the most urgently needed research. This chapter summarizes the major threats to coastally restricted forests and their attendant management strategies, and poses a series of questions for further research. Information is drawn from chapters in this book as well as from workshops[1] at two international symposia.

Management

Management concerns for coastally restricted forests are, for the most part, similar to those in all forested areas of the temperate zone. In addition, there are problems specific to

- Coastal regions, such as storms and climate change;
- Ecological islands, particularly the maintenance of genetic diversity;
- Poor competitors, where attention must be paid to the particular stressors to which each species is adapted; and
- Plants with a high moisture requirement.

The overriding problems facing coastally restricted forests are rooted in habitat loss coupled with regeneration failure. Habitat loss primarily results from logging and development and the subsequent alteration of hydrology and terrain. Regeneration failure is due to a combination of human and natural causes which encompass herbivory, interspecies competition, altered hydrology, fire exclusion, and the absence of effective regeneration programs after harvest or destruction. Sea level rise, climatic change, disease, and forest decline of unknown origins are also responsible for the disappearance of some coastal forests.

These multiple threats affect many species discussed in this volume, including *Chamaecyparis thyoides*,[2] *C. lawsoniana, C. nootkatensis, C. pisifera, Taxodium* and *Nyssa*, and *Sequoia sempervirens*.[3]

Management strategies for these threatened species and their ecosystems focus on two main themes: habitat preservation and effective regeneration measures. The authors' recommendations can be divided into several categories. First, maintain existing coastal forests and their adjacent buffer areas. Second, provide optimal natural regeneration conditions. Third, conduct genetic sampling to locate and protect unique stands. Fourth, develop artificial regeneration when natural regeneration fails. Finally, increase habitat restoration efforts and related research.

A few issues affecting the management of coastally restricted forests will be considered here. Detailed discussions specific to each system are presented throughout this volume.

Fire

It is now generally recognized that simplistic fire-suppression is not a successful strategy to protect sensitive systems for at least three reasons. The first reason is sadly apparent when uncontrollable fires rage through protected forests, feeding on many years' accumulations of flammable materials. The second is that some trees require fire or another disturbance for stand regeneration. The third is more subtle, for the absence of fire may shift the ecological balance to favor commoner plants, forcing specialized species out of their restricted habitats. The long-term impact of fire on the alteration of plant, animal, and microbial population balances requires further monitoring and research.

Prescriptions are both species-specific and site-specific. For example, while *Pinus serotina* requires fires every few years, under some conditions Stoltzfus et al. recommend intervals of six or seven decades between both fire events and logging to promote healthy stand regrowth of *C. thyoides*.

Maintenance of a Diverse Gene Pool

To locate and protect unique stands, Eckert[4] recommends genetic sampling. Why is it important to conserve something simply because it is at the edge of its range? An organism surviving at the limits of ecological stress may have unique genetic combinations. Its nearest relatives may, in the aggregate, possess all parts of the edge-species' genome, but only the individuals with particular genetic combinations can survive in the fringe habitat. Or, more rarely, new adaptive mutations arise. The adaptive combinations or mutations will tend to be lost in environments not selecting for them. The diversity conserved in edge-of-range sites allows for the adaptability of a species to future environmental change. Therefore, it is valuable to conserve not only many samples of species of concern, but it is vital to preserve multiple habitat samples as well.

The gene pool of an individual from another part of a species' range may survive in one section in the fringe environment, but not in another. Microsite conditions are of particular importance in a fringe habitat.

Many ancillary factors that affect a plant's hardiness near the limit of its range have particular management implications. For example, survival of a plant at the limit of its cold tolerance requires shelter from freezing winds, sufficient sunshine to promote good food reserves (a high sugar content in the cell sap delays freezing), and a situation that promotes gradual rather than sudden thawing (van Gelderen and van Hoey Smith 1993:353).

Restoration and Protection

Restoration of damaged lands, with emphasis on appropriate hydrology, is crucial to regaining healthy coastal forests. Alternative harvest methods may be able to mimic natural ecosystem processes, thereby retaining forest diversity and supporting more successful natural regeneration.

It must be recognized that coastally restricted forests require protected status. They are all, to varying degrees, under pressures that threaten to decimate their number and seriously alter the diversity and character of their associated communities. Those coastal forests that have been explored are known to harbor species recognized as rare, endangered, or possessing unique or highly prized values. Great promise lies in the unknowns, factors great or small that have not been examined or comprehended. These unknowns range from the uses of plant, animal, or microscopic materials—as pharmaceuticals, industrial chemicals, pesticides, or nutrients—to the forests' effects on the adjacent environments and climate worldwide.

Growing vegetation has profound influences far beyond the immediate surroundings. Until a hillside is stripped, it may not be apparent that the trees have been preventing erosion of the total soil layer. The quantity and quality of both surface water and the below-ground aquifer; the ambient temperature and air quality; the intensity and frequency of storms; the speed and intensity of coastal erosion; the quality and direction of beach deposition; the distribution and quantity of precipitation; and the health, diversity, and stocking of inland, inshore, and offshore fin- and shellfish are all intimately connected to an intact forest canopy and root mass.

In all situations, long-term monitoring is needed, both to track natural variation (such as sea level and climate change), artificial manipulation (of, for example, water regime or harvest), and the resulting ecosystem responses. Large-scale inventories tracking long-term patterns are essential for the evaluation of trends in coastal forests.

Research

Research needs are examined throughout this book. Although the most promising areas of research differ from species to species and from system to system, some common themes recur.

The major targets of basic investigation are:

- Ecophysiology
- Ecosystem functions
- Reproduction
- Genetic architecture, and
- Evolutionary biology.

The major practical aims are to:

- Develop high quality propagules for future harvest;
- Develop alternative harvest methods;
- Achieve the optimal balance between conservation and the utilization of forest products; and
- Determine what that balance should be.

A series of research topics has been chosen to more clearly determine the factors that select for and limit coastally restricted species. The rationale and background for each issue were presented in Chapter 1, Overview and Synthesis. The sequence of topics generally follows that chapter, first addressing the species, then their environments.

To highlight areas where research would be most valuable, a series of summaries were prepared. Tables 1.1–1.6 outline the data currently available on ecologically significant discriminants. Boxes 1.2 through 1.6 group the parameters according to whether or not they are shared by all or most of the species and systems.

Properties of the Species

The primary targets of species research are biological diversity, the functions of tree morphology, and the ecological strategies and properties of the dominants. In each category, a thorough collection of data is necessary prior to establishing the data's significance.

Biological Diversity

- Survey the forest canopy, soil, and water for all species, particularly microorganisms, fungi, and insects.

Morphology   Physical form may be the major key that fits a species to a restrictive environment. In some cases the structure reflects metabolic adaptations; in others, the form itself provides essential adaptive elements.

- Trees in saturated soil have shallow horizontal roots and no tap root. However, this is also true for *S. sempervirens* in well-drained, high-nutrient soils (Ornduff). What is the root-growth pattern of other coastal trees in well-drained situations?
- In a humid environment, what are the specific metabolic and ecologic functions of sclerophylly and xeromorphism, the water-conserving features common in desert plants?
- Do dissected leaves take in more moisture and nutrients than nondissected leaves of similar texture?

- Examine the hypothesis that small needles are advantageous under stressed conditions. Is there a pattern to the distribution of coastally restricted trees with larger needles—e.g., north-south, altitudinal, nutrient-related? Is the pattern correlated with ecotype (i.e., intra-specific), with different species, or both?

Ecological Strategies and Properties   Evergreenness seems to be a key property among coastally restricted trees. Work with various species in many environments suggests important factors to be examined to determine the correlation between soil nutrient availability and the evergreen habit of temperate-zone coastally restricted trees.

- Trace the translocation of carbon and nutrients between aging needles and young expanding ones.
- Trace the translocation of carbon and nutrients between needles and non-green plant parts.
- What are the changes in nutrient content during the life span of needles?
- Is the dropping of old leaves synchronous with, or sequential to, the start of growth of new leaves?
- Are the above factors uniform for each species in all its habitats, or are there significant within-species differences?
- What is the advantage of being a conifer if there is high nutrient availability and/or if moderate temperature permits year-round photosynthesis? This factor should be further explored to expose the dynamics differentiating these systems. See also the section on *Taxodium–Nyssa,* below.

Basin wetlands would be expected to be more dominated by the necessity for a water-conserving strategy than would flow-through systems, for the root region of trees in flowing water would have adequate moisture even in most times of drought.

- Is sclerophylly more pronounced in basin wetlands than in riverine sites?
- Is there selective or differential uptake of nutrients in any of the trees or their associated species? How does uptake vary with the nutrient supply or environmental conditions such as temperature, light, moisture, etc.?

Niche Breadth and Genetics

In the process of exploring the genetic capacities and expressed properties of each of the trees, the following aspects of each issue should be taken into account: the variation within the species (1) over time; (2) between individual trees within stands; (3) among neighboring stands; and (4) on transects north to south, in altitude, and from the coast inland.

- Does a single genome contain the capacity to thrive in a wide range of environments? Or are there a series of ecotypes, each selected for in specific microhabitats? That is, is there specialization and the concomitant loss of other genetic characters, resulting in many genetically diverse populations? (See chapters on *C. lawsoniana, C. thyoides, and C. nootkatensis.*)

The work of Hennon et al. and Russell on *C. nootkatensis* suggests a series of studies for each of the other coastally restricted species.

- In earlier times, did the species exist on a much greater range than at present?
- Where does the species rank on a specialist–generalist axis?
- Are species genetically highly variable, especially within populations?
- Are species phenotypically plastic (i.e., do they respond differently to different environments)?

Paul Hennon (personal communication, December 1991) suggested that avenues of inquiry would include exploring the genetic relationship between *Chamaecyparis* species and determining their ancestral epicenter, the biogeographic and evolutionary implications of the existence of a few outlier populations, the potential importance of reproductive strategies in relation to species distribution, and each species' sensitivity to specific soil conditions.

The work of Dunsworth, Hennon et al., and Russell, in three different parts of the range of *C. nootkatensis* prompts another set of questions.

- How different are trees growing in Alaska from those growing in British Columbia, Canada, and in northwestern USA? In what characteristics do populations in, for example, sea-level Alaska significantly differ from those in montane British Columbia?
- Are there temperature-compensatory responses that are replicated in altitude and latitude? For example, do the same adaptations found high on a mountain appear further north at a lower altitude?

Specialists and Generalists

- Which trees possess what r-selective (generalist) and K-selective (specialist) factors? (This follows from work of Eckert, Edwards 1992, Libby et al. 1968, Russell, and Zobel.)
- How are genotypes distributed between stands? (This follows from work of Eckert and Russell.)
- Are rare species poorer dispersers than common species?
- Collect data to determine if the extinction debt theory (discussed in Chapter 1) is applicable to coastally restricted systems. If so, what are the spatial scales to which it applies?

Secondary Chemicals

- Identify secondary chemicals produced by each species in plant parts, in water, and in soil. What are their functions in nature? What unusual metabolic properties of potential pharmaceutical, agricultural or industrial value do they possess?

Taxodium–Nyssa Forests

As the *Taxodium–Nyssa* ecosystem has many features that differ from the strictly coniferous, nondeciduous systems:

- Explore the possibility that its deciduous, relatively non-water-conserving features are related primarily to its existence in flowing-water, more eutrophic regimes.
- Compare *Taxodium* and *Nyssa* trees growing in (1) basin wetlands, (2) riverine wetlands, and (3) mesic sites for genetic variation and transpiration rates.
- What are the significant differences between the physiology, morphology, ecology, and/or life cycle of *Taxodium* and *Nyssa* and those of coniferous–nondeciduous systems?

If there are consistently perceptible correlations:

- Do *Taxodium* and *Nyssa* exhibit characteristics more similar to the southern coastally restricted species (such as *C. thyoides* var. *henryae, Pinus muricata,* and *P. radiata*) than to the northern ones (such as *C. thyoides, C. nootkatensis,* and *C. obtusa*)?
- Do *Taxodium* and *Nyssa* share significant qualities with the lowest altitude ecotypes of other coastally restricted species that they do not share with those species' high-altitude members?
- What are the significant features in the *Taxodium–Nyssa* environment that differ from those of strictly coniferous–nondeciduous coastally restricted systems? Examine climate, temperature, and altitude alone and in combination with the known substrate and hydrological differences.

Properties of the Environment

Geography   What physical parameters limit the geographic extent of each species' range? It is reasonable to assume that it is not the average temperature, precipitation, or humidity, but the extremes that are the significant factors (Alaback 1991, Weigand et al. 1992).

- To define the microclimatic variation and extremes, establish meteorological stations within stands of each species.

The Influence of Oceans on Nearby Land   What environmental properties limit or promote growth of coastally restricted species? If one property alone is not definitive, does a particular combination of factors track with these species? Take into consideration that some combinations of factors may have synergistic effects—the combination may be more effective than the sum of its parts—or that some co-occurring factors may cancel each other out. Especially note differences between forests with and without a coastally restricted tree canopy.

- Take measurements along transects from the coastline to 250 km inland, and vertically from the forest floor to above the canopy. Set up a standard data-collecting system for all coastal forests.
- Correlate atmospheric contents with soil components and properties: Does a property of the atmosphere, such as moisture or nutrient content, compensate for the lack of a soil component?

To test the hypothesis that the nutrient content in aerosol, precipitation, and fog drip is critical:

- Analyze and compare the ionic contents of coastal versus inland fog, aerosols, and precipitation. Examine for nitrogen, phosphorus, and minerals including trace elements.
- Measure uptake of nutrient along stems. (See also Ecological Strategies, above.)
- Measure wind velocities in, above, and near canopies.

Biogeography   What environmental factors led to the great distances between closely related species? In Europe, east Asia, and other intensively long-inhabited regions, it is generally assumed that the original coastal forests were extirpated (Poulinin and Walters 1985, Weigand et al. 1992), and that the geomorphology has been so drastically altered by human habitation that no trace remains of the primeval vegetation. This does not appear to be a convincing scenario for Canada.

- What factors on the Atlantic borders of Canada and Europe led to the absence of coastally restricted forests? Were there such forests in the past?

In the gap in *C. thyoides* distribution between the states of South Carolina and Florida, the only extant stands are in Georgia hillside seeps (Wilbur H. Duncan, personal communication, 1984).

- Are all the geographic outlier populations of each species found in particularly humid areas?

Fog

- Are all coastally restricted trees in fog belts? Are there similar sites near coastal temperate forest stands shrouded with the same fogs, but lacking these species? If so, are there definable, observably different conditions in those sites?

Correlations between Properties of Coastally Restricted Species and Their Environments

Table 21.1 and Figure 1.2 summarize the correlations between properties of coastally restricted species and their environments. To precisely identify the relationship between adaptations of the coastal forests and their environments, further research is required. Fruitful avenues include:

- Genetic investigations (following Eckert, Russell, and Zobel).
- Quantification of as many parameters as possible for both the species and habitats. Establish a series of graphs to be overlaid on each other to reveal multi-factor gradients of correspondence between each species property and each habitat characteristic. For example, compare the size of new leaves of trees in areas with differing water levels, total precipitation, and average humidity.

Climate Change, Human Disturbance

In addition to inquiry directed at the trees and their immediate ecosystems, there is great interest in the effects of both climate change and human disturbance, as well as the potential for minimizing and redirecting these factors.

Long-term Research Reserves

The preservation of large tracts in their native state is required for both applied and basic investigation. Long-term research reserves, appropriate to the longevity of the canopy dominants, must be established. The maintenance of long-term study plots is

Table 21.1 Correlations between Properties and Environmental Parameters of Coastally Restricted Species[a]

Species Property	Environmental Parameter
Native only to within 250 km of a marine coast (by definition)	Less than 250 km from marine coast (by definition)
High aerial moisture requirement	Frequent fogs, high humidity
Catastrophe-dependent	Major disturbances at irregular intervals
Broad-niche potential (wide ecological amplitude; capacity to thrive in a great range of environments) * thrive in protected plantations or amenity plantings far from the natural range and under very different conditions	A wide range of environmental types

Poor competitive ability
    — outside the favored niche
    — in favorable or "easy" environments
    — in much of its potential range
    * become dominants in special circumstances wherein they are well adapted to one or more restrictive factors
    * sensitive to saline groundwater and salt-laden winds[b]
    * do best in protected sites[b]
	Includes (but not restricted to) stressful environments

Needled evergreen	Low-nutrient substrate is common, never mandatory; often limited to particular topo-edaphic conditions within a range of climatically controlled regional forest types; salt and nutrient-laden fogs and aerosols

Discontinuous distribution
    * genetic and evolutionary effects found in ecological islands
    — founder effect
    — highly susceptible to change in the environment
    — greater variation between stands than from tree to tree within stands, due to reproductive isolation
	Markedly discontinuous sites

Moisture-conserving features (sclerophylly)
    — waxy cuticle
    — needle-shaped leaf
    — thick bark, deeply fissured
	Irregularly reccurring (may be rare) major disturbances: drought, fire, hurricane, clearcut; soil sometimes contains toxic elements

Paleoendemic	Geographic range previously greater than at present
Often, but not exclusively, monotypic stands	Often stressful; subject to episodic catastrophic events
Wind-pollinated; seed transported by wind	Frequently windy

Notes: * = subset of species property; – = modifier or example of species property.

[a]See also Figure 1.2, Table 1.1.

[b]Features that add to the picture of living "on the edge" in very difficult environments.

also necessary to accumulate baseline data in advance of the erratically occurring, but inevitable, catastrophic events that strike coastal forests (Conner).

Conclusion

Recognition of the value of an entire ecosystem is necessary to mobilize sufficient interest for its protection. Research provides insight into cause, interaction, and effect, so that successful management strategies can be devised. Research requires funding, which will only be invested when the intrinsic qualities are understood. The measurement of the immediate economic value of a forest must be integrated with its long-term importance.

It may be that there are few if any differentiating characters at the level of species components, and that the most significant qualities are to be found in the community as a whole. In some cases, data are available and simply need to be assembled in one place for comparison and evaluation. A combined database would facilitate and encourage research in appropriate areas. Part of the work already done on these species is not widely available due to language barriers: Japanese–English translations would be particularly valuable (Keiichi Ohno, Shin-Ichi Yamamoto, Donald Zobel, personal communications, 1990). A major purpose of this volume is to make material available to investigators and practitioners. Those who have additional information may study what others have found, add their data, and help to reshape and expand the understanding of these fascinating systems.

Acknowledgments   We wish to particularly thank both the volume authors and participants in the workshops for generously sharing their thoughts on management and research issues.

Notes

1. See Preface for details.

2. The common names of each coastally restricted forest dominant are listed in Table 1.1.

3. For further management and research information on *Chamaecyparis thyoides,* see Eckert, Phillips et al., Sheffield et al., and Stoltzfus and Good (Chapters 8, 10–12); on *C. lawsoniana,* see Greenup (Chapter 6); on *C. nootkatensis,* see Dunsworth, Hennon et al., and Russell (Chapters 3–5); on *C. pisifera,* see Yamamoto (Chapter 7); on *Taxodium* and *Nyssa,* see Conner and Day, and McWilliams et al. (Chapters 17, 19); and on *Sequoia sempervirens,* see Ornduff (Chapter 15).

4. Citations without dates refer to chapters in this volume. See note 3.

References

Alaback, P.B. 1991. Comparative ecology of temperate rainforests of the Americas along analogous climatic gradients. Revista Chilena de Historia Natural **64**:399–412.

Edwards, S. 1992. Foliar morphology of *Chamaecyparis* and *Thuja.* Four Seasons **9**:4–29.

Libby, W.J., M.H. Bannister, and Y.B. Linhart. 1968. The pines of Cedros and Guadalupe Islands. J. For. **66**:846–853.

Poulinin, O., and M. Walters. 1985. A guide to the vegetation of Britain and Europe. Oxford Univ. Press, Oxford.

van Gelderen, D.M., and J.R.P. van Hoey Smith. 1993. Conifers. Timber Press, Portland, OR.

Weigand, J., P.B. Alaback, A. Mitchell, and D. Morgan. 1992. Coastal temperate rain forests. Ecotrust, Portland, OR. Unpublished report.

Appendix A

Conversion Factors

Multiply	By	To Obtain
Metric to U. S. Customary		
millimeters (mm)	0.03937	inches
centimeters (cm)	0.3937	inches
meters (m)	3.281	feet
meters (m)	0.5468	fathoms
kilometers (km)	0.6214	statute miles
kilometers (km)	0.5396	nautical miles
square meters (m^2)	10.76	square feet
square kilometers (km^2)	0.3861	square miles
hectares (ha)	2.471	acres
liters (l)	0.2642	gallons
cubic meters (m^3)	35.31	cubic feet
cubic meters (m^3)	0.0008110	acre-feet
milligrams (mg)	0.00003527	ounces

grams (g)	0.03527	ounces
kilograms (kg)	2.205	pounds
metric tons (t)	2205.0	pounds
metric tons (t)	1.102	short tons
kilocalories (kcal)	3.968	British thermal units
Celsius degrees (°C)	1.8(°C) + 32	Fahrenheit degrees

U. S. Customary to Metric

inches	25.40	millimeters
inches	2.54	centimeters
feet (ft)	0.3048	meters
fathoms	1.829	meters
statute miles (mi)	1.609	kilometers
nautical miles (nmi)	1.852	kilometers
square feet (ft^2)	0.0929	square meters
square miles (mi^2)	2.590	square kilometers
acres	0.4047	hectares
gallons (gal)	3.785	liters
cubic feet (ft^3)	0.02831	cubic meters
acre-feet	1233.0	cubic meters
ounces (oz)	28350.0	milligrams
ounces (oz)	28.35	grams
pounds (lb)	0.4536	kilograms
pounds (lb)	0.00045	metric tons
short tons (ton)	0.9072	metric tons
British thermal units (Btu)	0.2520	kilocalories
Fahrenheit degrees (°F)	0.5556 (°F − 32)	Celsius degrees

Appendix B

Glossary

This list contains selected terms as used in this volume. The first time a technical term is used, it is defined in the text. The definitions given here are restricted to their usage in this volume. Many other meanings exist in other fields and in common usage.

adventitious.   In botany, plant parts growing from an unusual position e.g., roots from a leaf or stem.

aerenchyma.   Plant tissue containing large intercellular air spaces.

aerobic environment.   One in which oxygen is present.

aerobic organism   (such as bacteria). One requiring oxygen; an aerobe.

analog, ecological.   Ecological equivalent; ecological vicar. Different species that occupy similar niches in different places.

areolate.   Divided into small areas (areolae) by cracks or lines.

autotrophic.   Plant-like mode of nutrition. (Opposite: heterotrophic)

backswamp.   Wetland partially encircled by a river bend.

basin wetland.   Mire with no inflow or outflow; depends primarily on precipitation for water; bog. (Opposite: flow-through wetland)

biological diversity   (biodiversity). The variety of life forms, the ecological roles they perform, and the genetic diversity they contain.

board-foot   (bd ft). In the US, a common unit of measure for expressing inventory volumes for trees that currently are capable of producing saw logs (international 1/4-inch log rule). Defined as 1 ft by 1 ft by 1 in; the actual thickness is somewhat less.

bog.   A peat-accumulating wetland that has no significant inflows or outflows and supports acidophilic mosses, particularly *Sphagnum.*

canopy.   Top forest layer, usually tree tops.

cation exchange capacity (CEC). The ability of particles to absorb positively charged ions (cations), such as calcium and magnesium, in exchange for hydrogen ions.

CEC. See cation exchange capacity.

community. An association of interacting populations, usually defined by the nature of their interactions or the place in which they live.

constant companions; constant species. Plants or animals most frequently found growing together.

coppice. Group of small trees growing from root suckers or a cut tree stump. All members of a coppice form a clone (are genetically identical).

diversity. The number of different types of plants and animals in an area. Species richness.

dominant. a. Species having the most influence on community composition and form. b. Largest or most abundant species in the community.

ecological amplitude. Niche breadth. Also see niche. "Wide ecological amplitude" means broad niche breadth; the quality of persistence via tolerance and defense mechanisms rather than growth and reproduction. See also K-selection.

ecological equivalent. Ecological analog. Different species with similar properties occupying similar niches in similar, but geographically separate, environments.

ecotone. Boundary between two plant communities.

ecotype. Genetically differentiated subpopulation restricted to a certain habitat.

edaphic. Influenced by the soil.

endemic species. Species native and specific to a particular place.

epiphyte. Plant that uses another plant for its physical support, but does not draw nourishment from it.

ericaceous. Plants of the heath family (Ericaceae), e.g., *Rhododendron, Vaccinium* (blueberry).

extirpated. Locally extinct.

facultative. Refers to organisms growing in an environment without having a specific requirement for it.

family (synonym: open-pollinated family). In forest genetics, groups of individuals (siblings) with one or both parents in common. Often synonymous with progeny of one tree. See also population.

food chain. A representation of the passage of energy from a primary producer through a series of consumers at progressively higher trophic (feeding) levels. Thus, plant–herbivore–carnivore, etc.

food web. A representation of the various paths of energy flow through populations in the community, taking into account the fact that each population shares resources and consumers with other populations.

generalist. Organism that can change to fit different environments, that is phenotypically plastic or adaptable. See r-selection. (Opposite: specialist)

generation time. Time required to complete one full growth cycle. May refer to the life cycle of individuals (from birth to reproduction) or to the doubling time of cells.

heat sink. System that stores thermal energy.

heterotrophic. Animal-like (herbivorous or carnivorous) nutritive mode. (Opposite: autotrophic)

island, island phenomenon. In ecology, a relative term indicating a geographically isolated habitat, a habitat that differs so much from its surroundings that populations are reproductively isolated. Examples: a mountaintop, a pond, a forest fragment after logging.

kettle. In geology: a hollow; feature of glaciated landscape.

keystone species. A species that affects the survival and abundance of many other species in the community in which it lives. Its removal or addition results in a significant shift in the composition of the community.

K-selection. Evolutionary selection maximizing efficient utilization of resources. The strategy of highly adapted equilibrium species, specialists. Typical of stable environments. (Opposite: r-selection)

lapilli. Small stony or glass lava fragments.

latent heat. The quantity of heat absorbed or released when a substance changes its physical phase at constant temperature (e.g., from solid to liquid at the melting point, or from liquid to gas at the boiling point).

lenticels. Pores in the bark of a tree that perform gas exchange functions, similar to the stomata of leaves.

marsh. A frequently or continually inundated shallow wetland characterized by emergent herbaceous vegetation adapted to saturated soil conditions.

monoecious. Bears male and female reproductive structures on a single plant.

monotypic. An ecosystem containing only a single species at a particular level. Example: monotypic forest canopy; monotypic herbaceous cover.

msl. Meters above sea level.

mycorrhiza. A close physical association between the hyphae of a fungus and the roots of a plant which transfers nutrients between the two elements; is essential to the growth of some plants; actually a fungus-infected root.

neoendemic. An endemic species evolved from other species that remain nearby. See endemic, paleoendemic.

niche. The ecological role of a species in the community; the many ranges of conditions and resource qualities within which a species persists; ecological amplitude.

nitrification. Conversion of ammonia to nitrate, an aerobic process generally performed by aerobic (oxygen-requiring) bacteria.

obligate. Refers to organisms with a specific requirement for a given substrate or environmental condition.

paleoendemic. A relict plant that survives on substrates hostile to competitors and that is not surrounded by possible parent species. See endemic, neoendemic.

peat. Soil consisting of organic matter in various states of structural breakdown ranging from fibrous to mucky. Pure peat contains few nutrients available to plants, due to its high cation exchange capacity (CEC), the virtual absence of oxygen in its waters, and the binding of nutrients within undecomposed organic compounds. See bog.

phenotype. The characteristics of an individual resulting from the interaction of environment and genetics.

physiognomy. Superficial aspect.

podzol. Acidic, infertile, highly bleached soil low in iron and calcium, formed under cool moist conditions; has an ashlike upper layer depleted of colloids, iron, and aluminum, and a dark lower layer in which these substances have accumulated in toxic concentrations.

population. A group of interbreeding individuals. See also family. See Russell, Chapter 5.

radiation. In evolution, development and establishment of new species.

relative abundance. An aspect of diversity measured by Simpson's Index and Shannon-Weaver Index.

relict. Regional remnant of a biotic community otherwise extinct. In evolutionary biology, the persistence of an ancient organism.

rhizosphere. Root zone.

r-selection. Evolutionary selection for maximizing growth rate (r); an opportunistic strategy advantageous in rapidly changing environments. Strategy used by a generalist, opportunist, or colonizer. (Opposite: K-selection)

sclerophyllous. Describing vegetation with evergreen, firm, leathery leaves with thickened cuticle, structurally reinforced with lignin; an adaptation that conserves water and nutrients, discourages herbivory, typical of hot, dry regions.

secondary chemicals. Plant products that are not essential to self-maintenance.

serpentine soil. Very infertile, well-drained, reddish-brown soil with a distinctive vegetation cover; derived from ultramafic rock, high in magnesium, chromium, cobalt, and asbestos; low in calcium and plant nutrients; toxic to many plants.

slash. Residue of timber harvest.

snag. Standing dead wood.

specialist. Species in which genetic differences can be detected among populations from relatively small environmental gradients. The greater the differences are *among* populations as compared to *within* populations, the more specialized they are—i.e., niches are greatly subdivided. See K-selection. (Opposite: generalist)

species richness. Species abundance; the number of different types of living forms.

stomata. Pores in the exterior of plant leaves through which gases and water move. Singular: stoma.

substrate. 1. The physical base, such as sand, peat, rock. 2. The substance on which an organism feeds, or on which an enzyme acts.

swamp. Wetland dominated by trees or shrubs (in USA).

taxonomic clustering. The numbers of related taxa (life forms) (e.g. species per family, families per class, classes per phylum) in an area. A community with many closely related forms (i.e., a high degree of clustering) reflects a relatively stable environment with strong environmental pressures, which tends to give an advantage to particular adaptations.

tephra. Volcanic material such as ash or lapilli.

translocation. Movement of minerals and compounds within a plant.

transpiration. The loss of water vapor from a plant to the outside atmosphere, via stomata in leaves and lenticels in stems.

ultramafic rock. Serpentine rock. Igneous rock with <45% silica, composed principally of ferromagnesian minerals, forming serpentine soils with a distinct vegetation cover.

vegetative reproduction. Asexual process involving unspecialized plant parts—e.g., root, stem, leaf.

vicariants. Ecological vicars. Species filling equivalent niches.

vicariousness. Degree to which a given species (or taxon) is found in the same environment elsewhere. Ecological equivalence.

wetland. Areas neither strictly terrestrial nor completely aquatic, where the water table is seasonally or usually at or near the surface, or where the land is covered by shallow water.

xeromorphic, xeric. Concerning plants whose morphology is adapted to dry conditions. See sclerophyllous.

Glossary Sources

Allaby, M., ed. 1992. The concise Oxford dictionary of botany. Oxford Univ. Press, New York.

Botanical Latin to English definitions. 1993. The Write Co., Sheldonville, MA.

Brooks, R.R. 1987. Serpentine and its vegetation. Dioscorides Press, Portland, OR.

Cowardin, L.M., V. Carter, F. Golet, and E.T. LaRoe. 1979. Classification of wetlands and deepwater habitats of the United States. U.S. Fish Wildl. Serv., Biol. Serv. Program, FWS-OBS, 79/31.

Harvey, P.H., and M.D. Pagel. 1991. The comparative method in evolutionary biology. Oxford Univ. Press, New York.

Isaacs, A., J. Daintith, and E. Martin. 1984. Concise science dictionary. Oxford Univ. Press, New York.

Margulis, L., and K.V. Schwartz. 1988. Five kingdoms. Freeman, New York.

Mitsch, W. and J. Gosselink. 1992. Wetlands. 2nd ed. Van Nostrand Reinhold, New York.

Moore, P.D. and D.J. Bellamy. 1974. The peatlands. Springer-Verlag, New York.

Ricklefs, R.E. 1993. The economy of nature. W.H. Freeman, New York.

Specht, R.L., ed. 1979. Heathlands and related shrublands of the world. Elsevier, Amsterdam.

Stearn, W.T. 1983. Botanical Latin. David and Charles, London.

Wilson, E.O., ed. 1988. Biodiversity. National Academy Press, Washington, DC.

Wilson, E.O. 1992. The diversity of life. Belknap Press of Harvard Univ. Press, Cambridge, MA.

Index

When both the scientific name and common names of species are mentioned in the text, they are listed in the index under both terms. However, descriptive subentries appear only under the scientific name. Common name entries include the same page numbers, but they are undifferentiated (i.e., they include no subentries). To facilitate finding cross-references within the text, the index contains the name of each chapter's senior author followed by the first page of the relevant chapter.

The letter *a* after a page number indicates information in an appendix; *b* indicates information in a box; *f* indicates information in a figure; and *t* indicates information in a table.

326   INDEX

endangered, 158
fire and, 17, 307
fog and, 28
genetics, 171–182
growth, 11, 22t, 47t, 48, 120–122, 149–151, 171
habitat, 21t, 49, 168–169
harvest, 117–118, 120–122, 156–163, 172
herbivory, 161, 169
inventory, broad-scale, 111–122
in Mashantucket Pequot Cedar Swamp, 137–139
migration, postglacial, 172, 181–182
morphology, 23t
in New Hampshire and Maine, 171–182
in New Jersey Pinelands Biosphere Reserve, 142–154
in North Carolina, 156–169
plantability, 168
pure stands, 116–117
range, 22, 25, 40, 40t, 171
reproduction, 11, 22t, 29, 148, 153–154, 160–169
seed production, 148
size, 39, 40t
soils, 25
succession, 151–152
susceptibility to loss, 171–172
timberland, 115t, 116t, 117–118, 119t
value-added conservation, 162–163
vegetation patterns, 125
Chamaedaphne calyculata (leatherleaf), 136–137, 138, 147t
cherry, black *(Prunus serotina)*, 127
chestnut, American *(Castanea dentata)*, 125, 135, 138
China, 4
chi-square analysis, *Chamaecyparis thyoides,* 174
Chosenai arbutifolia (Korean willow), 208
cinquefoils *(Potentilla* spp.), 127
Cladium mariscoides (twig-rush), 135
Clean Water Act, 160
Clethra alnifolia (sweet pepperbush), 127, 147, 153
climate, 25, 42f, 127
    Chamaecyparis nootkatensis and, 67, 73, 78
    global climate change, 29–30
clovers *(Melilotus* spp.), 127
clubmosses *(Lycopodium* spp.), 127
clumps, *Pinus pumila,* 210b
coastal temperate rainforest (CTRF), 25
Coccoloba floribunda, 245, 246t
codominants, 15
Combretaceae, 248
competition, 9b, 12, 14
Compositae (composites), 127
cone production
    Chamaecyparis nootkatensis, 73–74
    Pinus pumila, 208–217
    Sequoia sempervirens, 222–223
Connecticut, 124–139, 172
Conner, William H., 271, 278
Conocarpus erecta (botoncillo), 240t, 241
Cornus spp. (dogwoods), 127
cottonwood
    Populus deltoides, 282
    Populus heterophylla, 286

cotyledons, 60
cranberry, bog *(Vaccinium oxycoccus),* 188
crayfish *(Procambarus clarkii),* 287
cress, marsh *(Nasturtium officinale),* 127
Croton niveus, 245, 246t
crowfoot, water, *(Ranunculus trichophyllus),* 132, 133
Cryptomeria japonica (Japanese red cedar), 102
Cupressaceae, 47
Cyperaceae, 135
cypress. *See also Taxodium–Nyssa* (cypress-tupelo) forests
    Alaska yellow-. *See Chamaecyparis nootkatensis*
    bald. *See Taxodium distichum*
    hinoki. *See Chamaecyparis obtusa*
    pond *(Taxodium distichum* var. *nutans),* 257, 262, 266, 267t
    Sawara. *See also Chamaecyparis pisifera*
    Taiwan. *See Chamaecyparis formosensis Taxodium distichum* (bald cypress)
    Taiwan red. *See Chamaecyparis taiwanensis*
Taxodium spp., 5b, 11, 12, 115, 119, 120t

Dalbergia glabra, 239, 240t, 245, 246t
dams, wetland loss and, 281
dangleberry *(Gaylussacia frondosa),* 147, 153
deciduous trees, 5b, 134–135, 137
Decodon verticillatus (swamp loosestrife, water willow), 135
decomposition, 287–288
deer, white-tailed *(Odocoileus virginianus),* 161, 162, 165, 169
Dennstaedtia punctilobula (hayscented fern), 127
dewberry *(Rubus hispidus),* 147
dicotyledons, 44t
discontinuous distribution, 9b, 10, 17, 31f, 314t
dispersal ability, 14–15
diversity, 19b, 307–308, 309
dogwoods *(Cornus* spp.), 127
dominance, 9b, 12, 15, 227, 237
drainage, 62
driads *(Dryas integrifolia),* 132
drip-tip feature, 11
Drosera rotundifolia (round-leaved sundew), 148
drought, 13–14, 18, 29
Dryas integrifolia (mountain avens or driads), 132
Dryopteris, 137
    novaboracensis (New York State fern), 127
    simulata (Massachusetts fern), 148
Dunsworth, Glen B., 70

Eckert, Robert T., 171
ecological amplitude. *See* niche
ecological strategies, 12–16, 310
ecosystems, 3–4, 5, 21t
ecotone, 27
elderberry, common *(Sambucus canadensis),* 127, 138
electrophoresis, 174
elk, Roosevelt *(Cervus elaphus roosevelti),* 231
elm
    American *(Ulmus americana),* 137
    Japanese *(Ulmus davidiana),* 189, 191–197
endangered species, 158